Rite Of Passage.
A Photographic Introspective of the Canadian Tree Planting Industry.

Published by DJ Bolivia Inc.
P.O. Box 6074, Sackville, New Brunswick, E4L 1G6.
jonathan.scooter.clark@gmail.com

ISBN 978-1999016807

Hardcover edition, printed by IngramSpark.

Inquiries: www.jonathanclark.ca

Disclaimer

This book is presented solely for educational purposes, and to allow the public to gain a greater understanding of the tree planting industry in Canada. The author and publisher are not offering it as professional, legal, financial, physiological, medical, human resources, or any other type of advice. While best efforts have been used in preparing this book, the author and publisher make no representations or warranties of any kind, and assume no liabilities of any kind with respect to the accuracy or completeness of the contents, and specifically disclaim any implied warranties of appropriateness or fitness of use for any particular purpose.

Neither the author nor the publisher shall be held liable or responsible to any person or entity with respect to any loss or incidental or consequential damages of any kind caused, or alleged to have been caused, directly or indirectly, by the information contained herein. No warranty may be created or extended by website links, photos, videos, or written materials related to or referenced by the author and publisher. The content of any websites referenced in this book may change at any time.

The contents of this book should not be construed as a suggestion or recommendation, explicit or implicit, for any person or persons to seek employment within the forestry sector in general, nor specifically within the tree planting or reforestation sectors. Tree planting is extremely difficult work, and most people are not well suited for employment positions within the industry.

Acknowledgements

Many thanks to all of the people who appear in the various photographs within this book. Without the photographs of people at work, which help demonstrate the human-interest aspect of the tree planting industry, this book would have been pretty boring.

The photos in this book were compiled from two separate years of managing a planting camp in Western Canada. I tried to take photographs of everyone in both camps, for inclusion in this book. My apologies to approximately half a dozen of the people in these two camps for whom I unfortunately didn't seem to capture a good photograph. It was not my preference to omit anyone from this book.

The information in this book may present a good overview of the Canadian tree planting industry. For more detailed information, both with respect to finding a tree planting job and learning the specific techniques necessary for tree planting, look for another of my books called "Step By Step: A Tree Planter's Handbook," which is available on Amazon.

The majority of these photos were taken with a Canon 5D Mark IV, using one of several different lenses. Predominantly, I used the EF 24-105, the EF 17-40, or the EF 200 fixed focal length lens. A handful of photos were taken with a Canon T5i body.

Preface

This book is a project that I've wanted to tackle for quite a while. For years, the most popular part of the Replant website was the collection of photo galleries. I had always thought it would be nice to have a coffee-table style photo book that would feature a solid collection of high quality planting photos. Such a book would be a good way to show non-planters what the tree planting industry is all about, at least as much as it's possible to do that without actually forcing them to experience a summer of planting.

While my original vision was for the book to be a "best of" compilation from Replant, I quickly decided against that approach, because I wanted the photos to be "new" to everyone reading the book. Unfortunately, my job as a camp supervisor keeps me fairly busy, so I can't spend as much time on photography as I'd like. After a full season of collecting photos that were specifically intended for this book, I decided that I only had about half the number that I needed. For that reason, the photos in this book were taken during two consecutive planting seasons. Due to this timeline, some of the people in this book didn't actually work with each other, but in other cases, you'll see two photos of the same individual that were taken in different years.

My original intent was to also include a number of photos from BC's coastal planting industry in this book. However, that industry is so different from planting everywhere else in Canada that I opted to give the coastal photos their own separate book. That book is called, "From Our Footsteps, Giant Forests Grow."

I try not to get my camera too wet, so you'll find a disproportionate number of photos that were taken on sunny days. This "weather bias" may unfortunately promote a slight misrepresentation of what tree planting can really be like, but to be fair, it's impossible to really understand planting from an outsider's perspective. To understand it, you have to do it.

Despite the weather bias, I hope that any non-planters reading this book will still walk away with a much better understanding of the types of challenges that planters deal with on a daily basis. It's a job that many planters love to hate, but for better or for worse, one that they'll never forget for the rest of their lives.

www.jonathanclark.ca

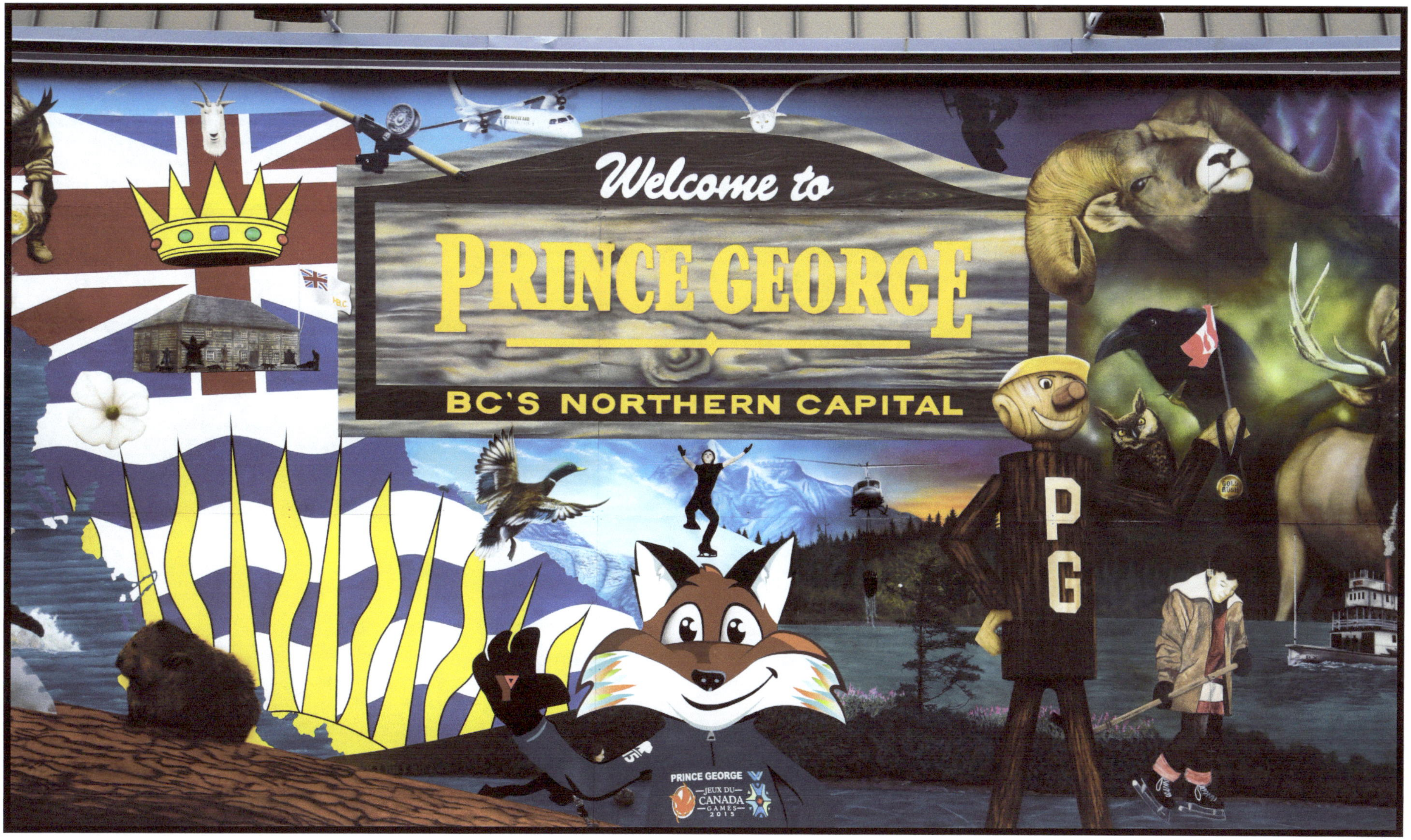

For planters in western Canada, Prince George is one of the centers of industry. Many first-time planters fly into the Prince George airport on their way to join their crews. This mural is the first thing that people see when they walk out of the Prince George airport. Mister PG, seen in the painting, is a real-life statue which is currently located at the intersection of highways 16 and 97. These are the two main highways running through northern British Columbia.

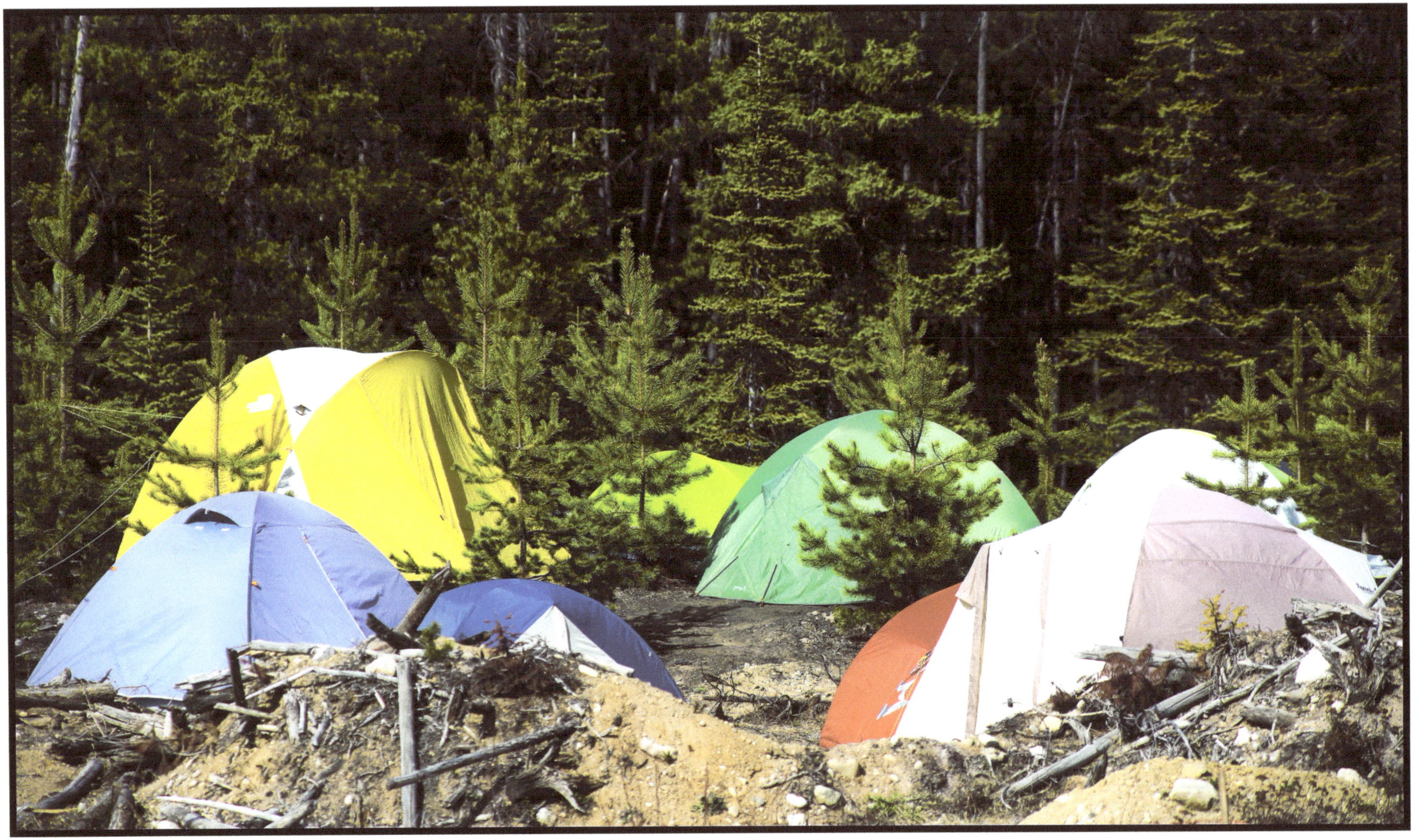

Although some crews are based out of motel accommodations, a large portion of the industry works out of tent camps. It is estimated that there are approximately seventy to eighty tent-based planting camps working in British Columbia alone, representing perhaps as much as 75% of BC's provincial tree planting labour force. Remote tent camps are also common for planting operations throughout the rest of Canada.

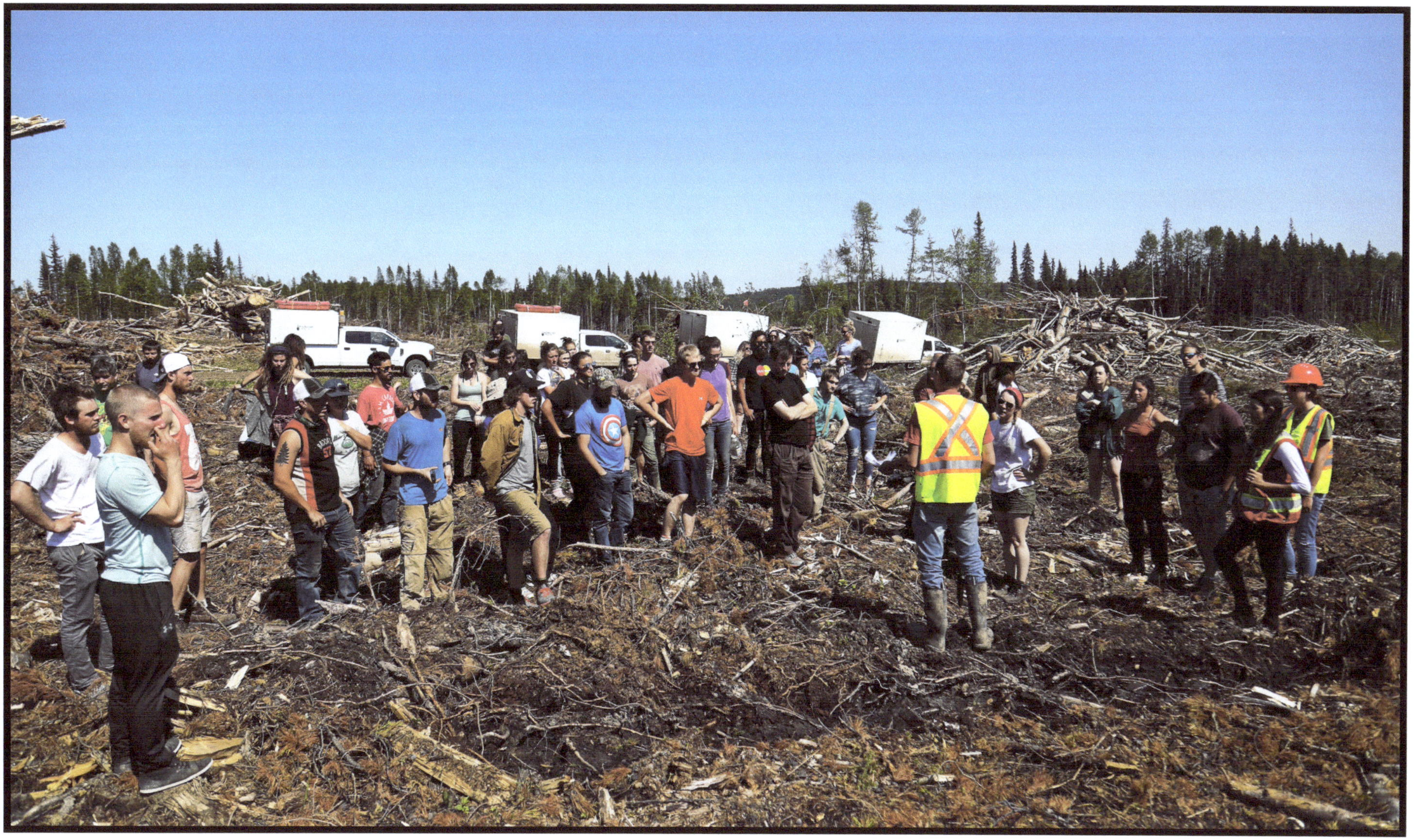

At the start of each separate "contract," all of the planters will typically have a meeting with the forester, to allow the forester to explain their standards for planting quality, density, stock-handling, and other contract requirements. This meeting is called the "pre-work conference." Planters may typically work on anywhere from a few to half a dozen or more separate contracts during any given planting season.

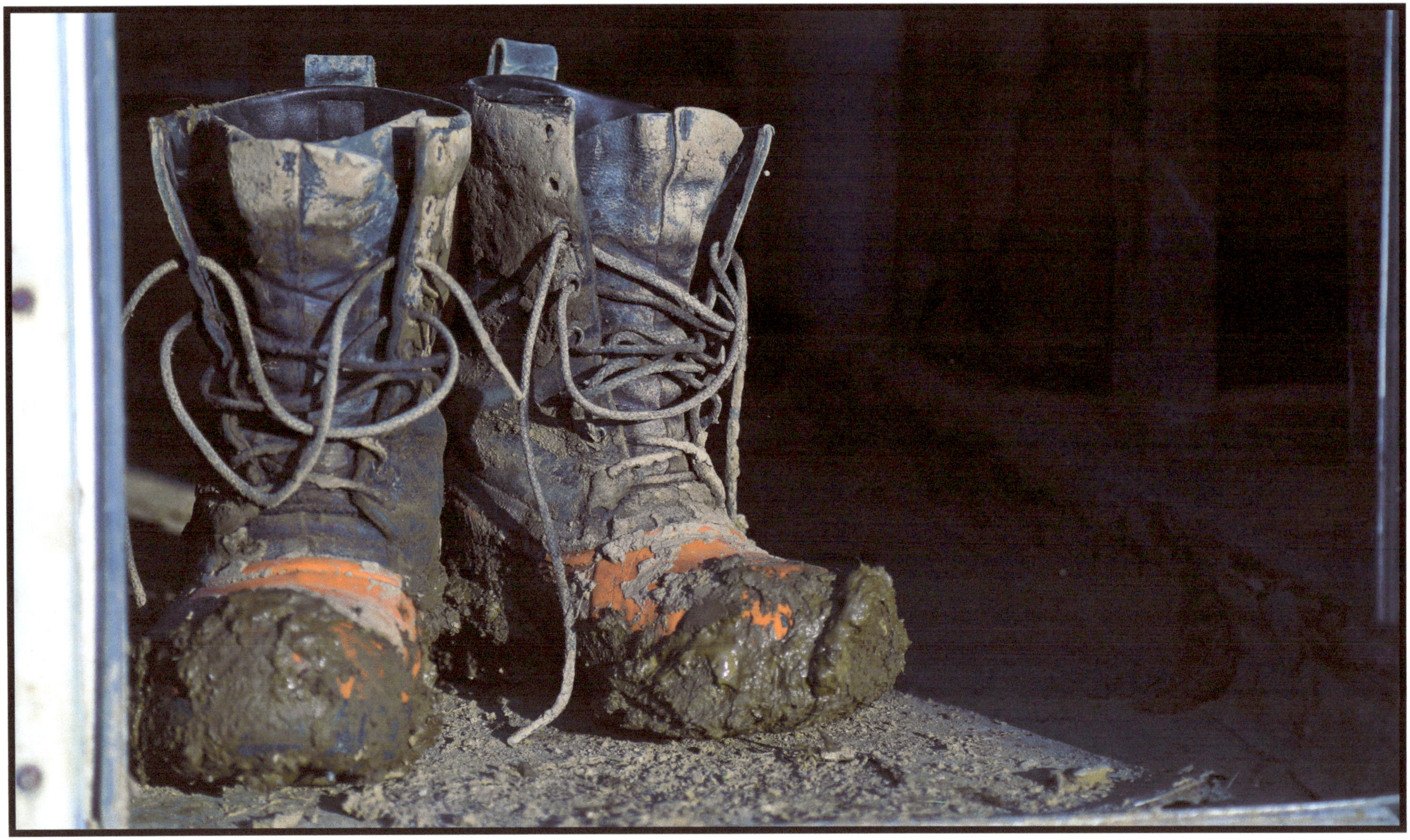

If you're a tree planter, be prepared to get dirty. You're going to be living in a world of mud.

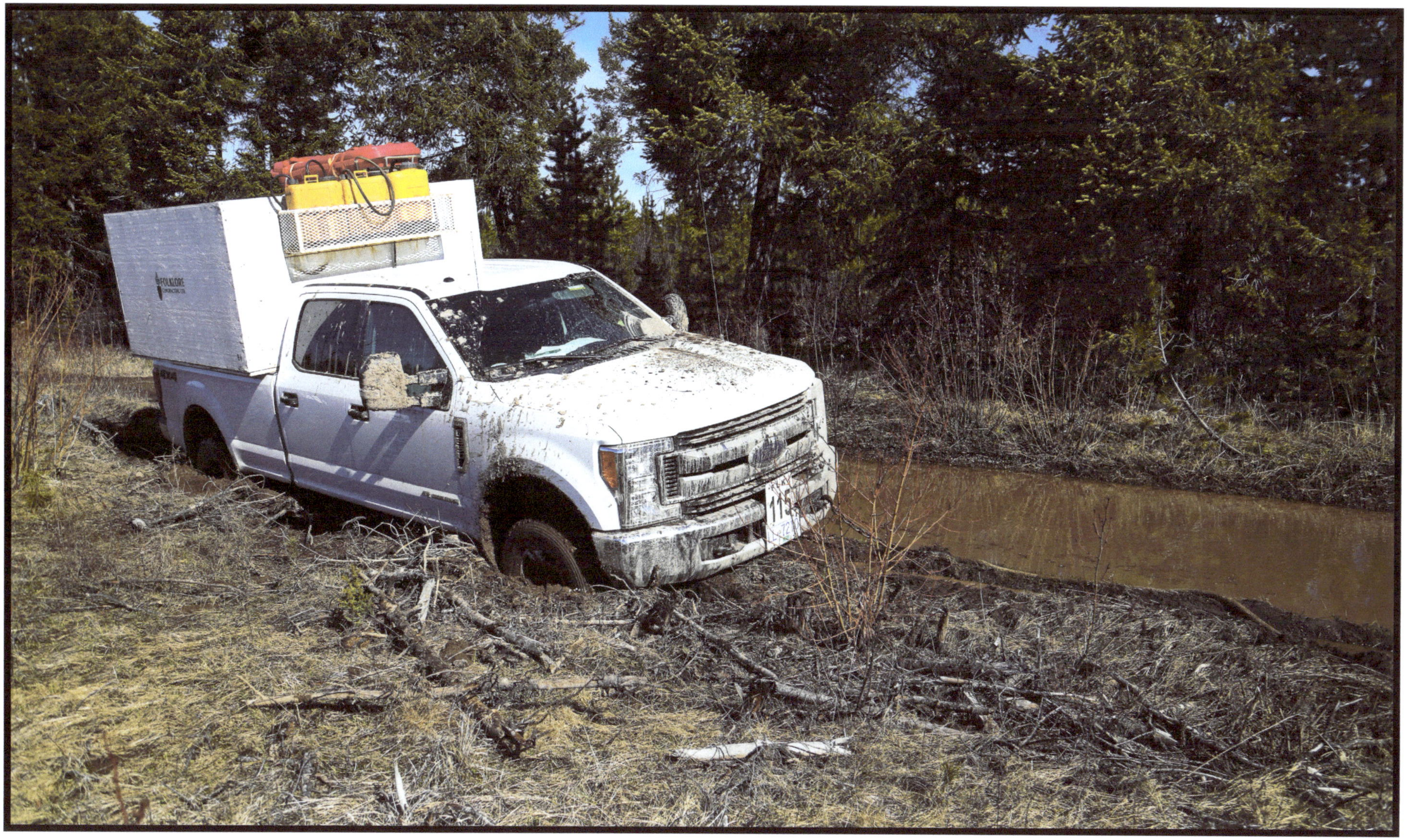

Roads? Where we're going, we don't need roads! Actually, based on this photo, it's obvious that roads can be quite helpful. Sometimes we waste a lot of time dealing with stuck trucks.

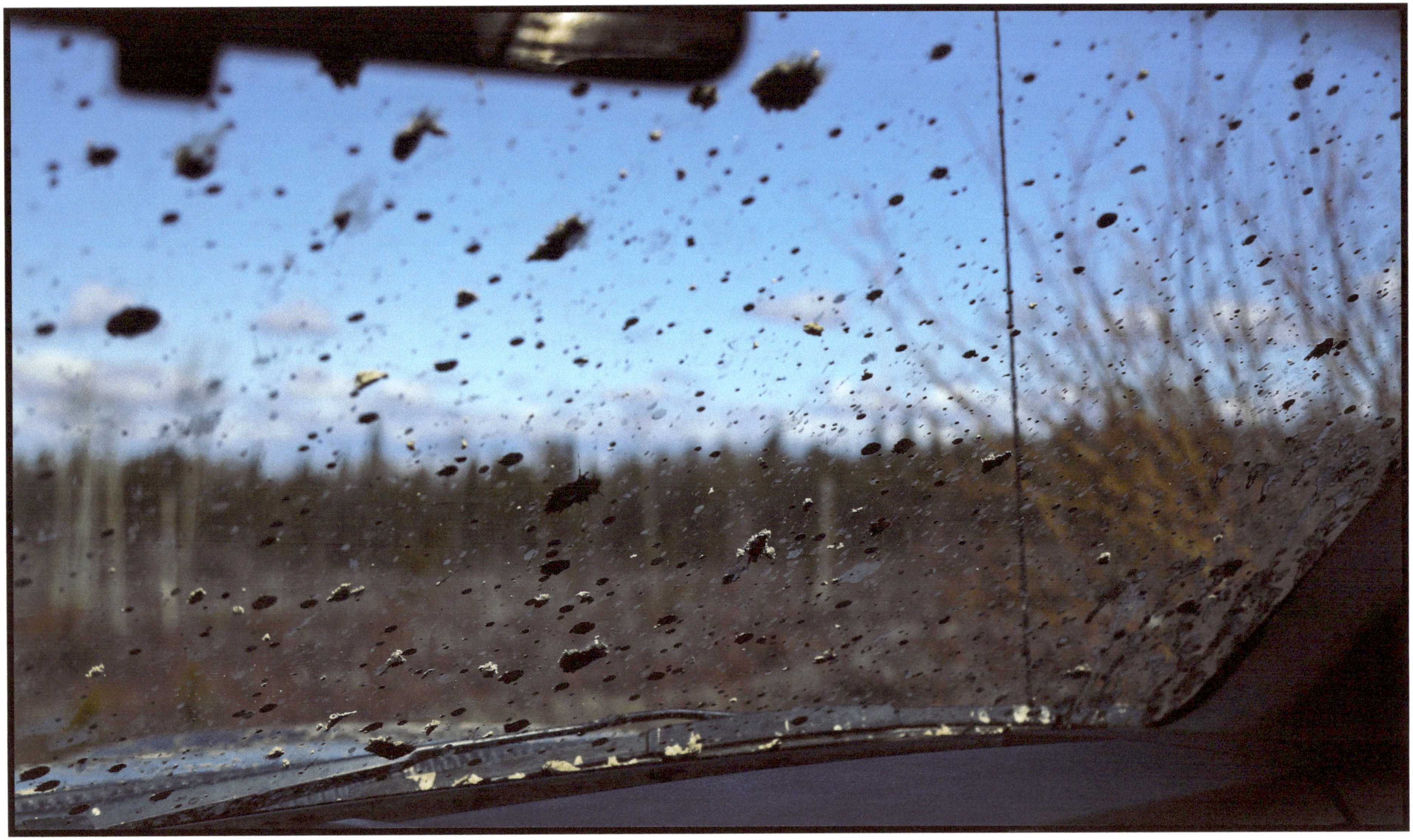

On a positive note, we're proud supporters of the windshield washer fluid industry.

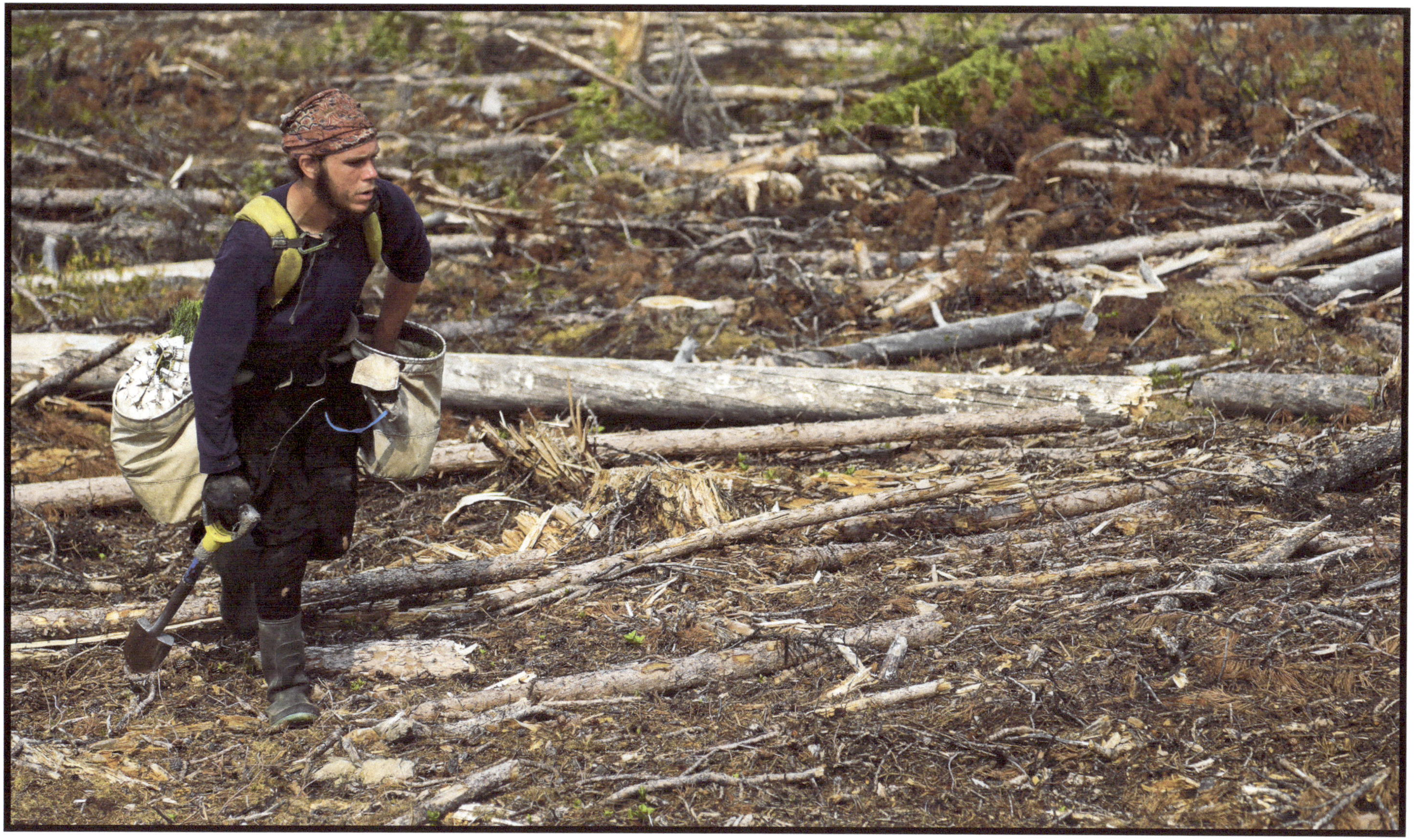

This is Aidan. In the season that this photo was taken, Aidan was the top planter in our camp. He planted slightly over 200,000 trees during that season. That season lasted for three calendar months (May/June/July) and there were about sixty-six full planting days during this time.

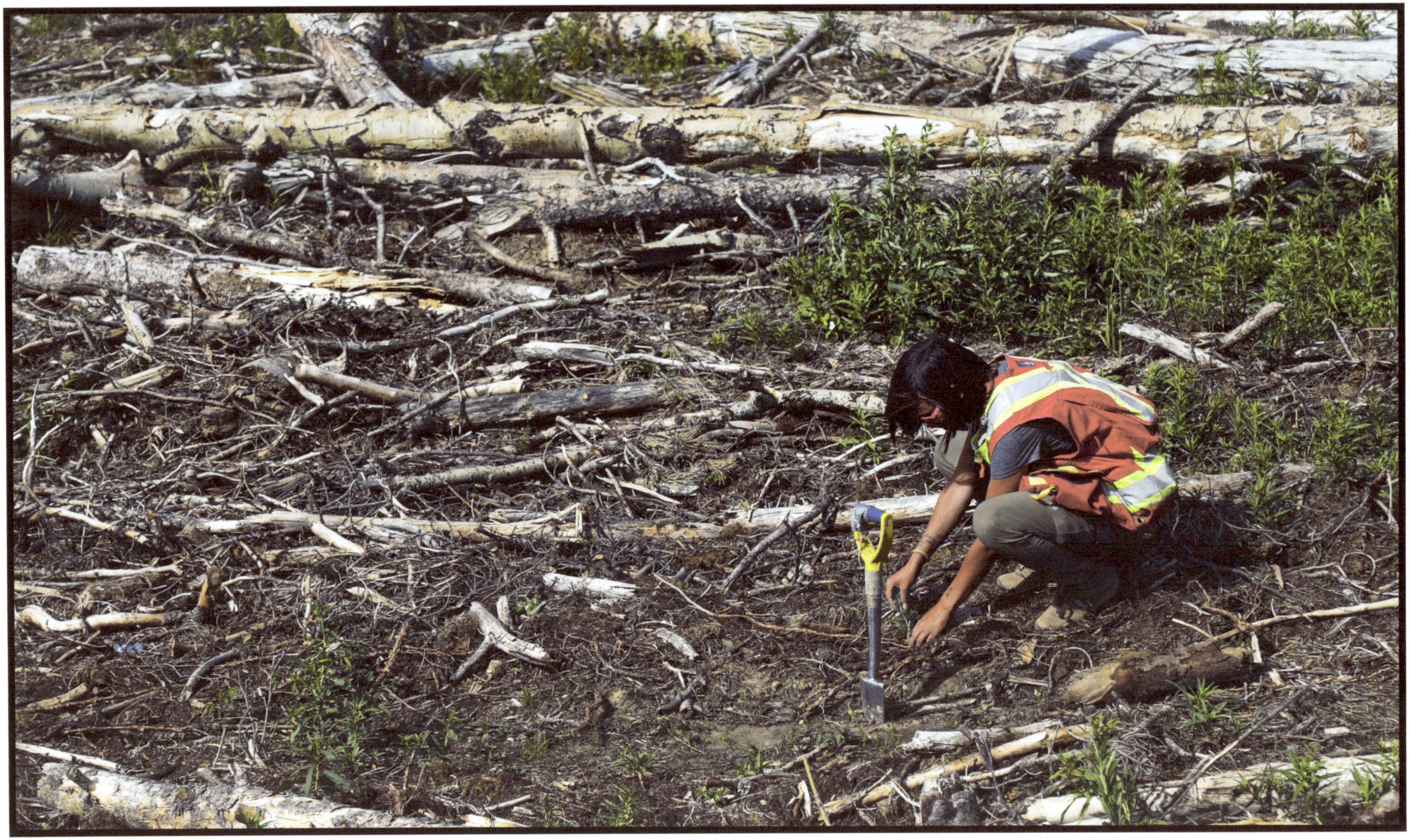

A lot of people can learn to plant several thousand trees per day. However, the trick is to plant them according to the quality standards that the foresters ask for. Here we see a "checker" who is hired by the local forester to ensure that the planters are meeting the required quality and density specifications.

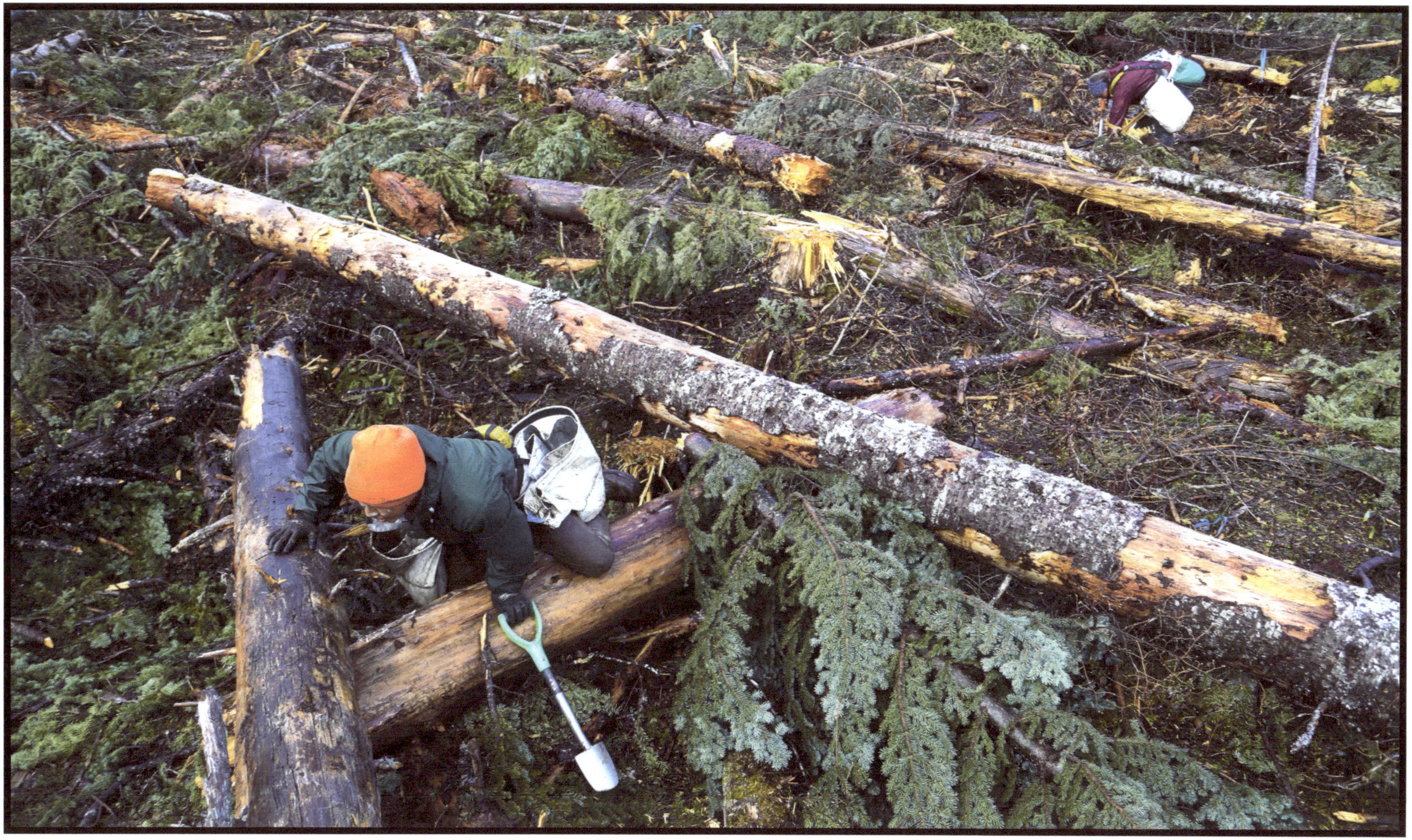

A lot of first-time planters assume that they'll be working in farmers' fields. However, many cut-blocks are covered with slash, which is a mix of garbage wood and branches. It seems inconceivable that nice logs like these would be left behind, but some of them aren't valuable enough to justify the high cost of being trucked to a local mill. On a positive note, the slash will eventually decompose and act as fertilizer for the future forest.

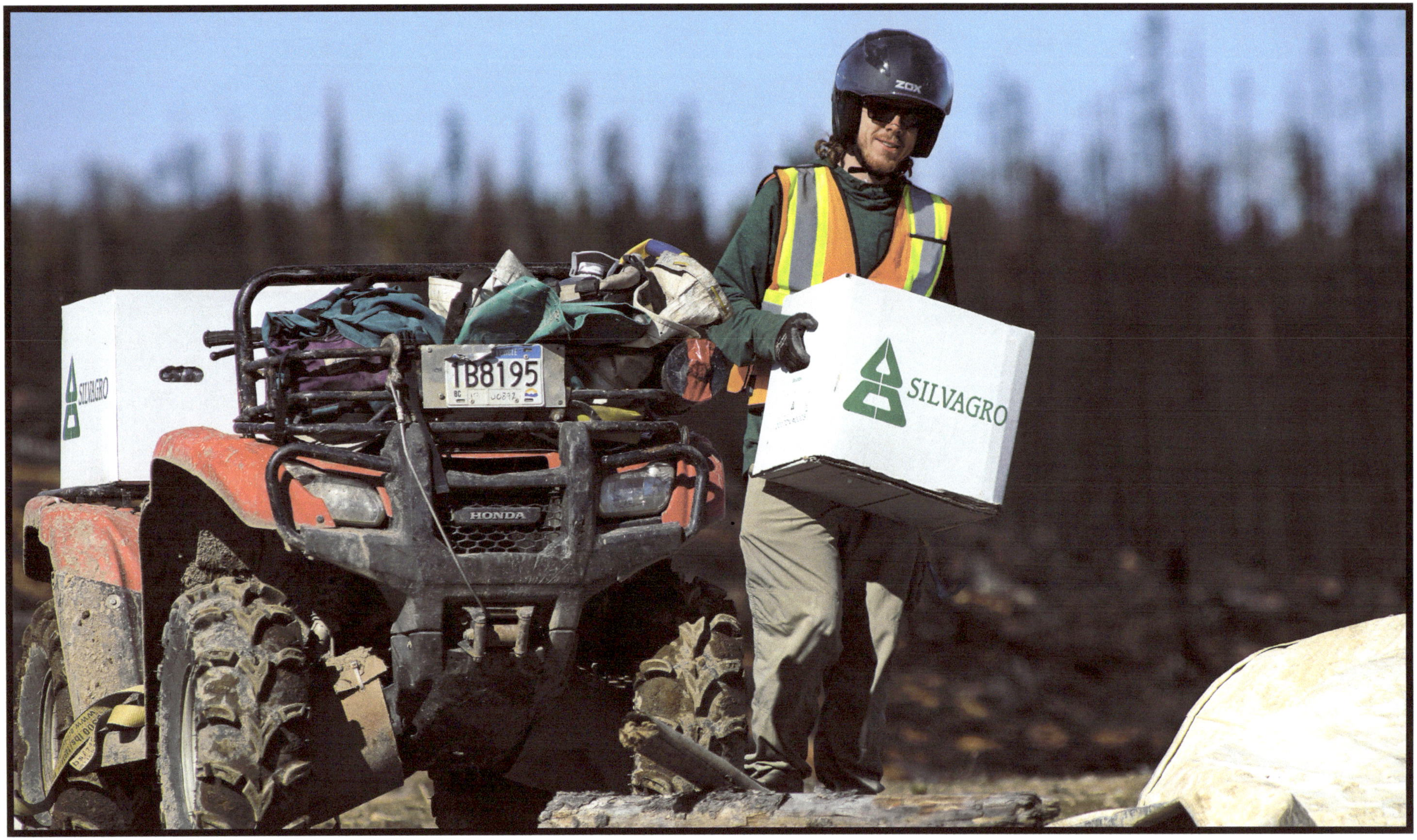

Many cut-blocks don't have roads that are suitable for truck traffic. It's common for field staff to use ATV's to deliver trees to otherwise inaccessible parts of the block. These ATV's are often referred to as "quads" because they have four wheels. Quads are light enough that they can usually travel effectively on the temporary roads or skidder trails in the blocks, although it is rare that they can travel directly across harvested ground without skidder trails.

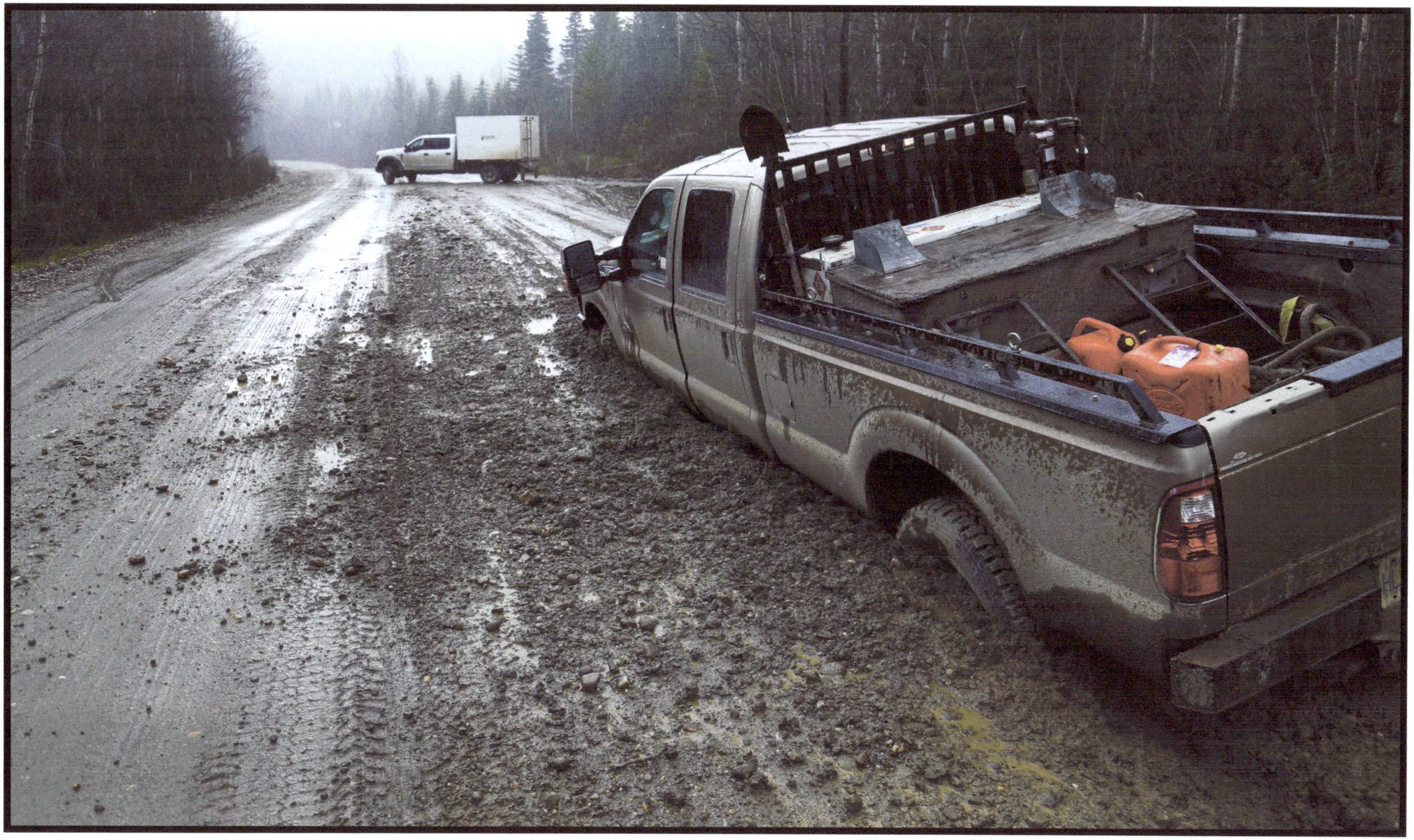

One day, one of my crew leaders (Jess) borrowed my truck for half an hour. I didn't see her again for four hours. She finally came running up to me on the block, out of breath, and said, "the road ate your truck." She had just jogged almost ten kilometers, uphill. When she said that my truck was stuck far away on the main logging road, I thought she was joking. She wasn't.

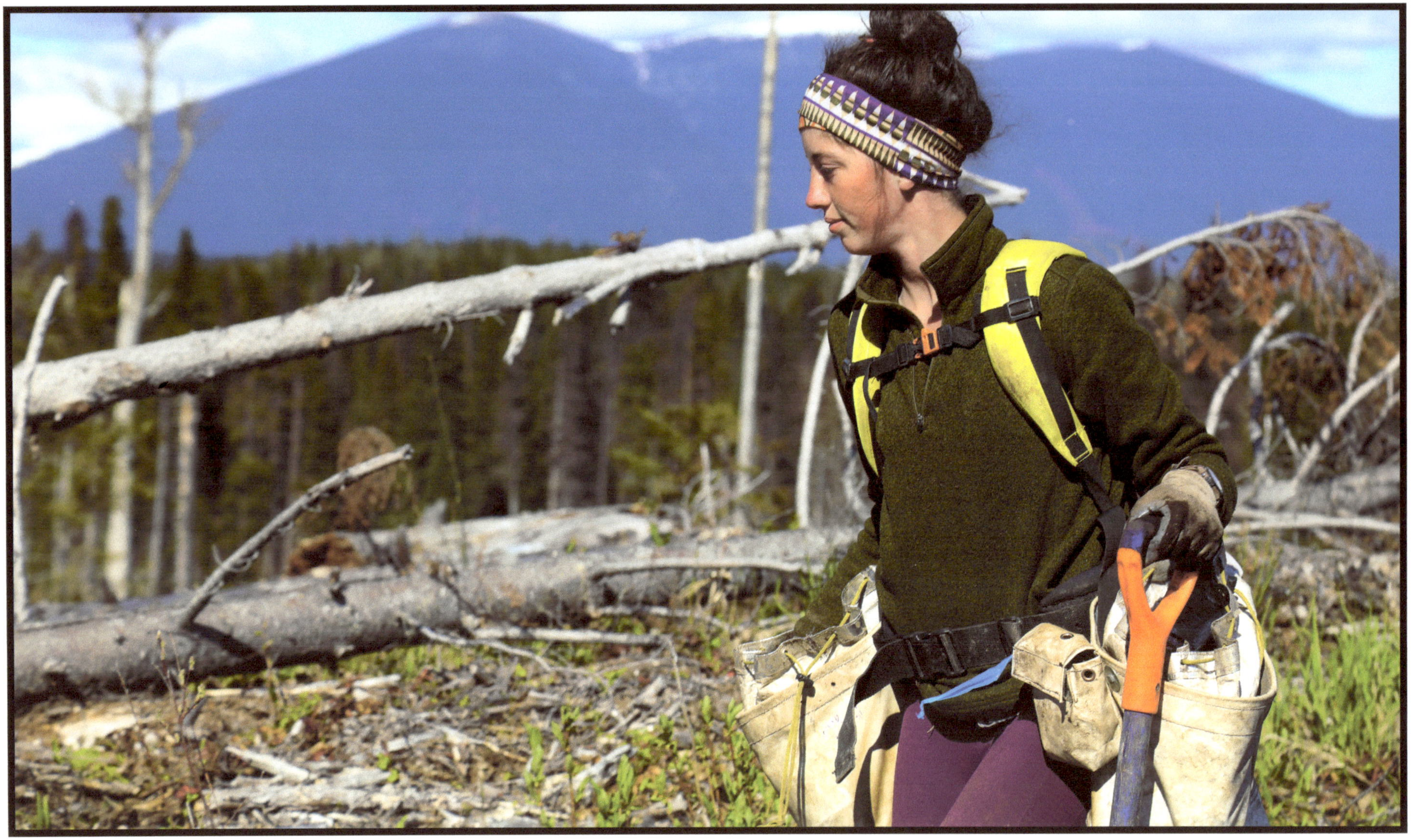

You'll notice that many of the photos in this book were taken on sunny days, in great weather. I don't like to get my camera wet. Not every day is this pleasant. We work in the rain. Sometimes we even work in light snow.

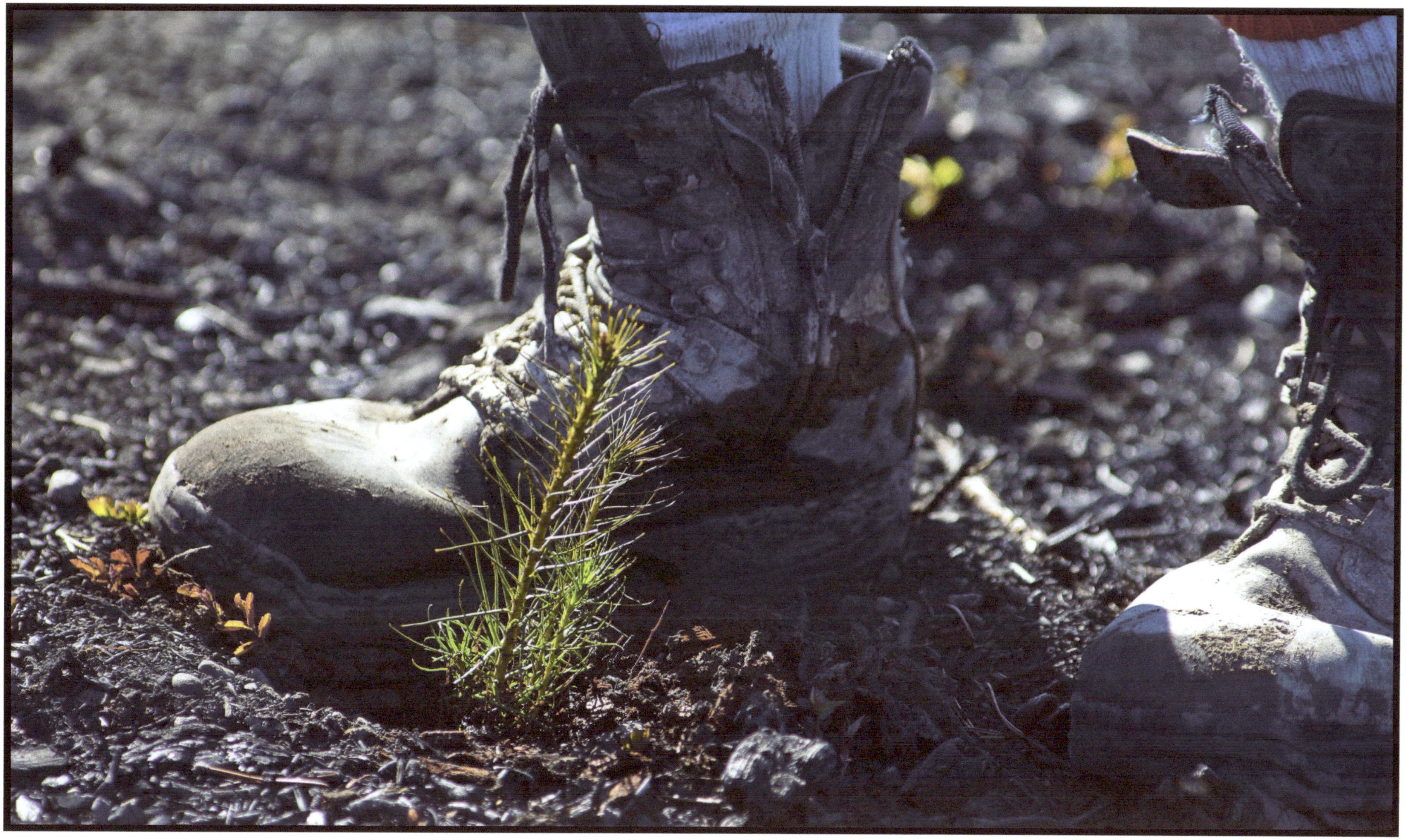

This is a baby pine tree seedling. Several types of pine trees are planted throughout Canada. This particular tree is a lodgepole pine, which is especially common in western Canada. Probably almost half of the trees planted in Canada are pine trees. Several spruce species are also planted frequently, as well as a dozen other diverse species (mostly coniferous). This is probably a typical size for a seedling planted in Canada.

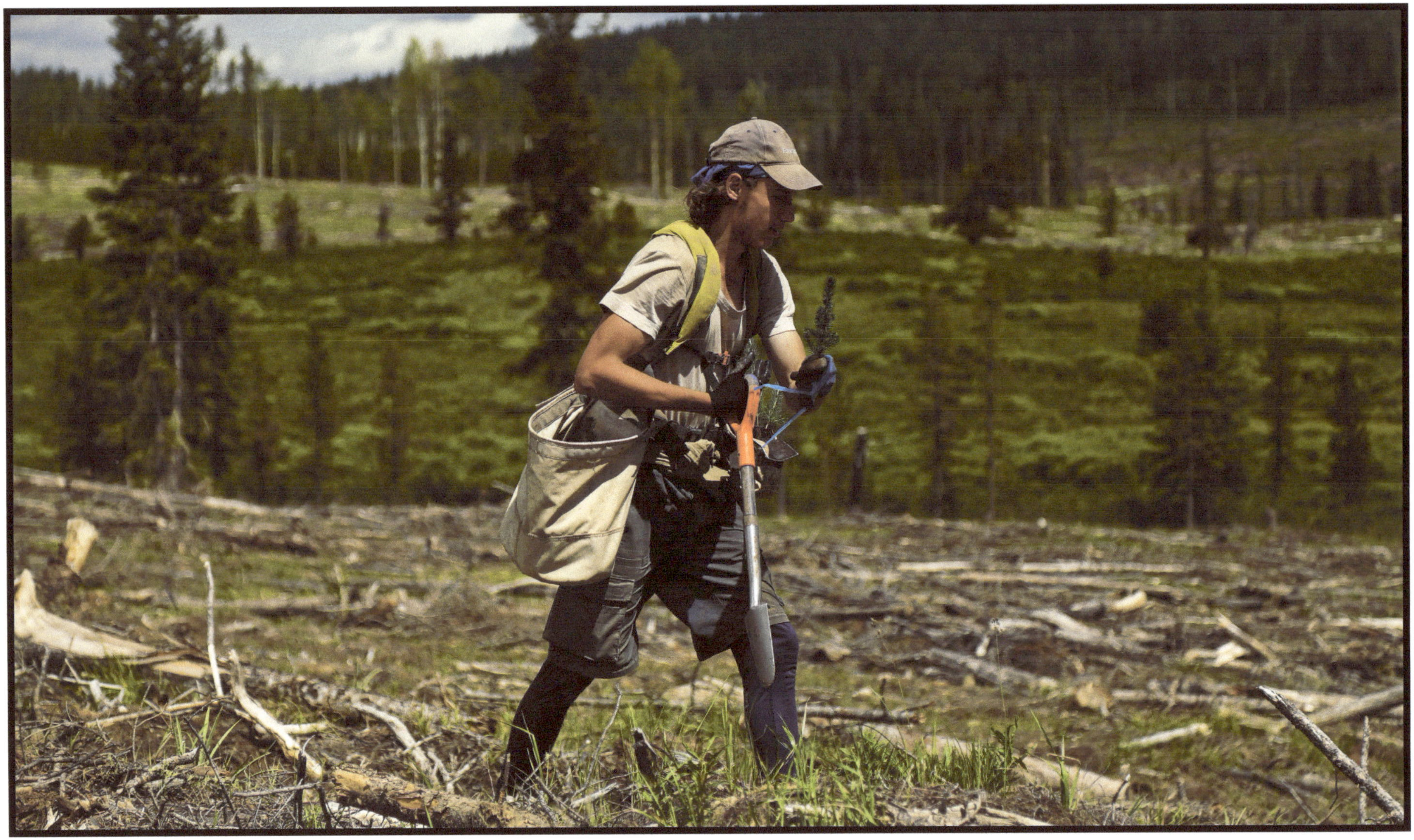

This is Cam, from Ottawa. Cam knows that in order to be successful at planting, you need to put in as much effort as possible every day. You have to work just as hard on the tough days as you do on the easy days. Cam says that he learned this wisdom from Aidan, and adds that a planter should, "Always try your hardest because working hard is a reward in itself." If you're not prepared to give 100% while planting, you're not going to succeed as a planter.

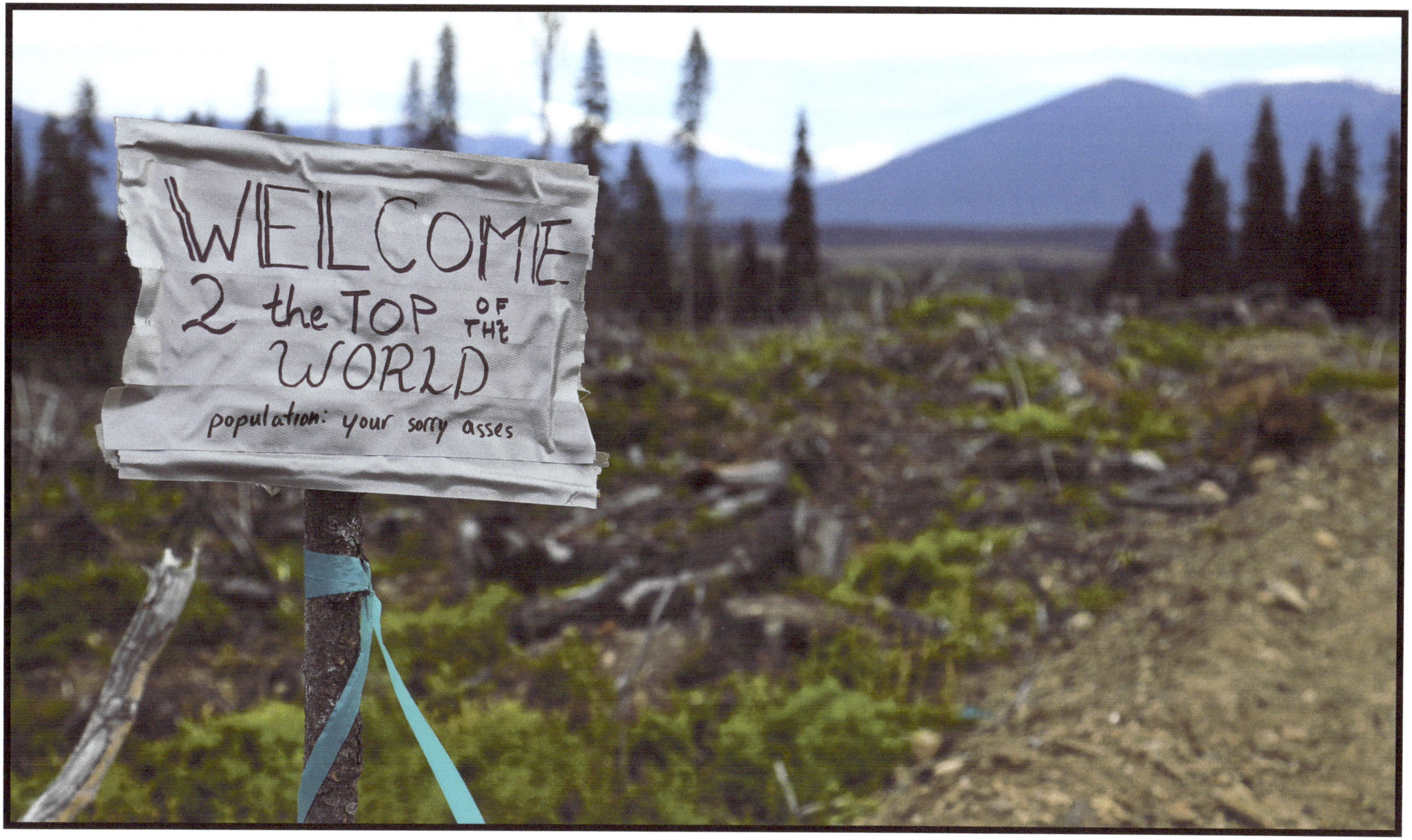

My interpretation of this sign was that one of the planters likes it better on the bottom.

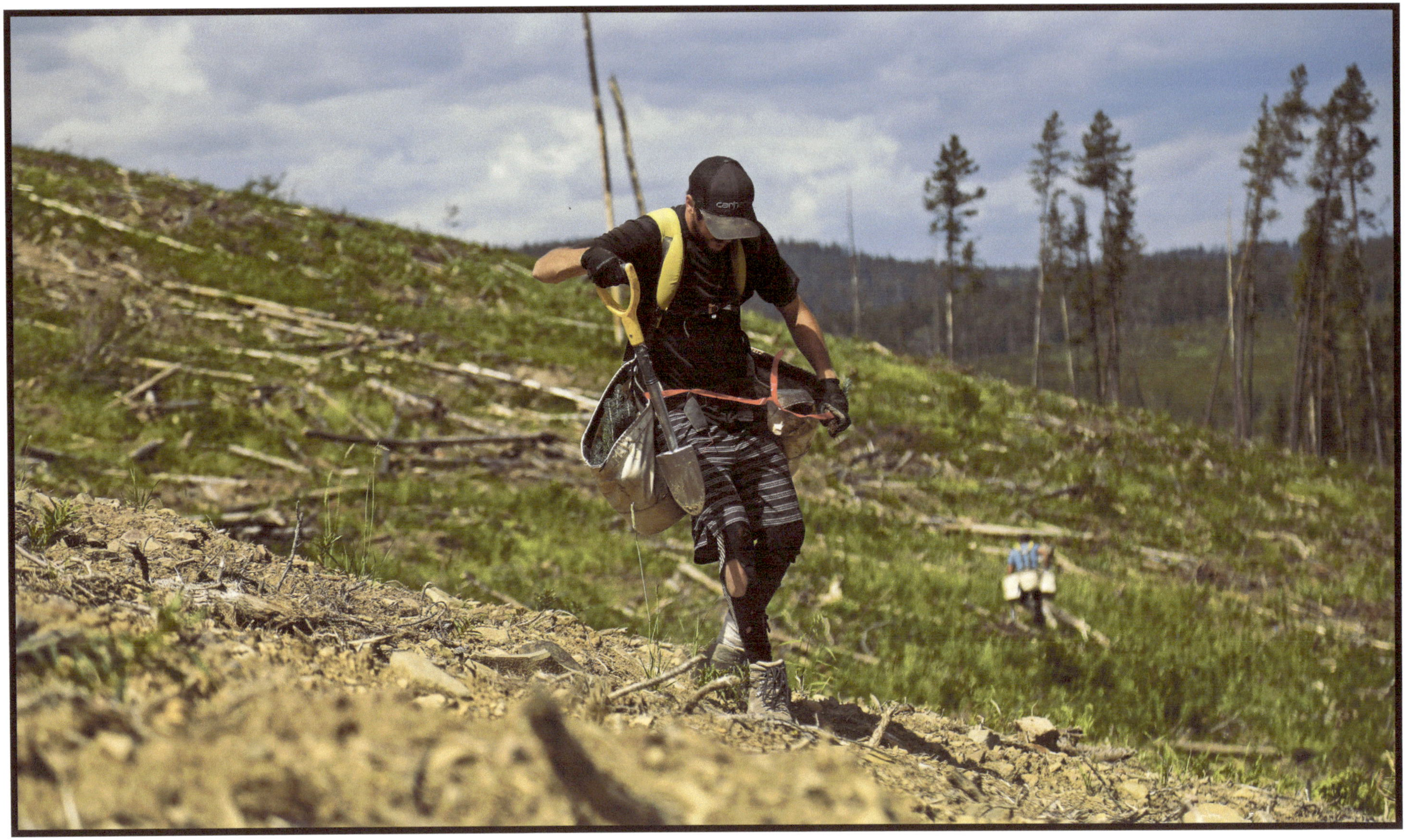

Blocks are rarely flat.

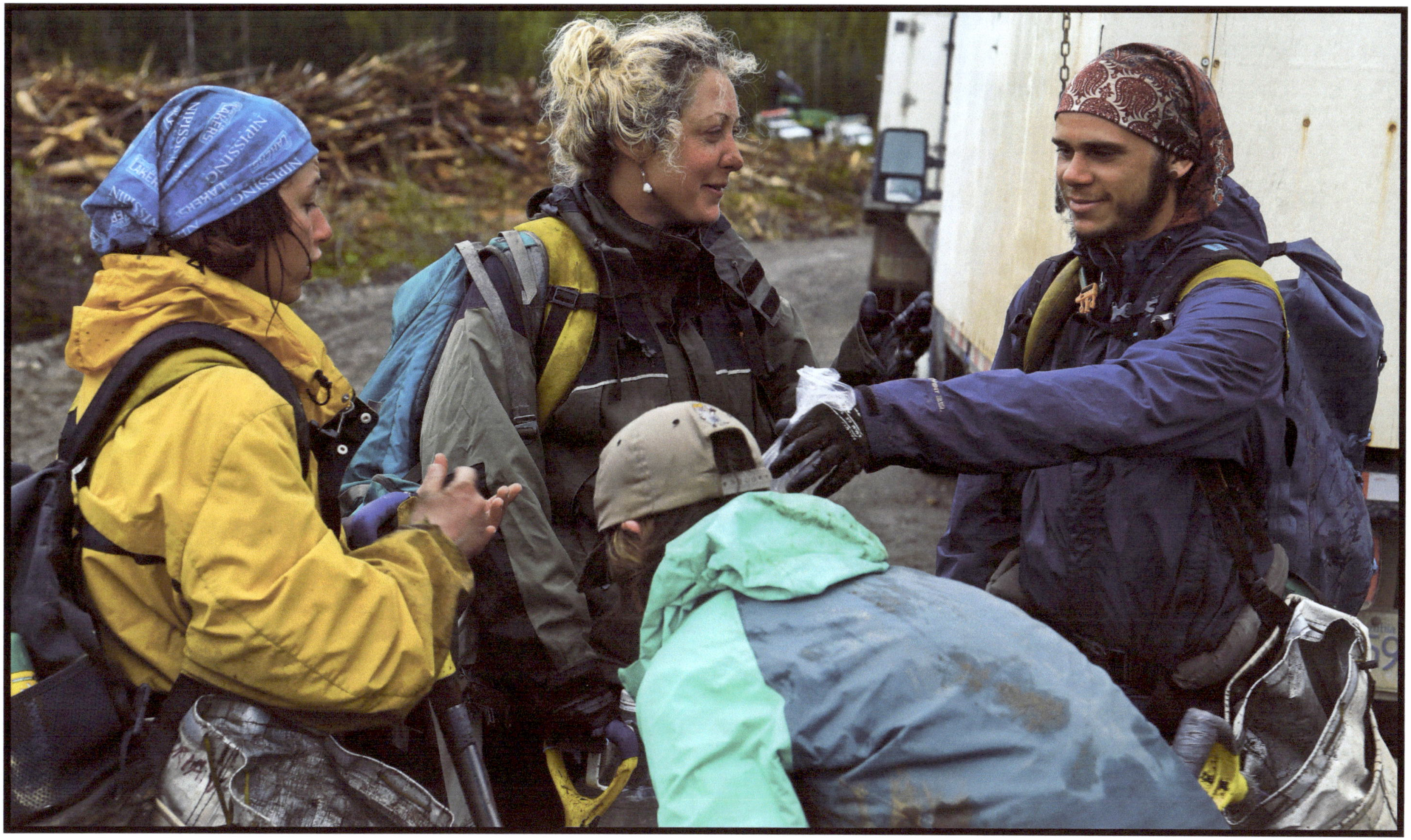

Here we see Aidan sharing a bag of gummy bears at the end of the last day of a shift, before driving "home" to camp. Tree planters typically don't pay attention to a regular seven-day weekday-based schedule. It's more common to do a 3&1 or 4&1 rotation, where the crews work for three or four days then take a day off, regardless of what day of the week it may be.

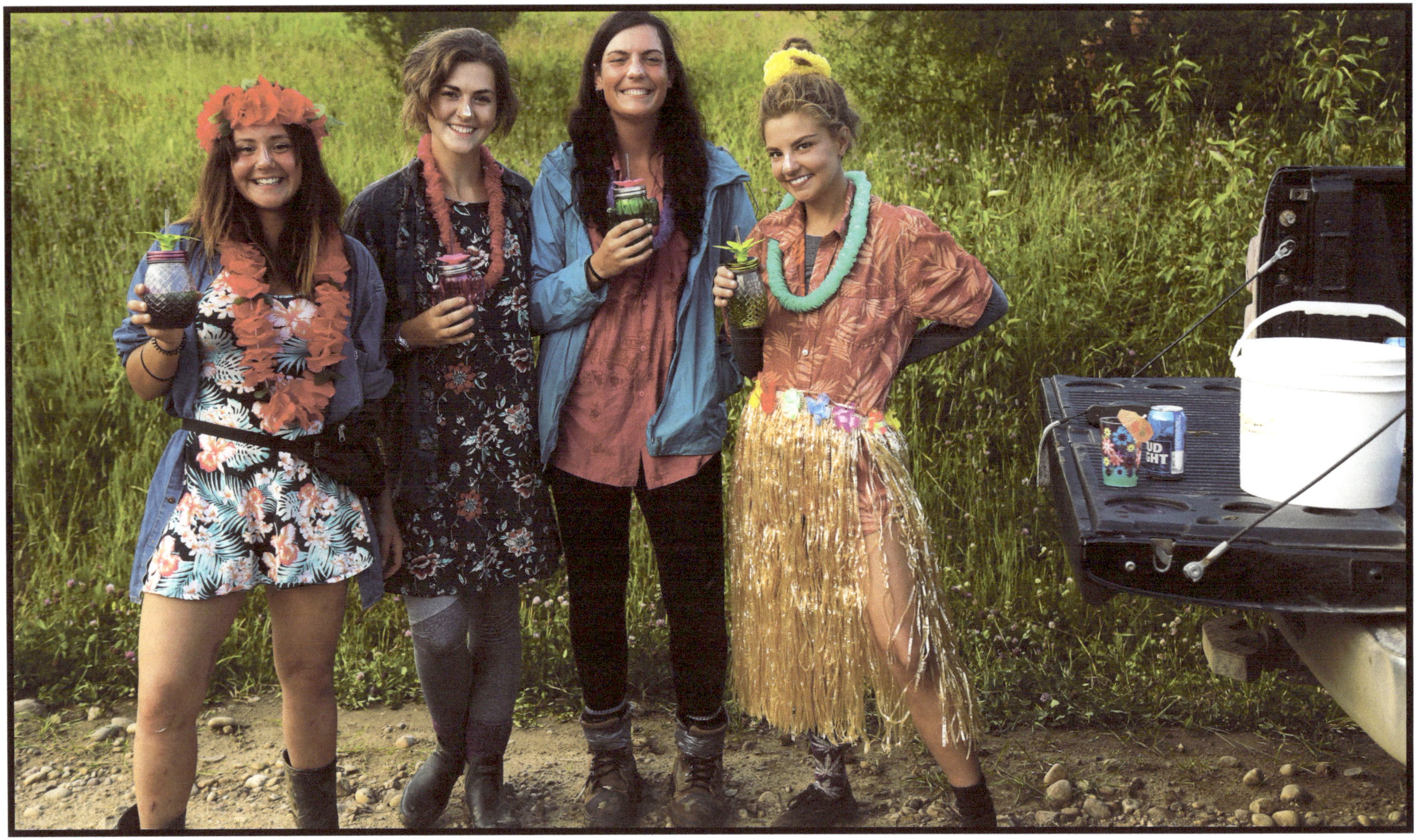

On the "night off," which is actually the night before the day off (rather than the night at the end of the day off), it's common for planters to socialize around a campfire, or to have theme parties in camp. This timing allows everyone to sleep late for a few hours on their day off, to recover.

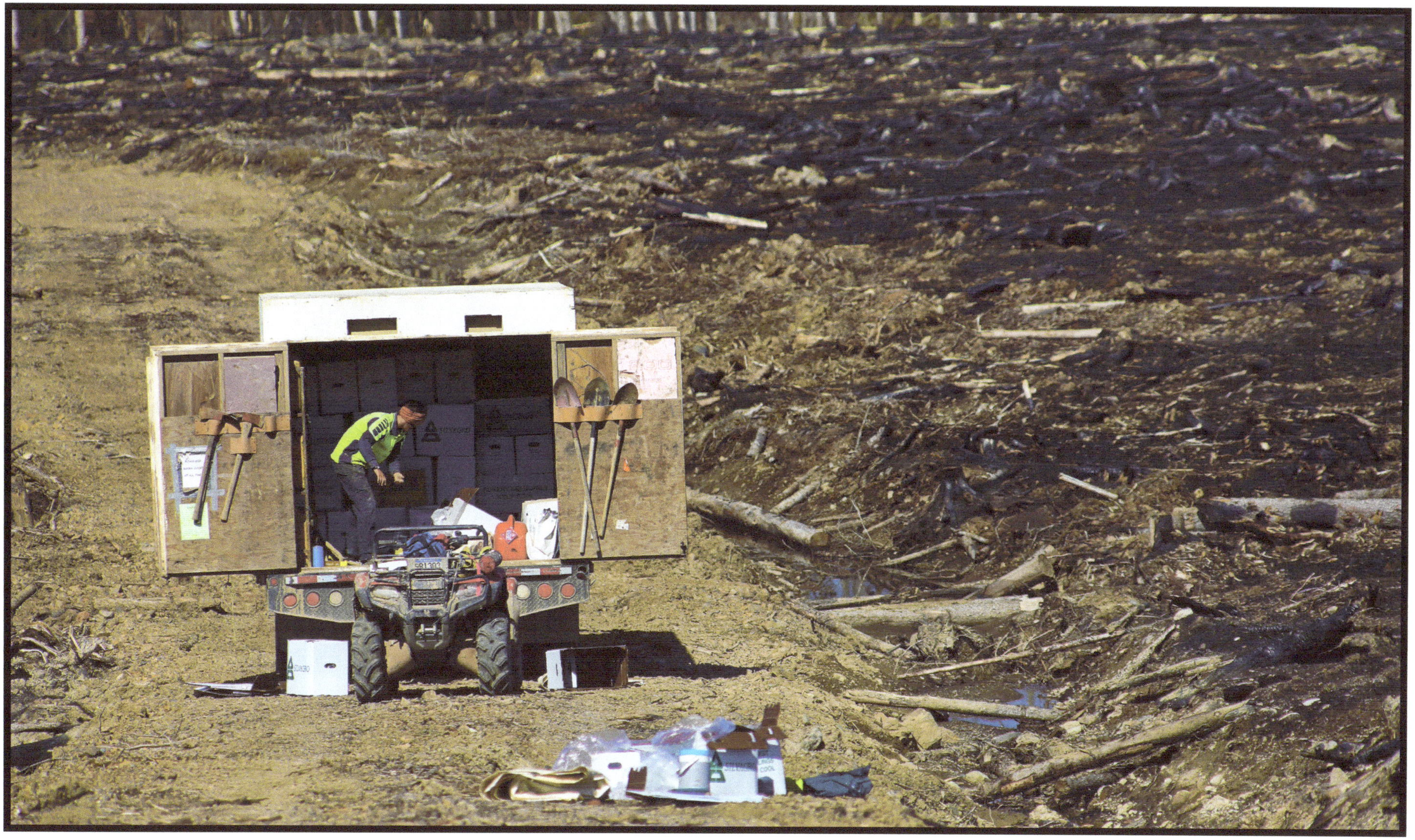

Sometimes, planters work on contracts designed to reforest areas that were burned by wildfires, instead of just cleaning up after logging activity. It's usually easier to plant on burned blocks than on "regular" post-harvest blocks, because a lot of the slash and grass will have burned off.

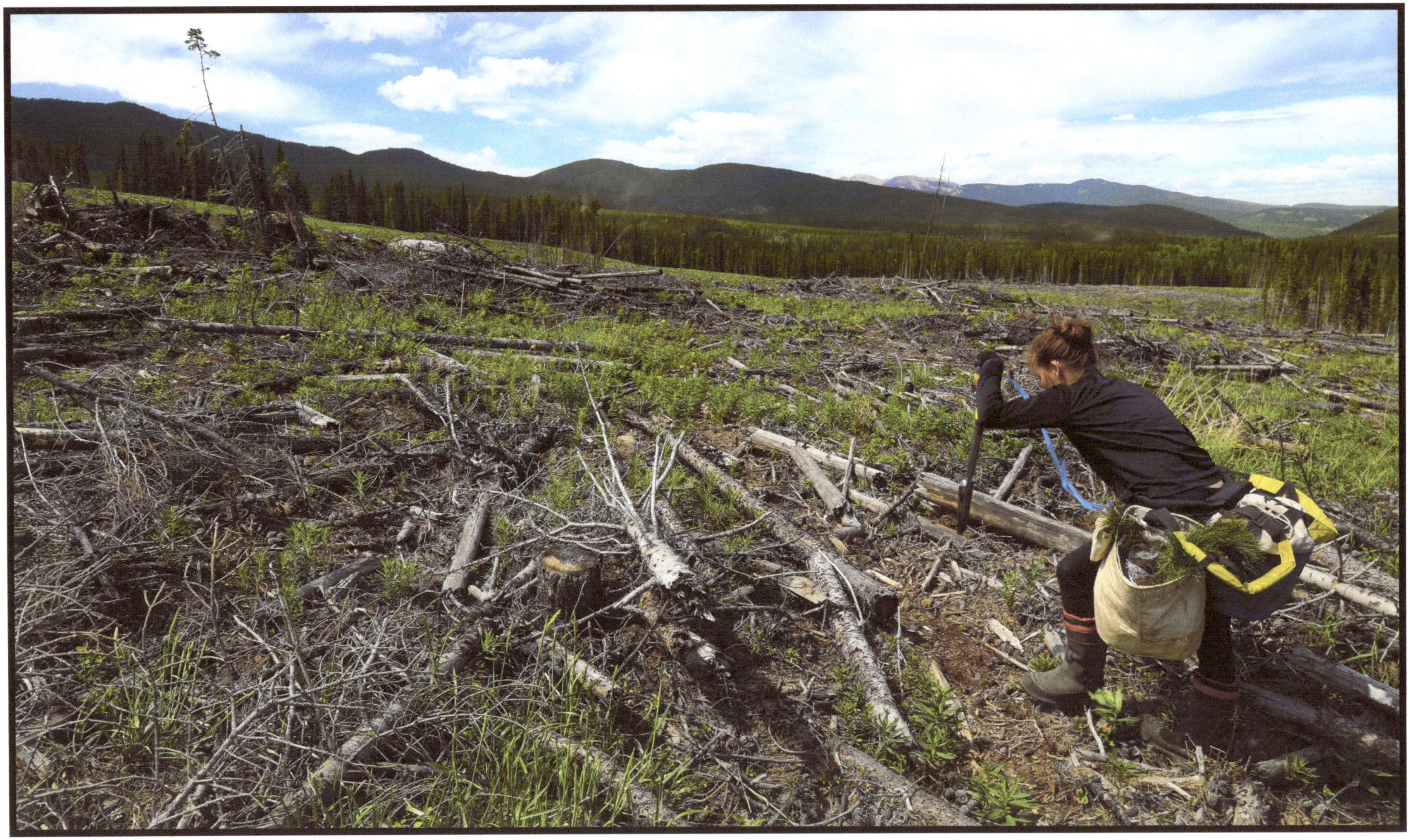

Zola is from BC, but she studies geography at the University College London (in the United Kingdom). Zola loves travelling and cooking, and in the long term, is planning a career in sustainability and environmental-related entrepreneurship.

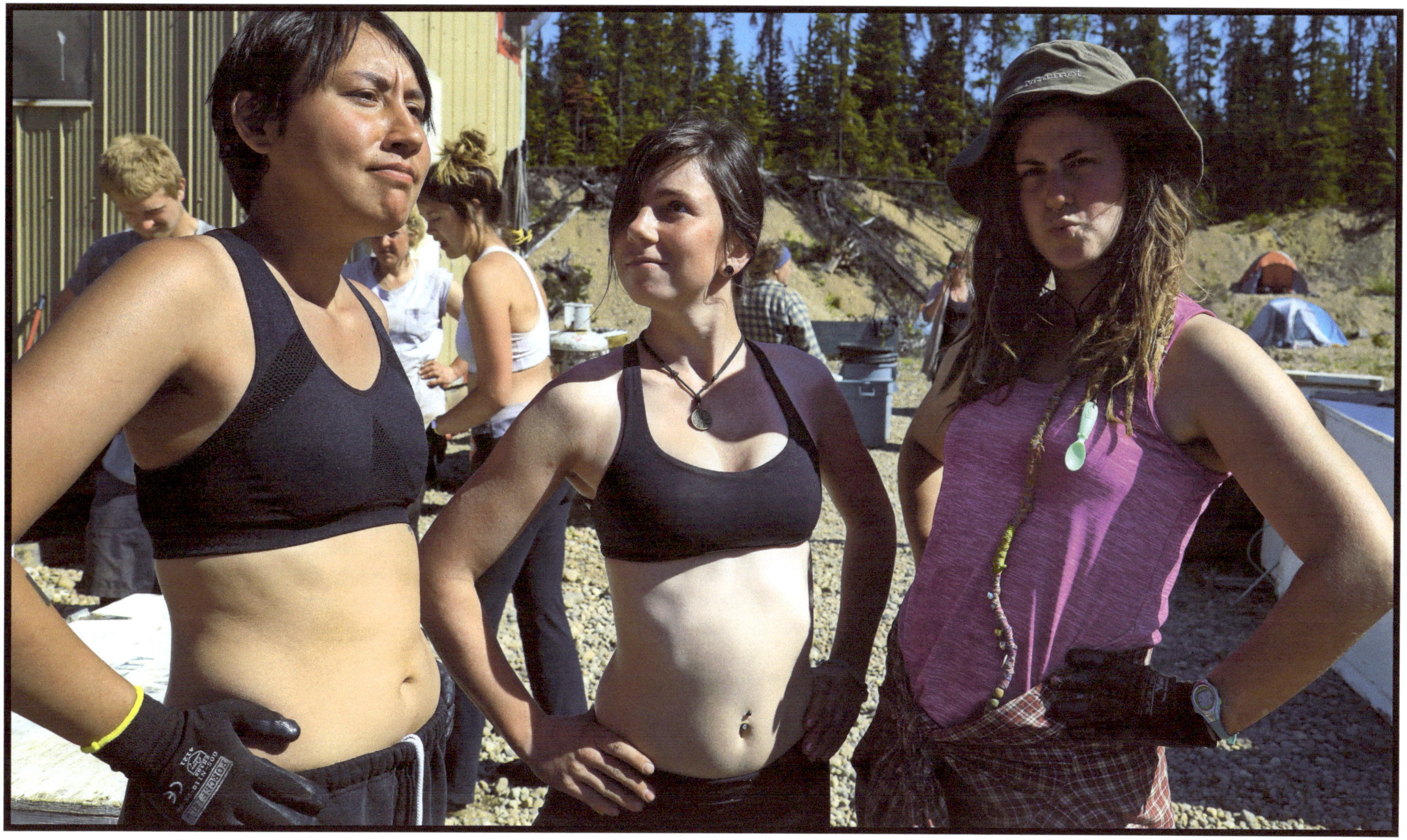

Even in early May, temperatures can sometimes get pretty warm. If you're working in a location that doesn't have a lot of mosquitoes or black flies, you might have a great opportunity to get some sunshine while you're at work. This photo was taken during a camp move, rather than on a planting day.

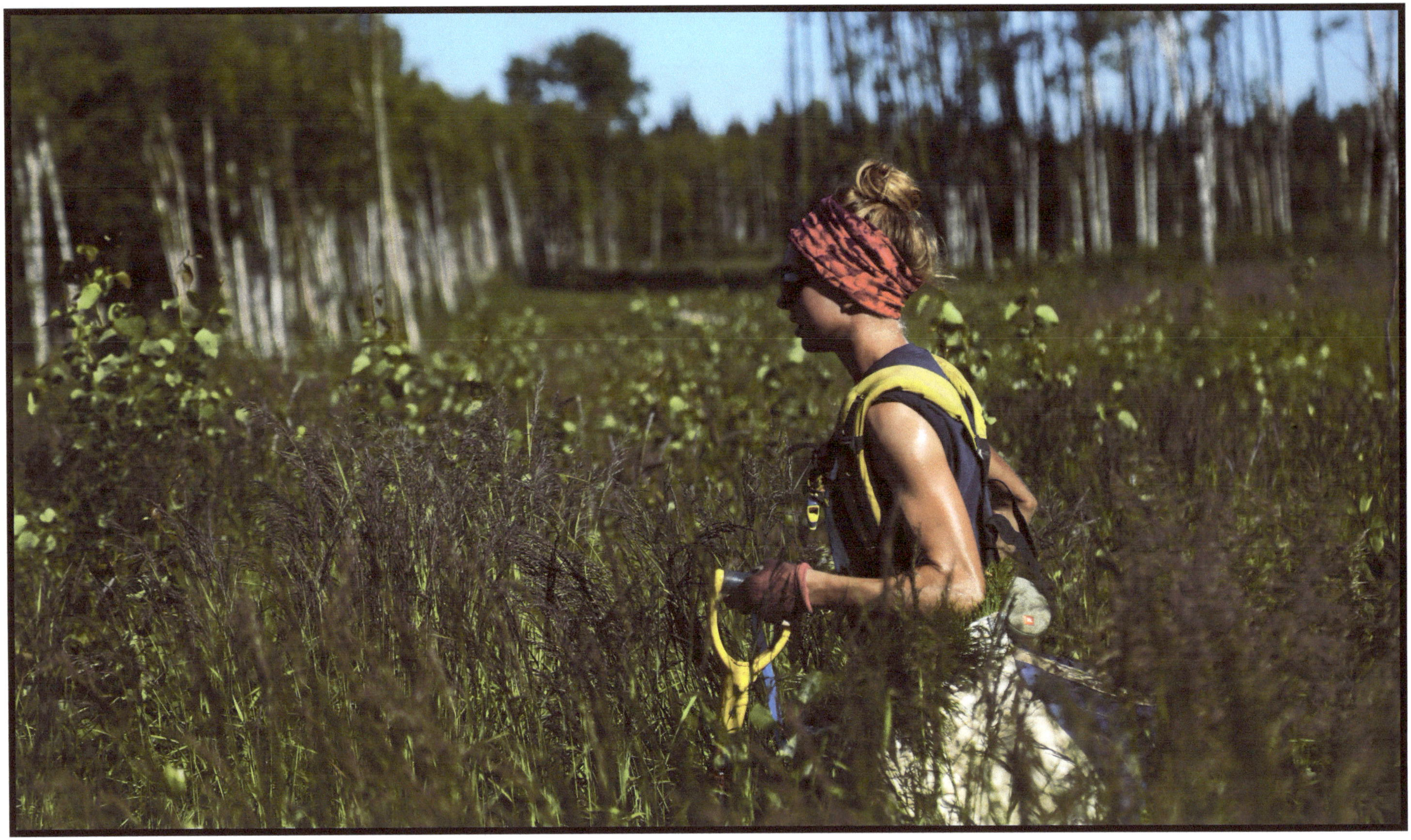

Later in the season, you may wish that there hadn't been so much sunshine for the past few months, because you may end up working on blocks that have become quite grassy. Nobody likes a grassy block.

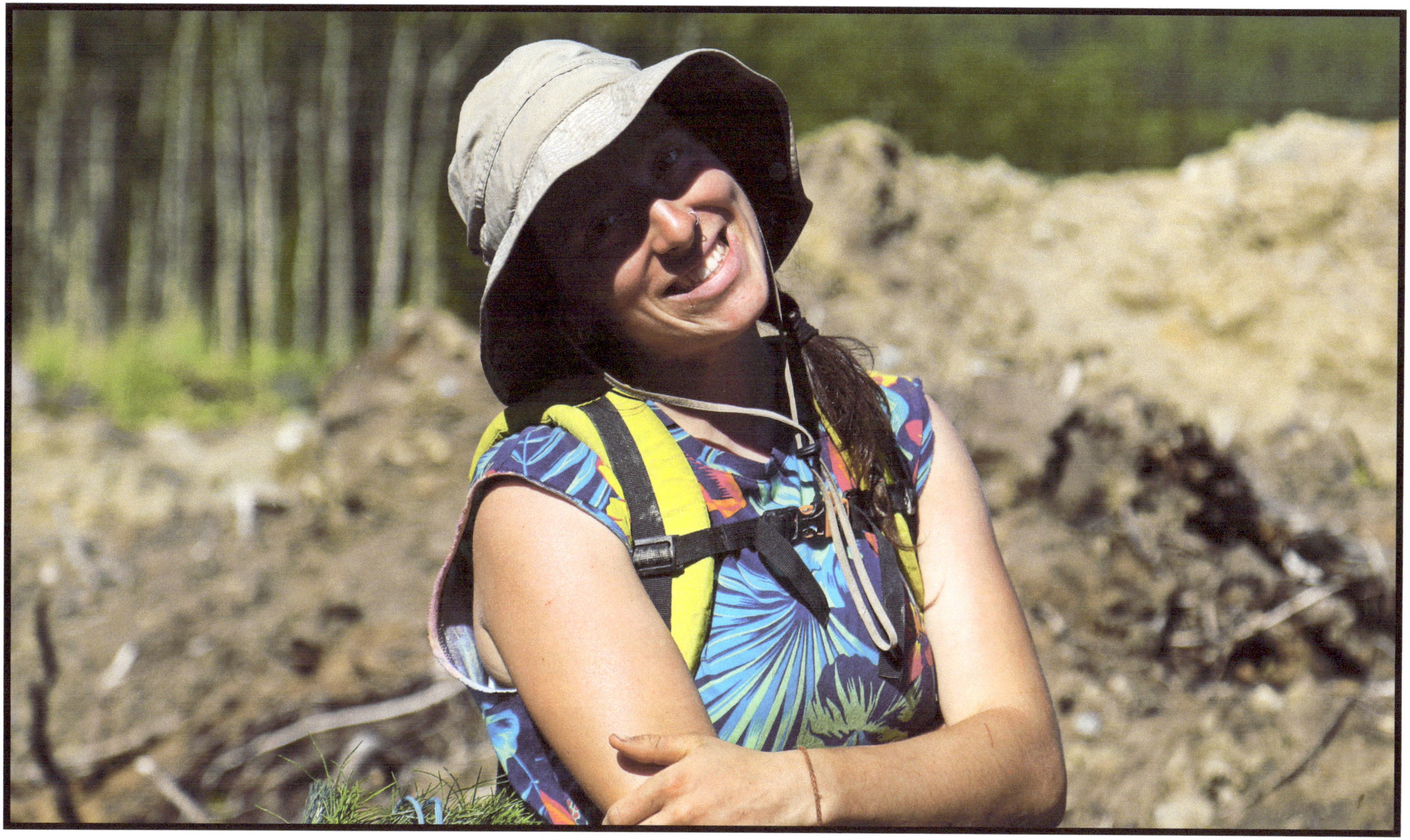

Kat is from the UK. Although the majority of planters working in Canada are Canadian citizens, you'll occasionally see a few non-Canadians working in the industry (especially from other Commonwealth countries). Kat is from Farnham, Surrey. She attended both the University of Bristol in the UK and the University of Salamanca in Spain, and has a Masters in Chemistry. Kat loves hiking, and is currently travelling around the world.

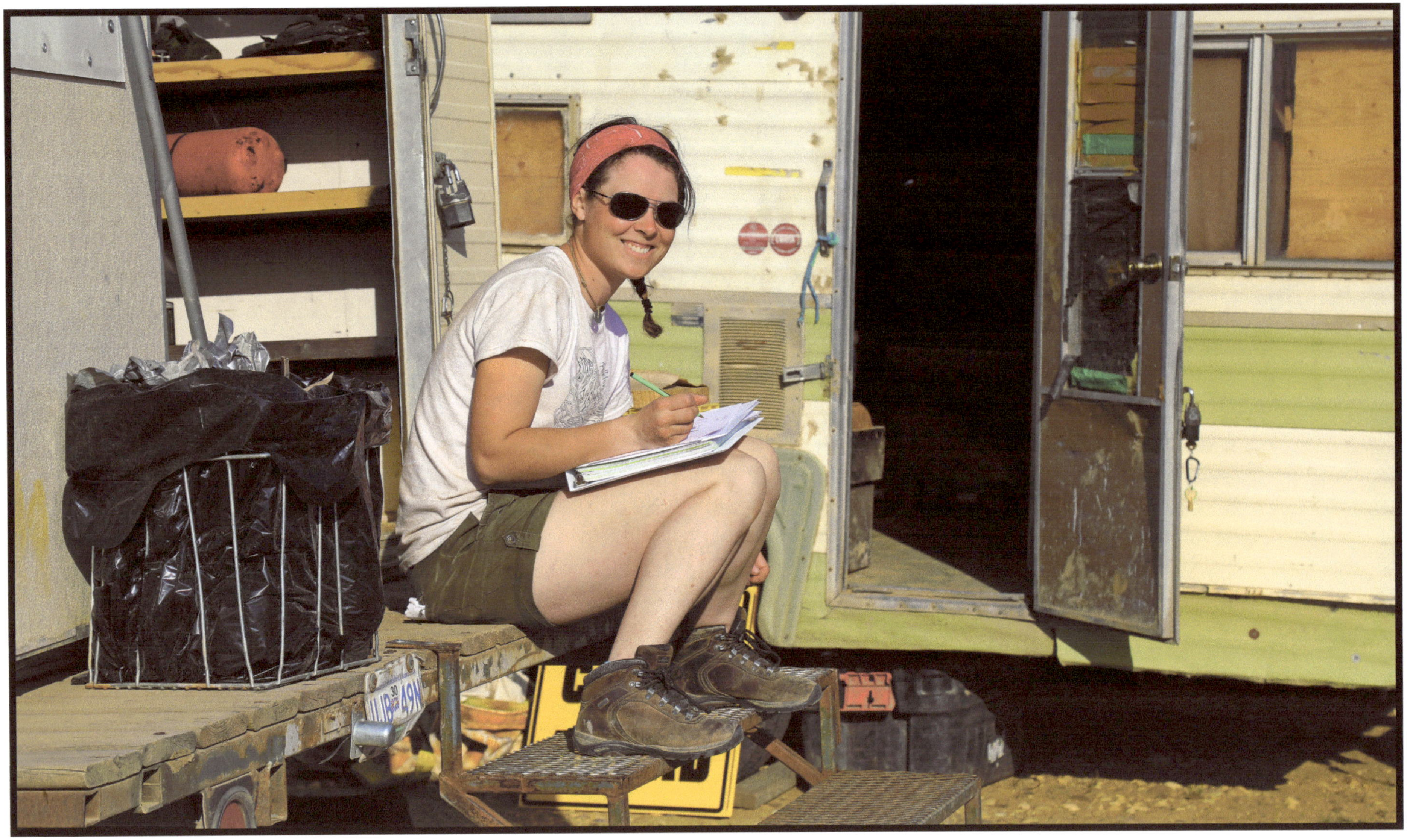

When most people think about tree planting, they only have visions of working on the blocks during the day, or living in a tent at night. However, there's a lot of paperwork and documentation that needs to happen behind the scenes on any given contract. Regular documentation and internal inspections are key components of due diligence and liability management, with respect to everything from safety to camp standards to project metrics.

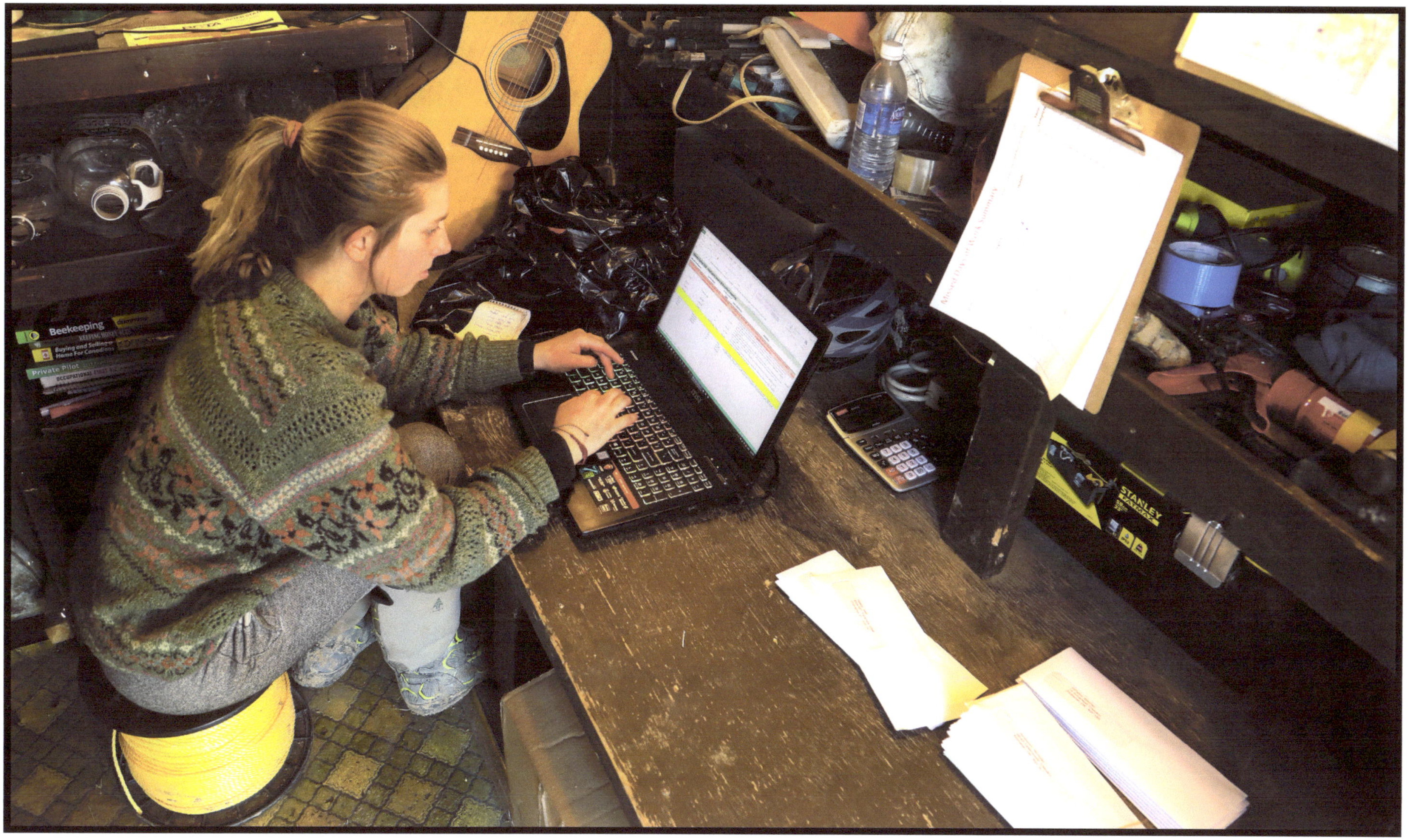

Over the past decade, the industry has becoming increasingly reliant on using technology to share data, whether it be spreadsheets about daily production, incident reports, stock monitoring reports, or dozens of other types of data that is tracked regularly. Many camps that operate in remote areas with no cell coverage will install satellite-based internet systems to allow camp management to share data with clients on a daily basis.

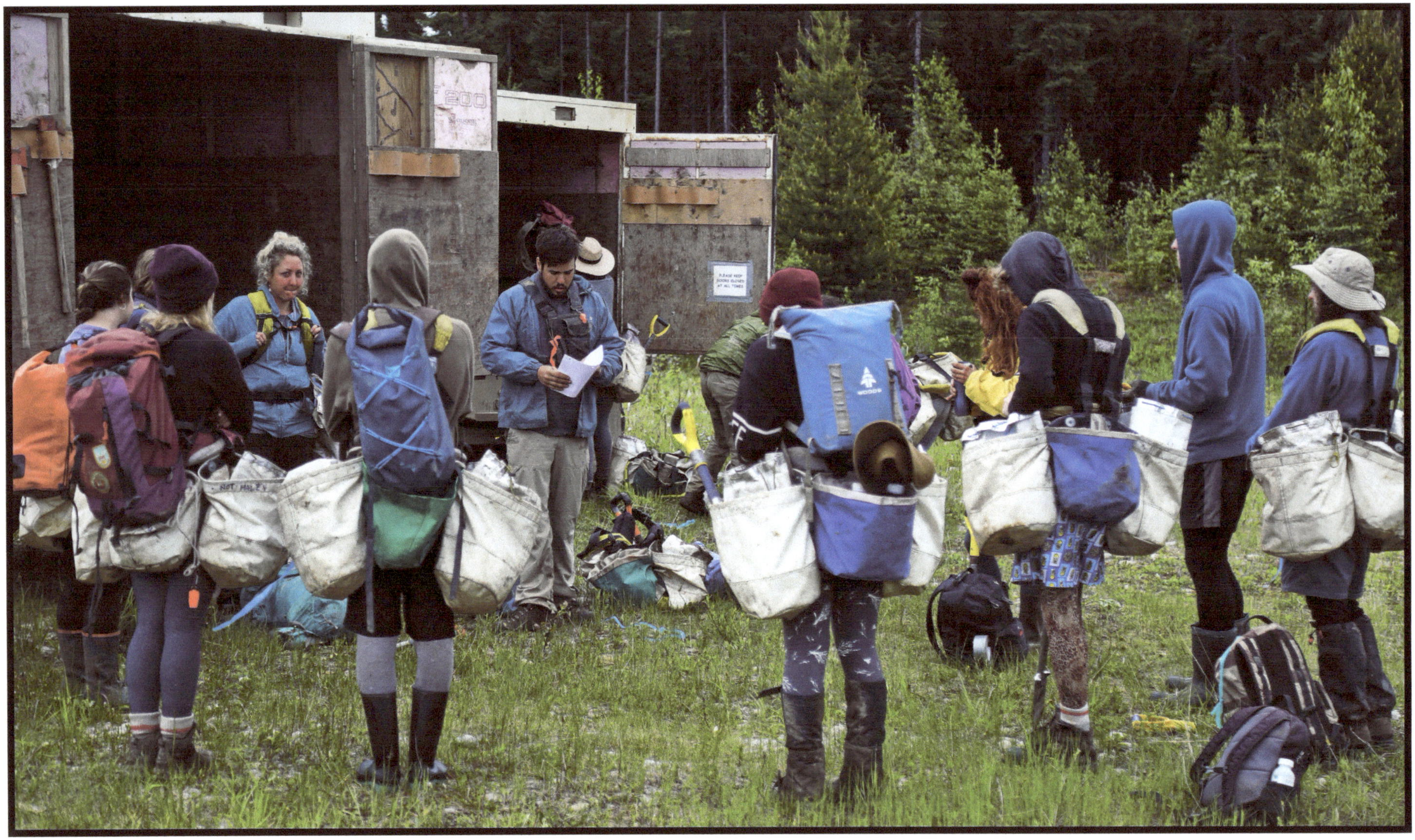

A key component of any company's safety program will be the daily tailgate meeting. Every crew should have a brief meeting in the morning to allow the foreperson and crew members to discuss safety and potential hazards, and to make sure that everyone understands the work plan for the day. This meeting usually only takes three to four minutes, but is invaluable in keeping everyone on the same page and maximizing the crew's efficiency during the day.

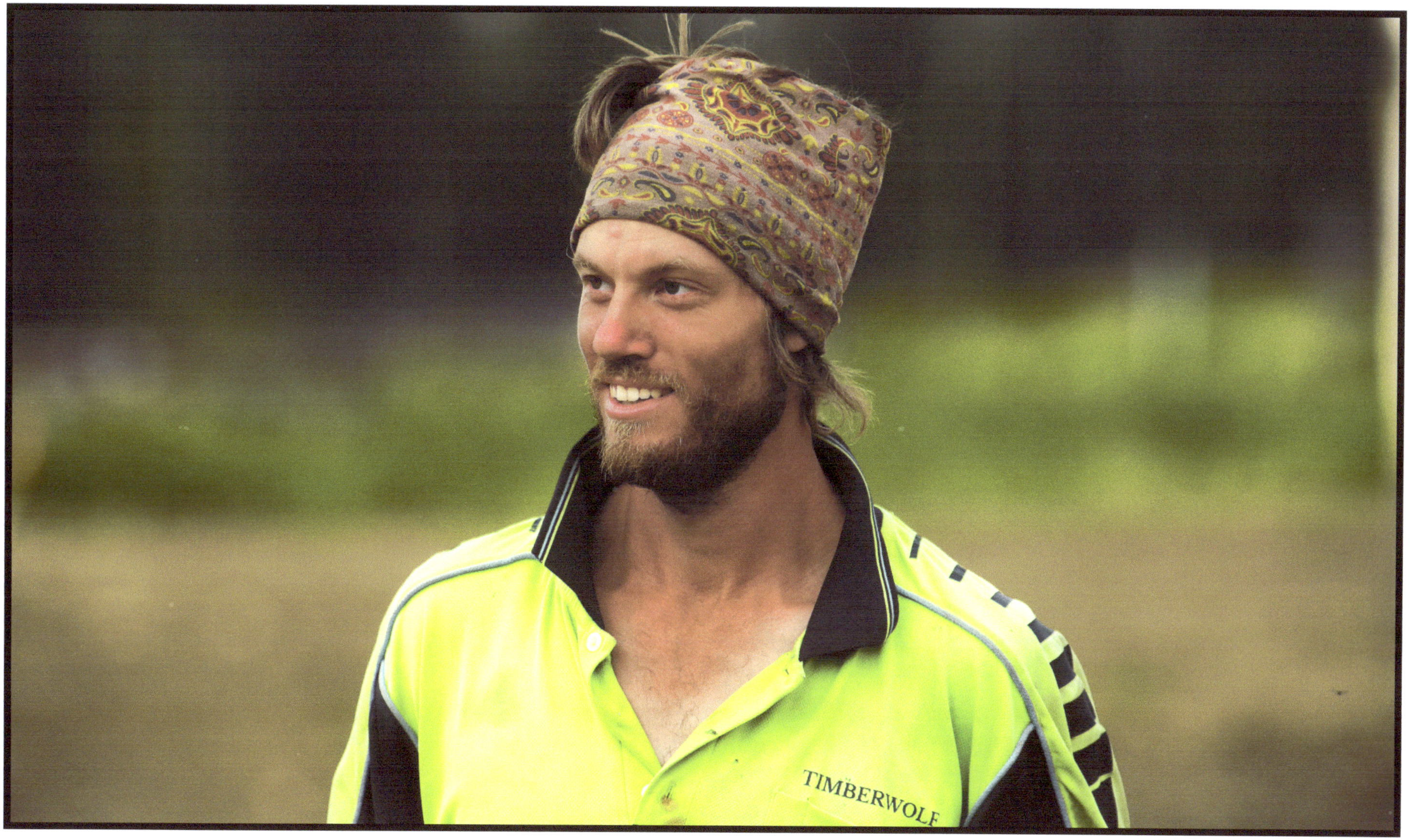

Sebastian is a crew leader. He has experience planting in both Canada and Australia, and loves touring Canada in the off-season to pursue his geocaching hobby.

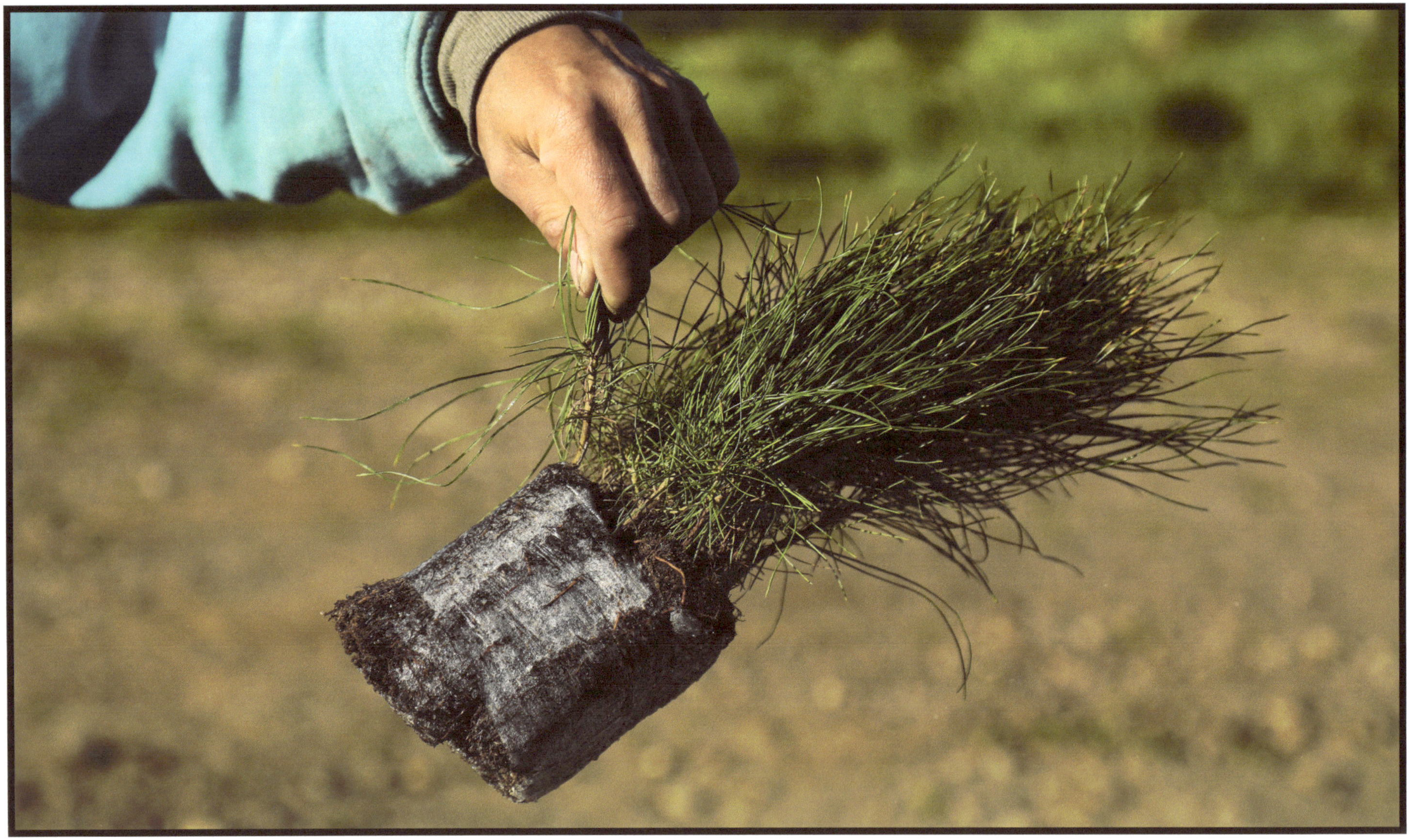

Despite all the best planning, things still go wrong on a regular basis. This shipment of trees was completely frozen when it arrived on site, due to ongoing management issues at the forest nursery in Quesnel. Although it looks like this bundle of trees is held together by a plastic bundle wrapper, the wrapper has been removed to show solid ice. It took several days for this shipment to thaw out before the crew was able to plant the seedlings.

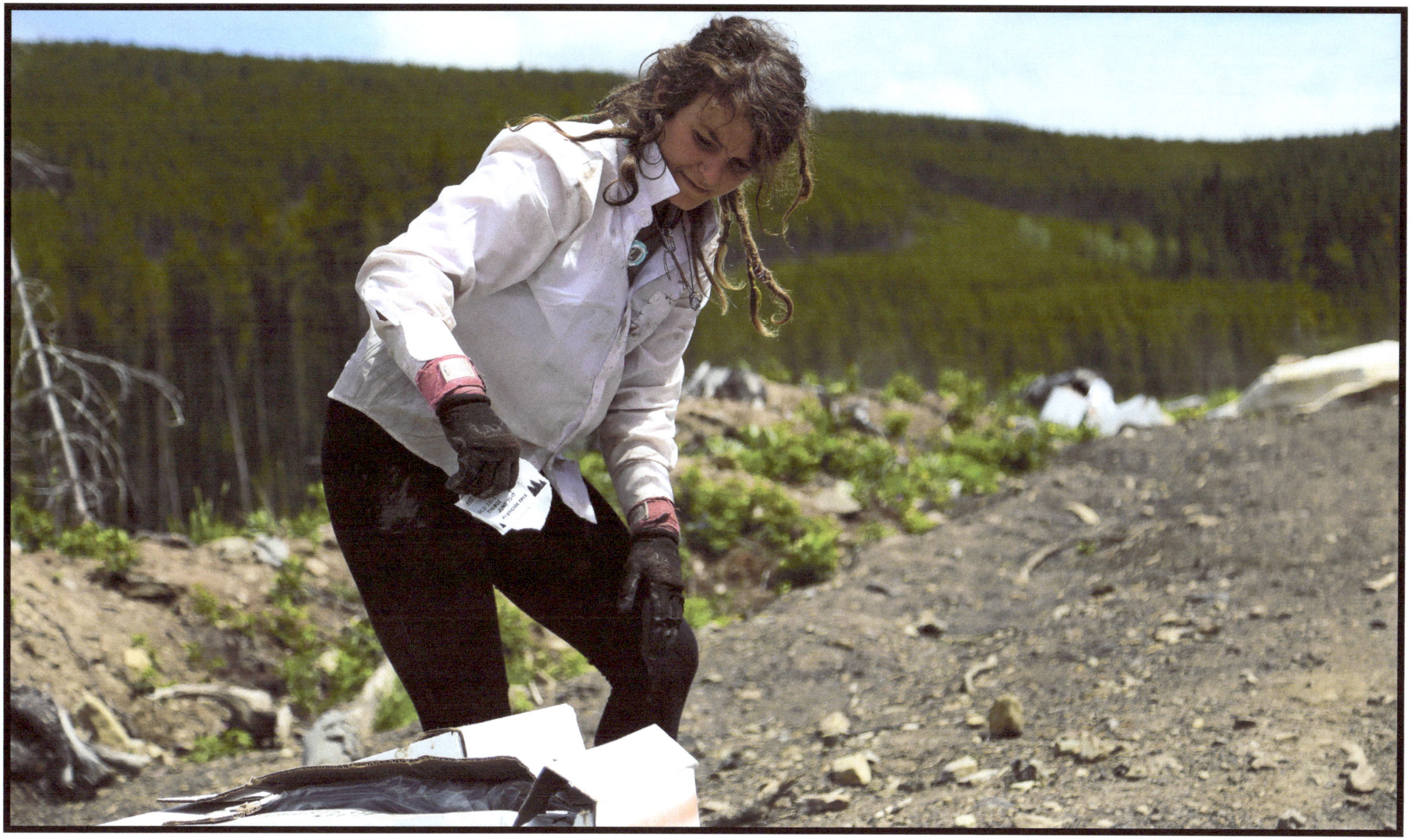

Laura, a fine arts graduate of Mount Allison University in New Brunswick, started planting in the Maritimes before she transitioned to the industry in western Canada. Laura is a very talented artist. You can see some of her work on Instagram at @laurarose_art

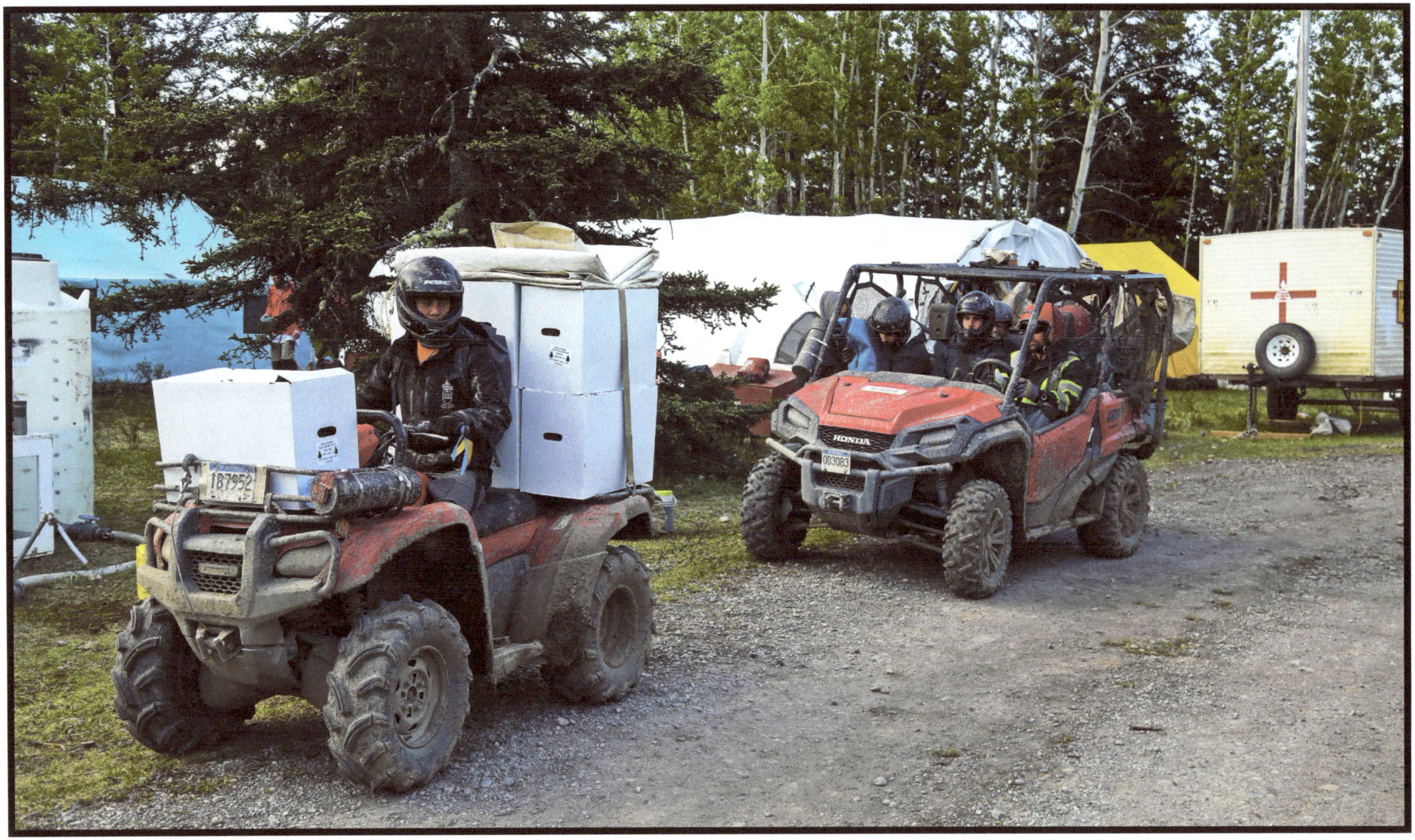

In addition to quads, some companies also use larger ATV's that are able to carry passengers safely. We refer to these ATV's as "Side-by-Sides." They aren't suitable on all projects.

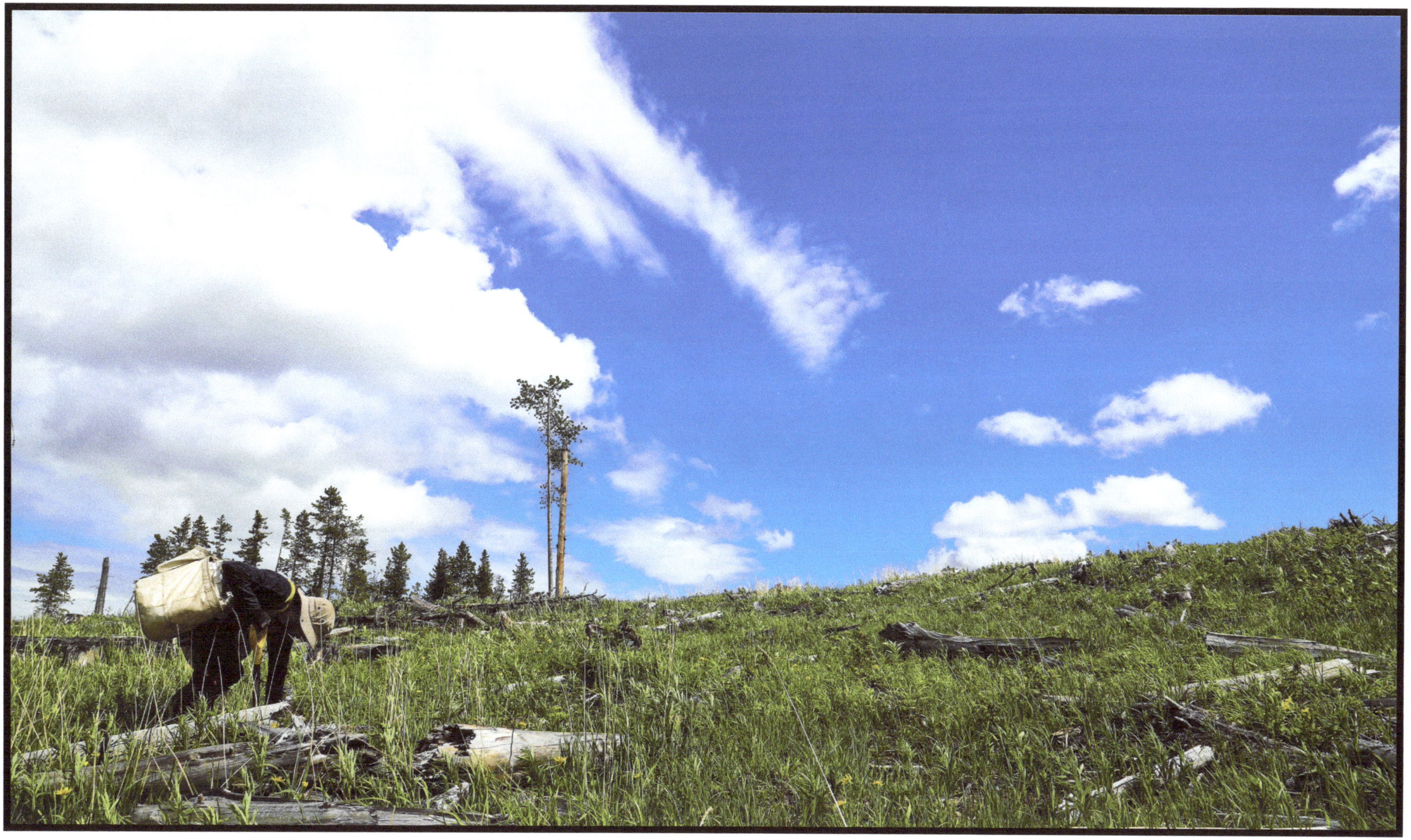

After a season of planting, most people forget about the rain days and the mud, and remember only the sunny days and good times. Selective memory is a wonderful thing.

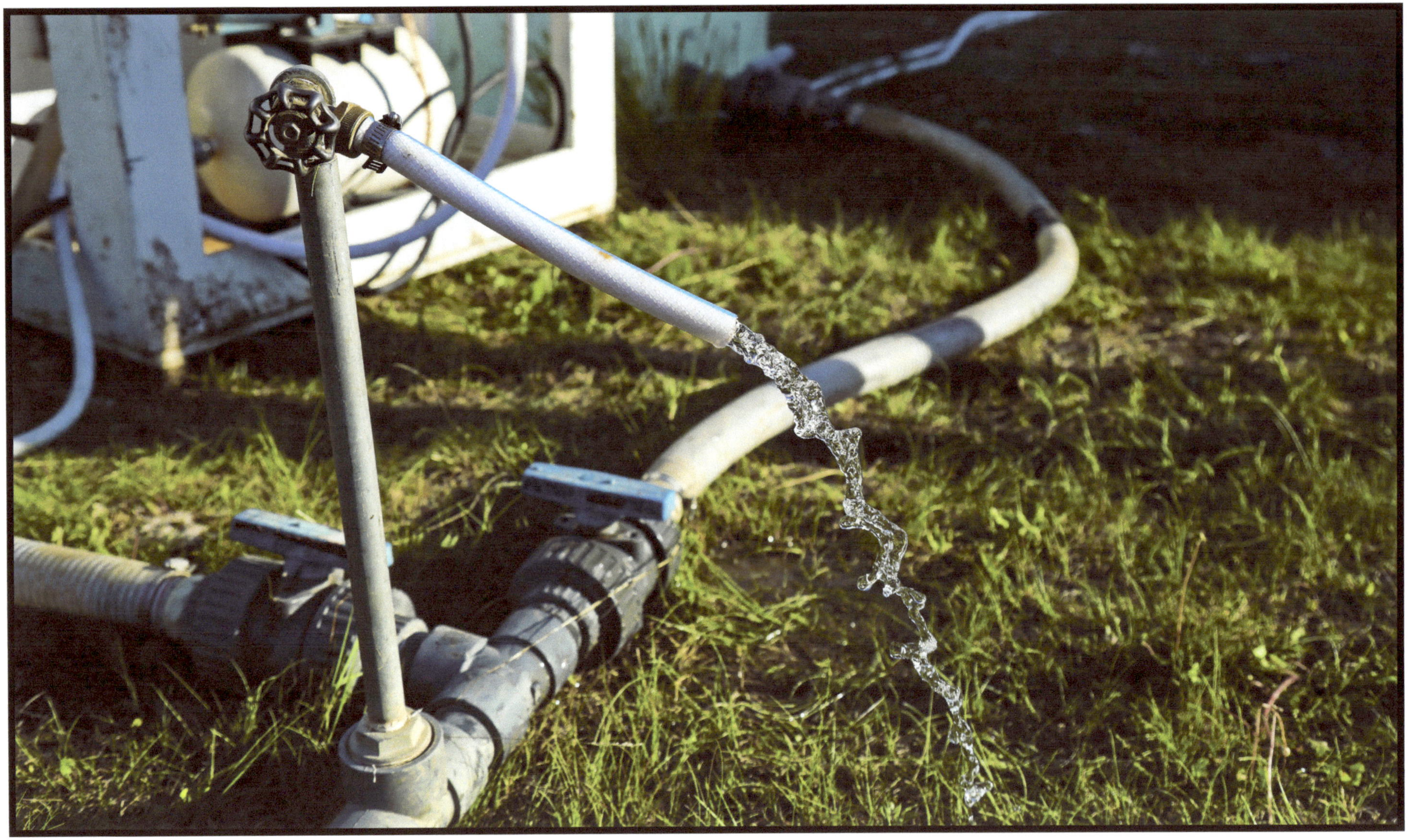

Water is just about the most important thing for planters to bring to the block. Most camps pay a lot of money to truck in large quantities of clean, safe, potable drinking water. During hot, sunny weather, it's common for a camp to use as much as two thousand litres of potable water per day, just for drinking water and kitchen use.

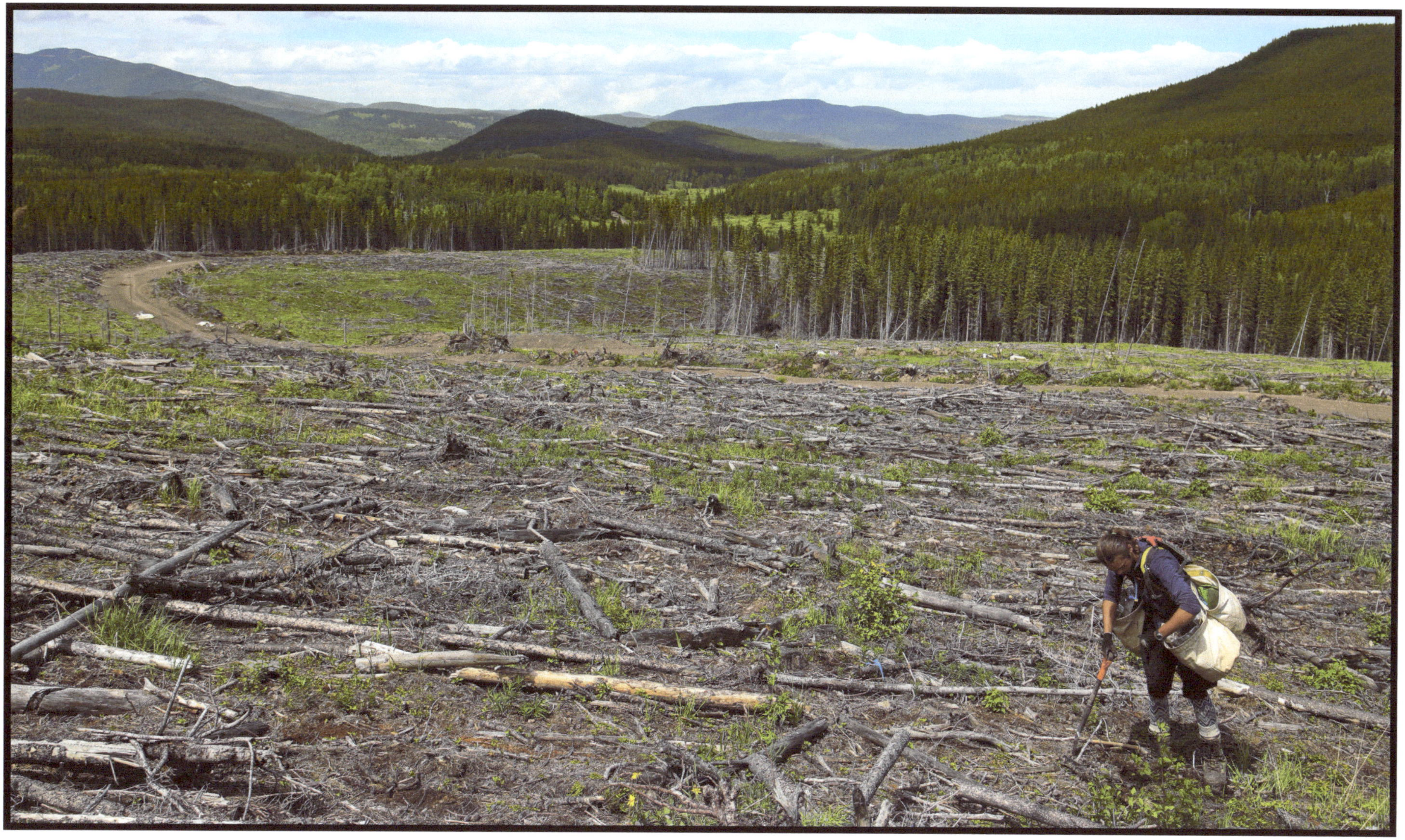

Ty is from Ontario. During the off-season, he studies Audio Production and Live Lighting & Sound, and also plays bass guitar in a few different groups. He also creates music using Cubase audio production software, and as a hobby, reads a lot of science fiction.

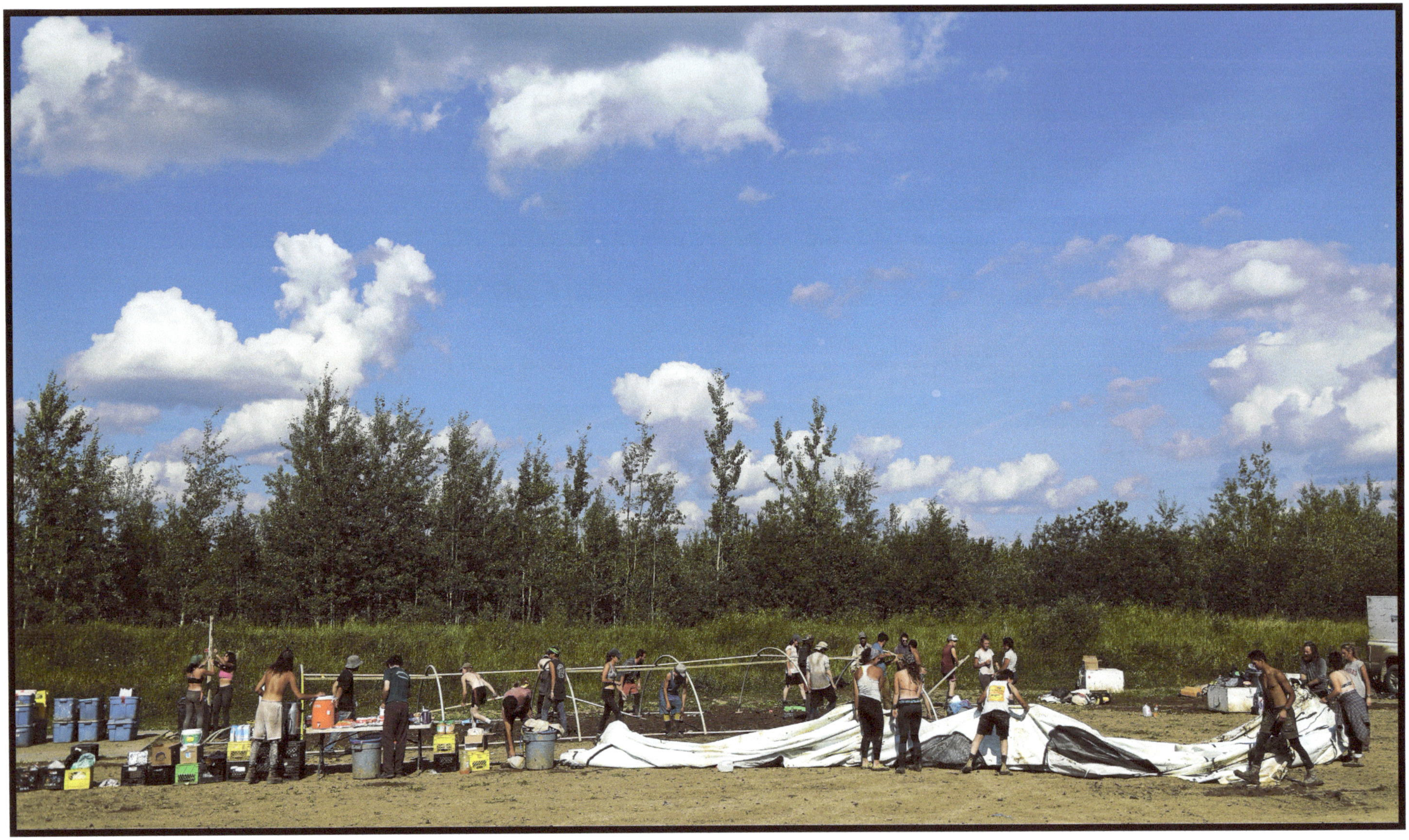

Moving a camp is a very complicated and logistically challenging process. However, with participation from a large group of planters, it's easy to tear down and pack up a camp in not much more than two hours, and to set up in about three hours. The process of moving a camp is very challenging and expensive, but it's often worth it to be staying much closer to the work sites than would be possible when working out of motel-based accommodations in "nearby" towns.

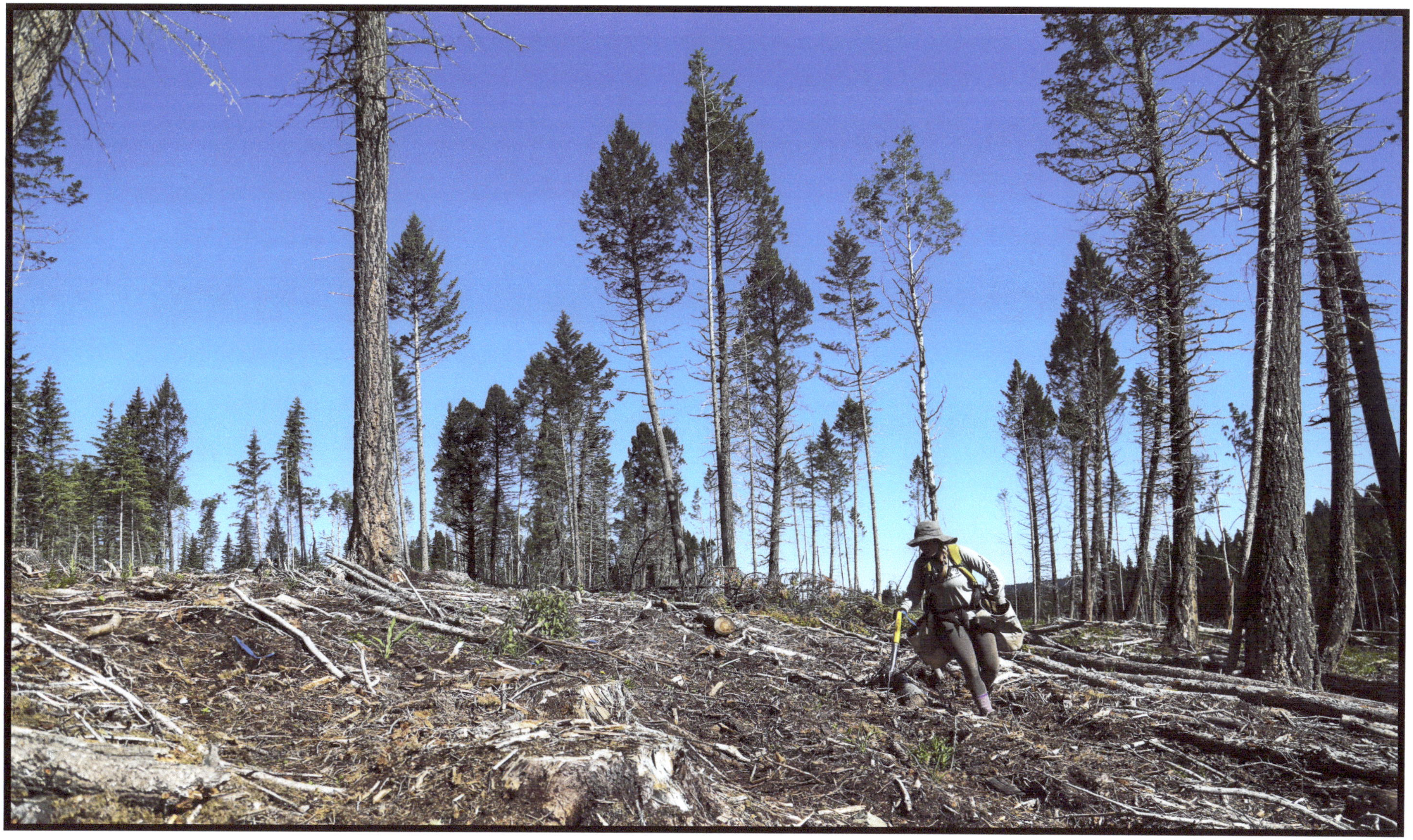

Hayley started planting in Ontario (her home province) before making the transition to the industry in western Canada. Hayley is a graduate of Western University in London.

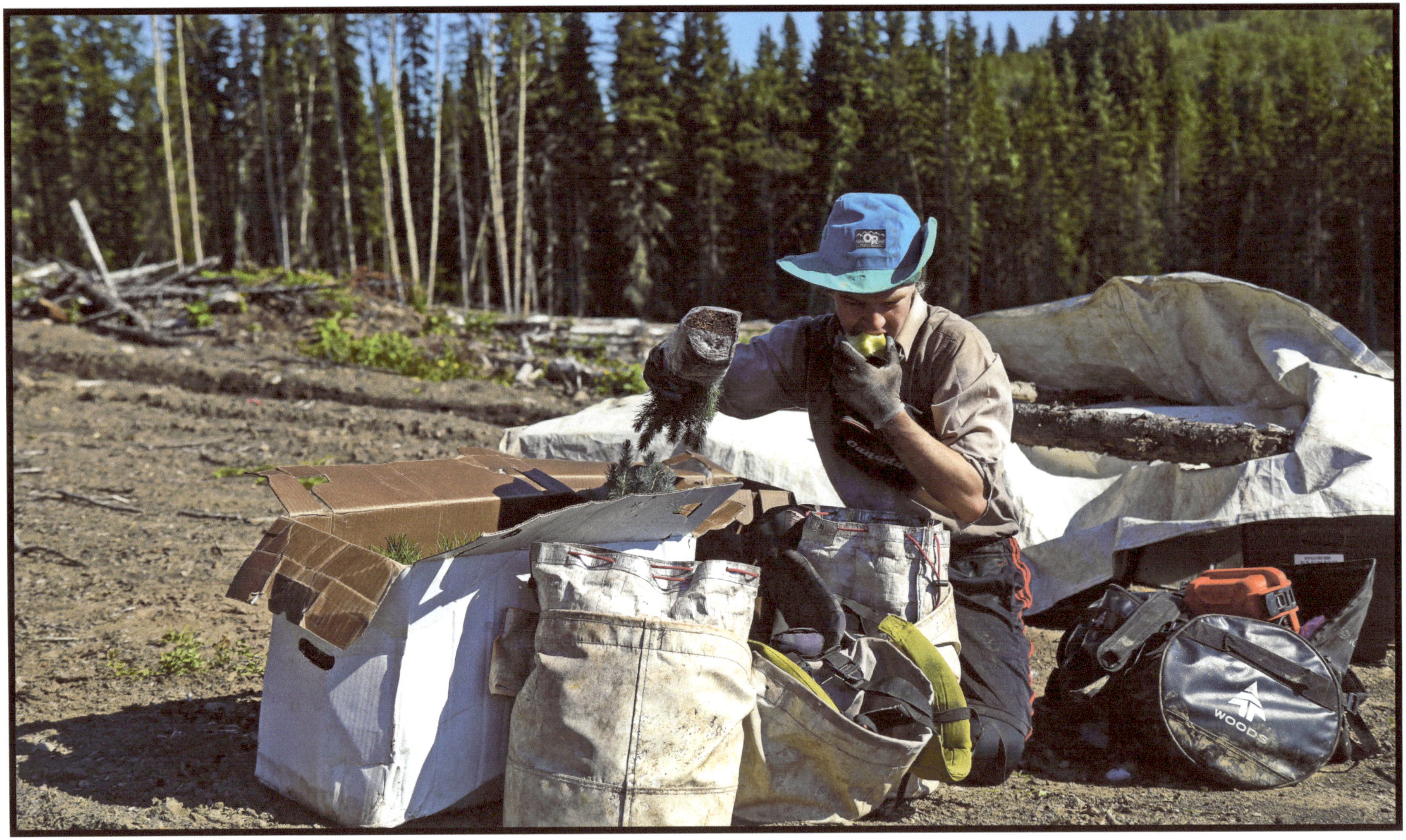

If you can multi-task, you can save time and ultimately make more money. Even something as simple as eating while bagging up, which only saves a couple of minutes, makes a difference if you do it every day throughout a full season of planting.

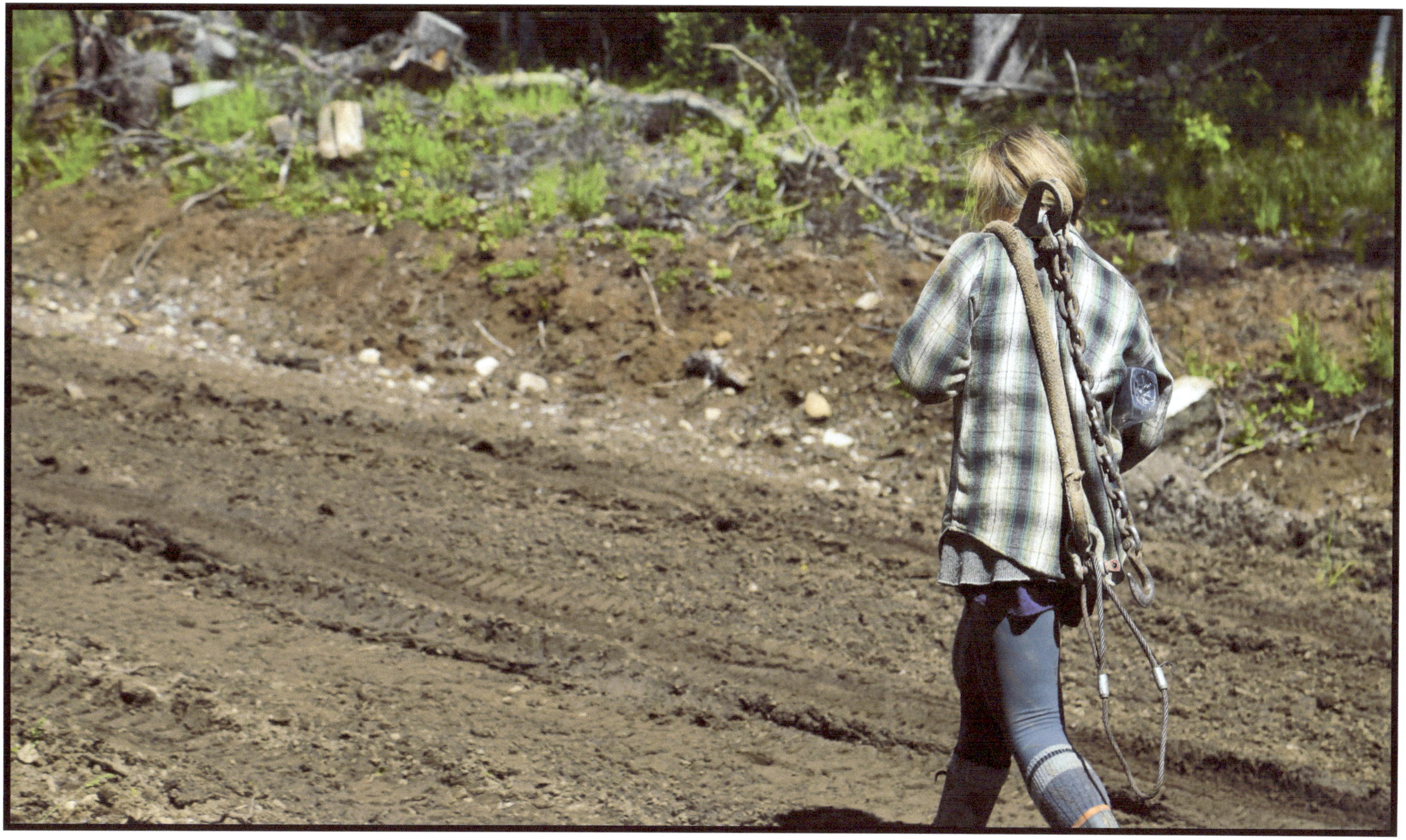

Trucks get stuck frequently enough that it's smart for each vehicle to carry a tow rope.

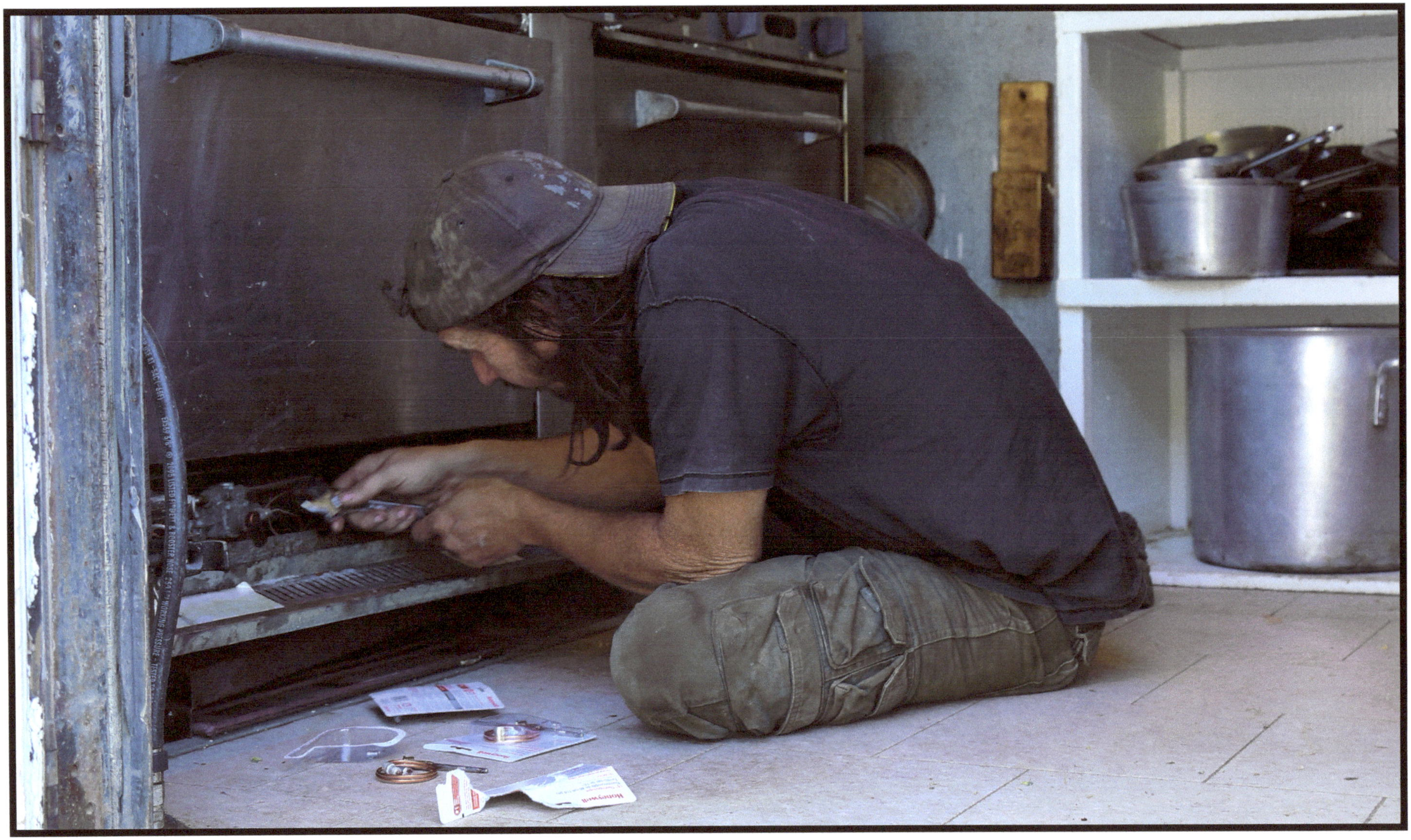

As a camp manager, my job description is quite simple: "Do everything." In this photo, I'm replacing a dead propane thermocouple on a propane oven. No matter whether you're a planter, a crew leader, or a camp manager, you learn to get very good at improvising while on the job, in order to get things done efficiently. Expect the unexpected.

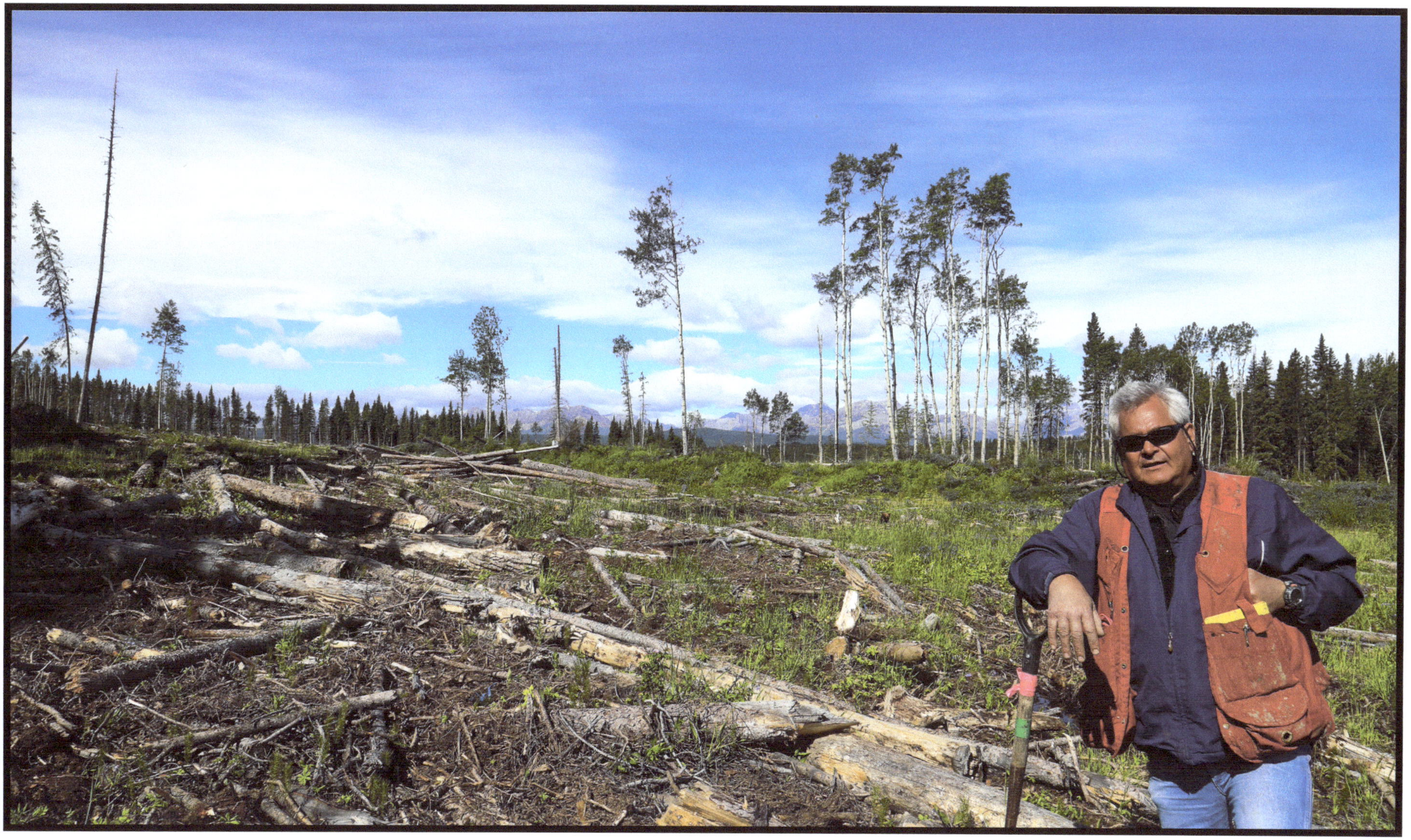

Darrell is a forester in Alberta. Darrell has decades of experience, and he knows that what works in terms of academic theory doesn't always work when put to the test under local field conditions. Darrell loves to come out to the bush to visit the planters on the blocks, and to be available to answer any logistics or planning questions that arise.

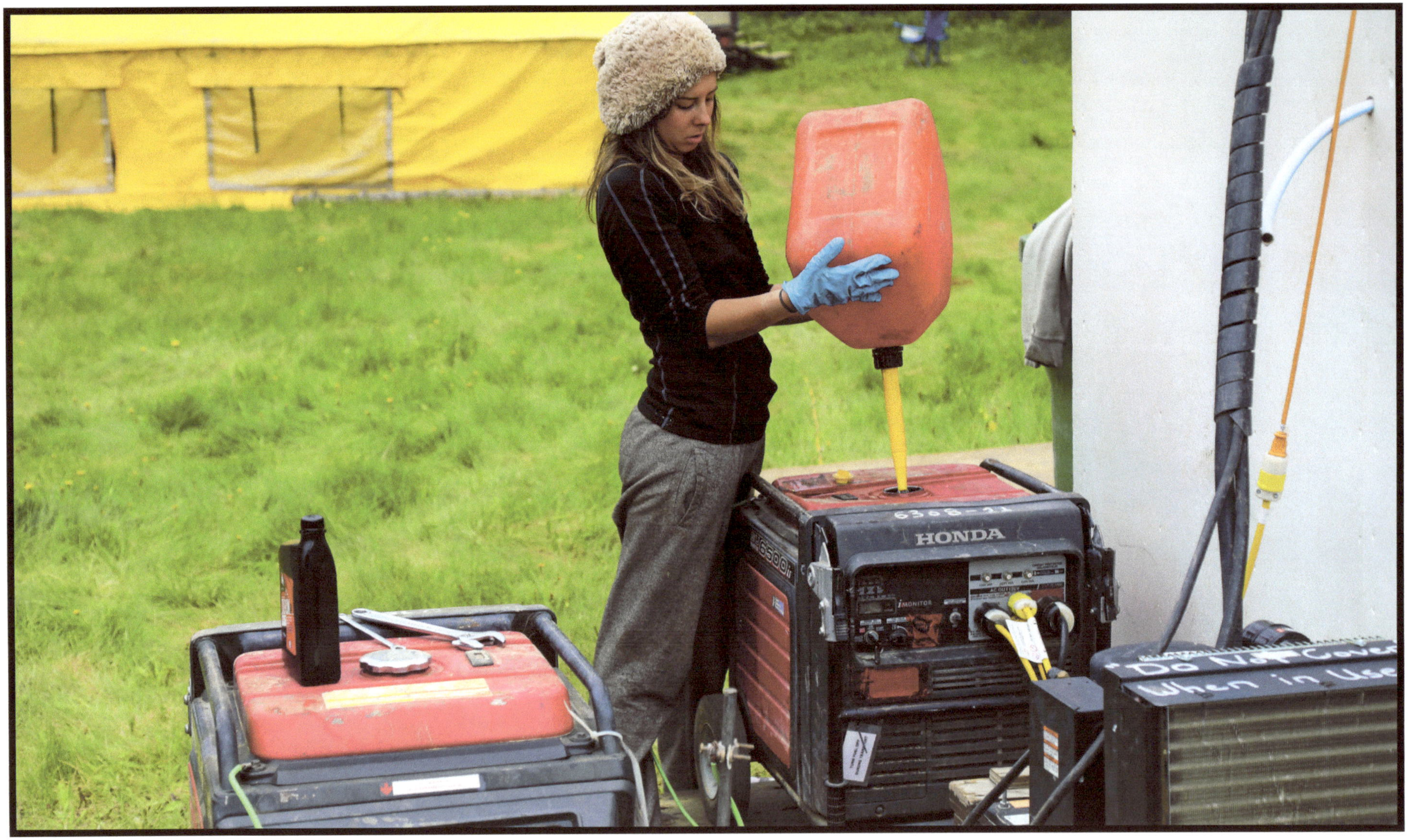

A planting camp typically has several gasoline-powered generators, which provide electricity to the kitchen and other key camp structures. These generators don't provide electricity to the planters' tents. Keeping the camp powered is a challenge, because the generators sometimes burn upwards of 150 litres of gasoline per day.

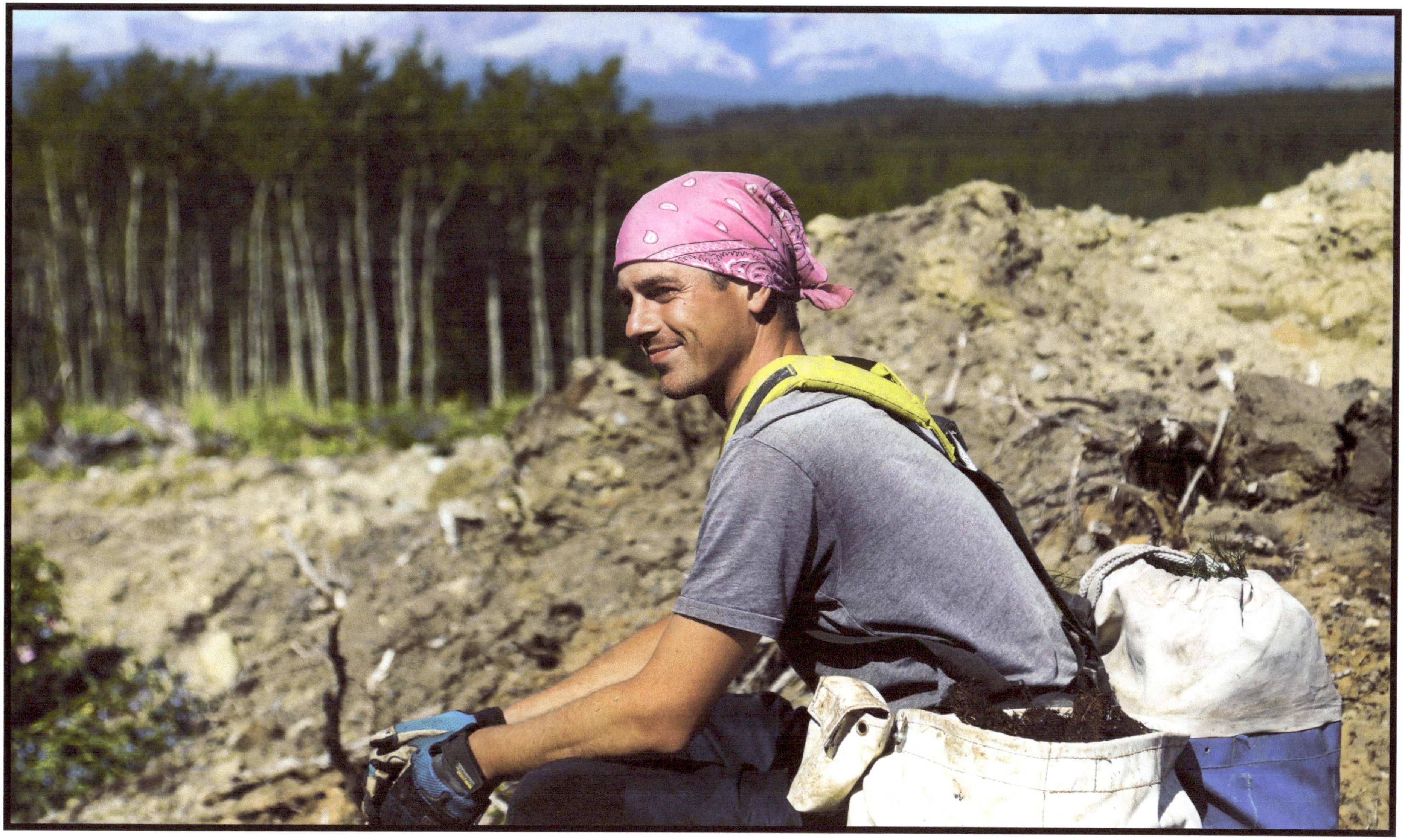

Mitch started his planting career working for an Ontario-based company before migrating to work in British Columbia. Mitch describes himself as an ex-military nomad who originally hails from Hampton, New Brunswick. It's always amazing to see how many people from all across Canada (including the Maritimes) work for companies in British Columbia. British Columbia tends to attract the cream of the crop when it comes to tree planters.

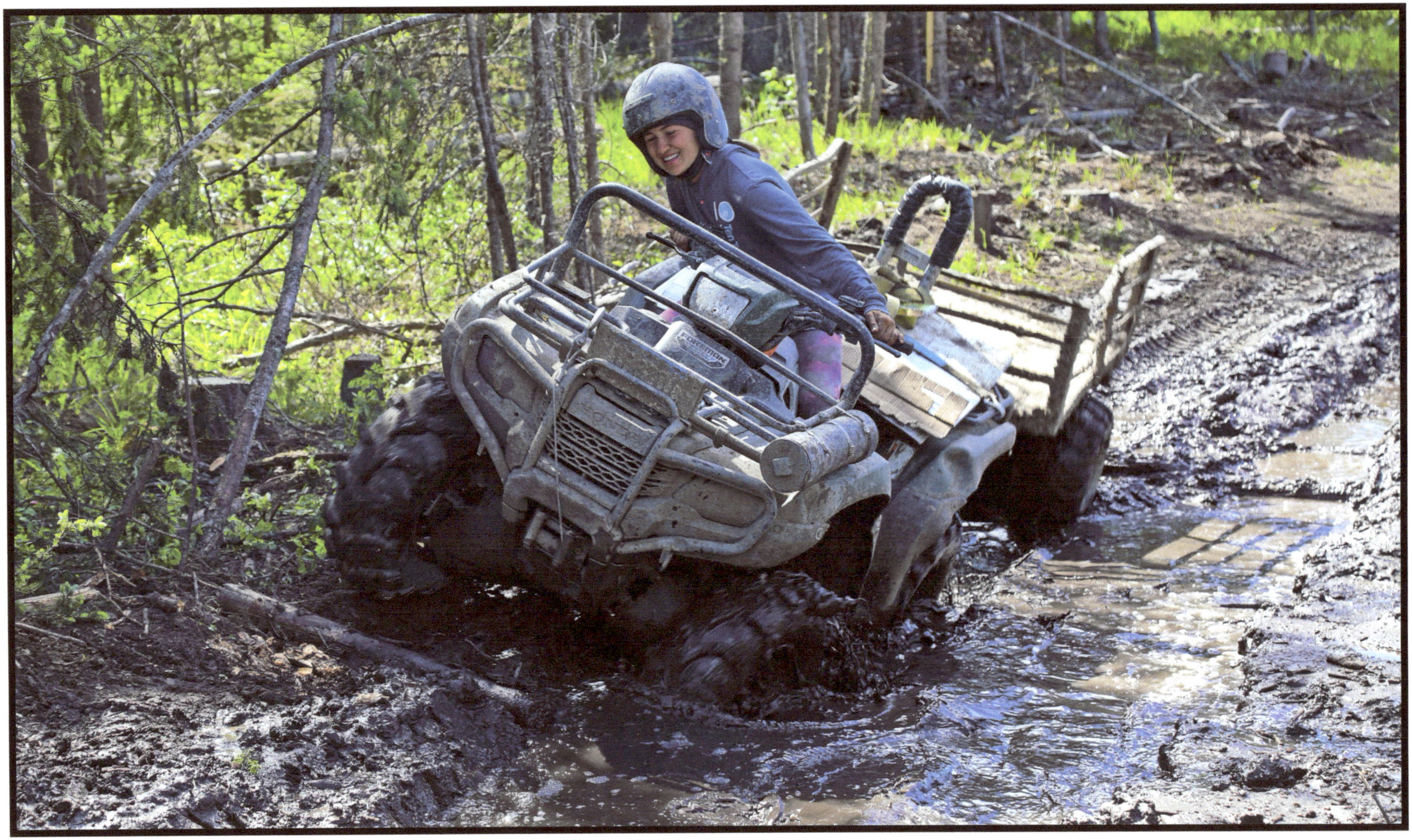

Even the quads have trouble getting through some muddy areas. This quad has a small trailer attached, to help with tree delivery.

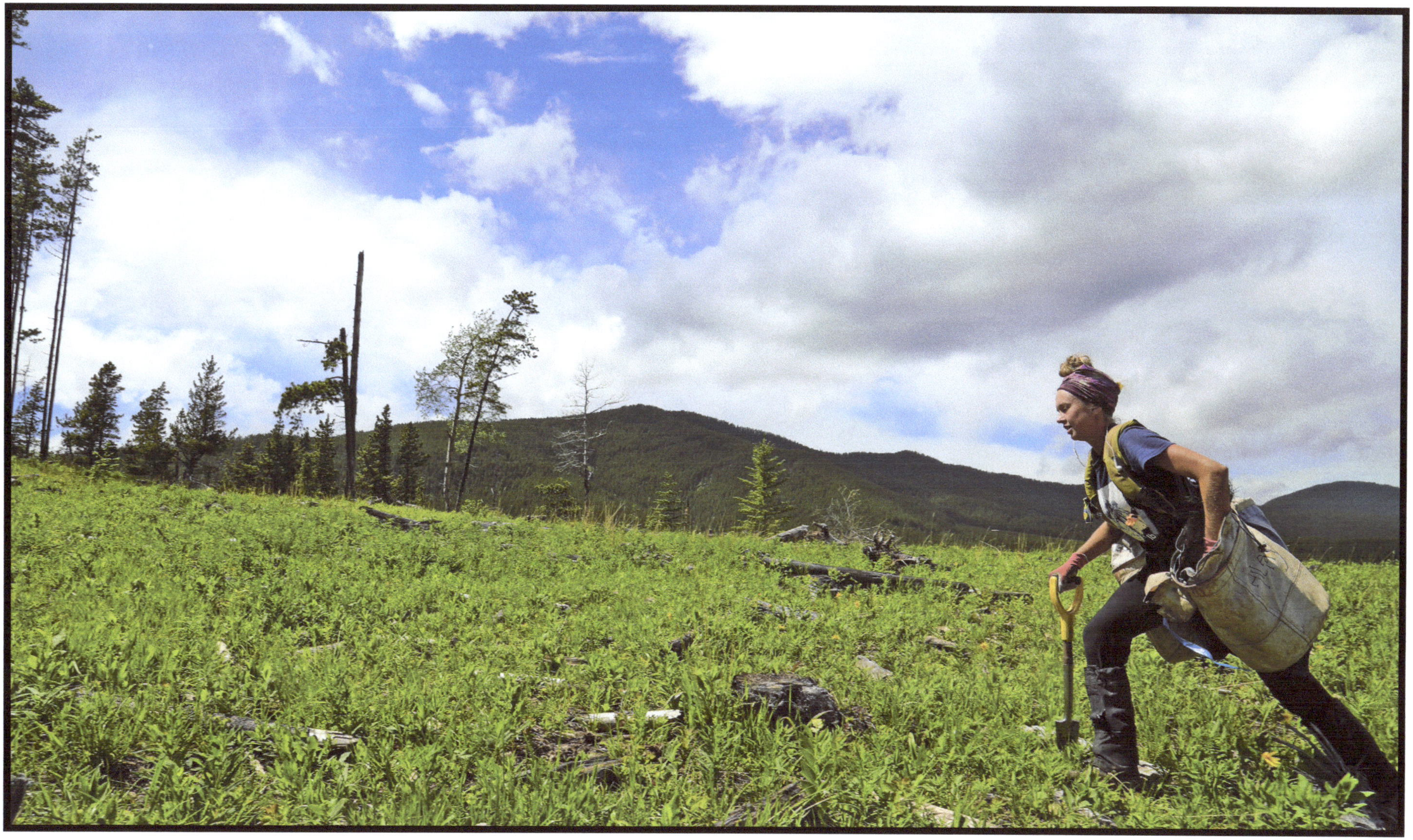

Hilary is from Ontario, and holds a Bachelor of Fine Arts degree in Dance from Ryerson. Hilary really enjoys yoga, and plans to get her yoga teacher training certification. She says that, "Movement is something that always holds my interest. Studying dance, anatomy, and movement has definitely helped me with planting at times." We hope that Hilary doesn't move to a different career path.

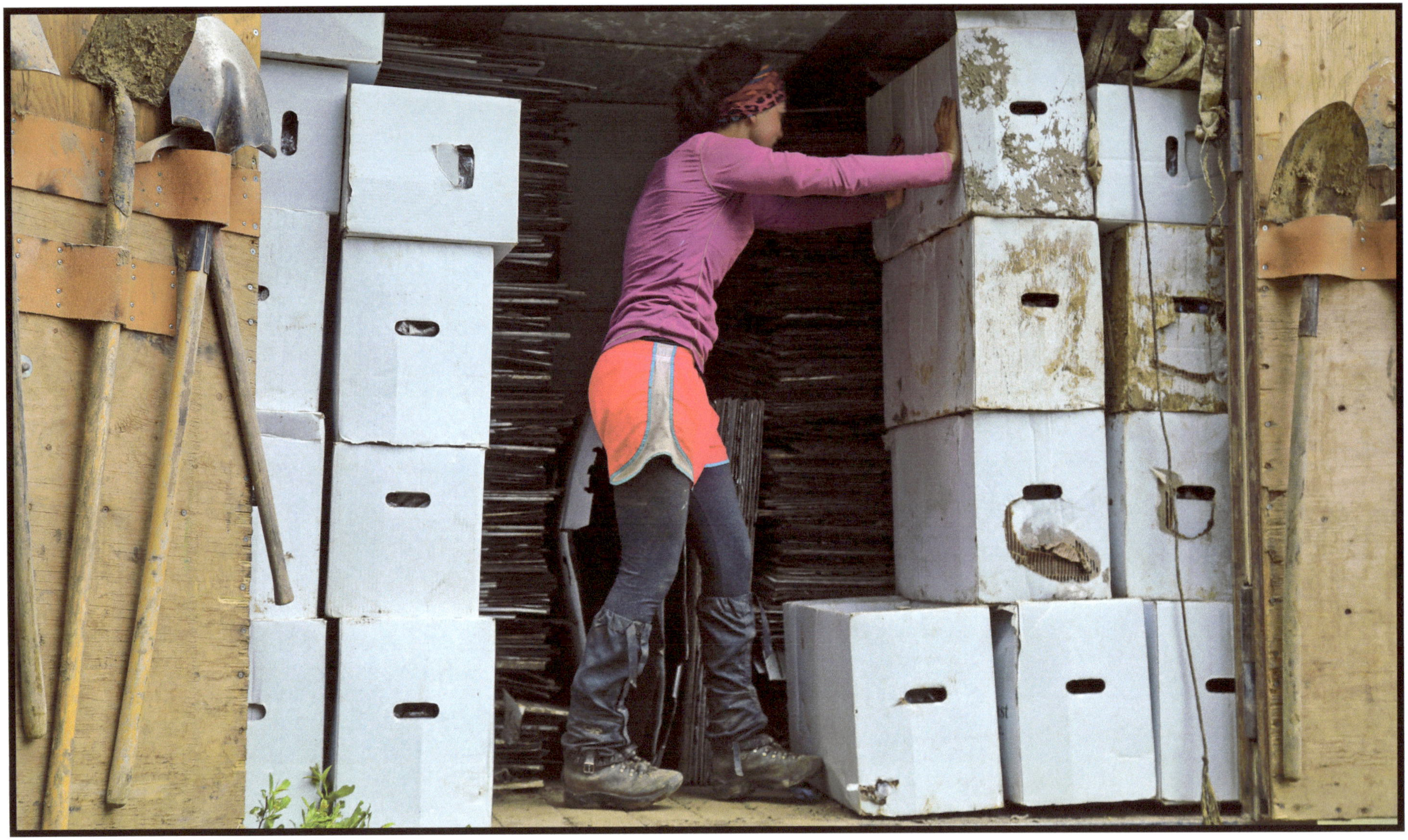

When a trip to the local landfill and back is a six-hour adventure, it's important to pack as much garbage into the truck as possible, to minimize the number of trips required. A full-size planting camp generates about one large pickup load of garbage like this for every two work days. We recycle this cardboard when possible, but due to the high wax content, many recycling centers won't accept it. This is one of the most frustrating things about our industry.

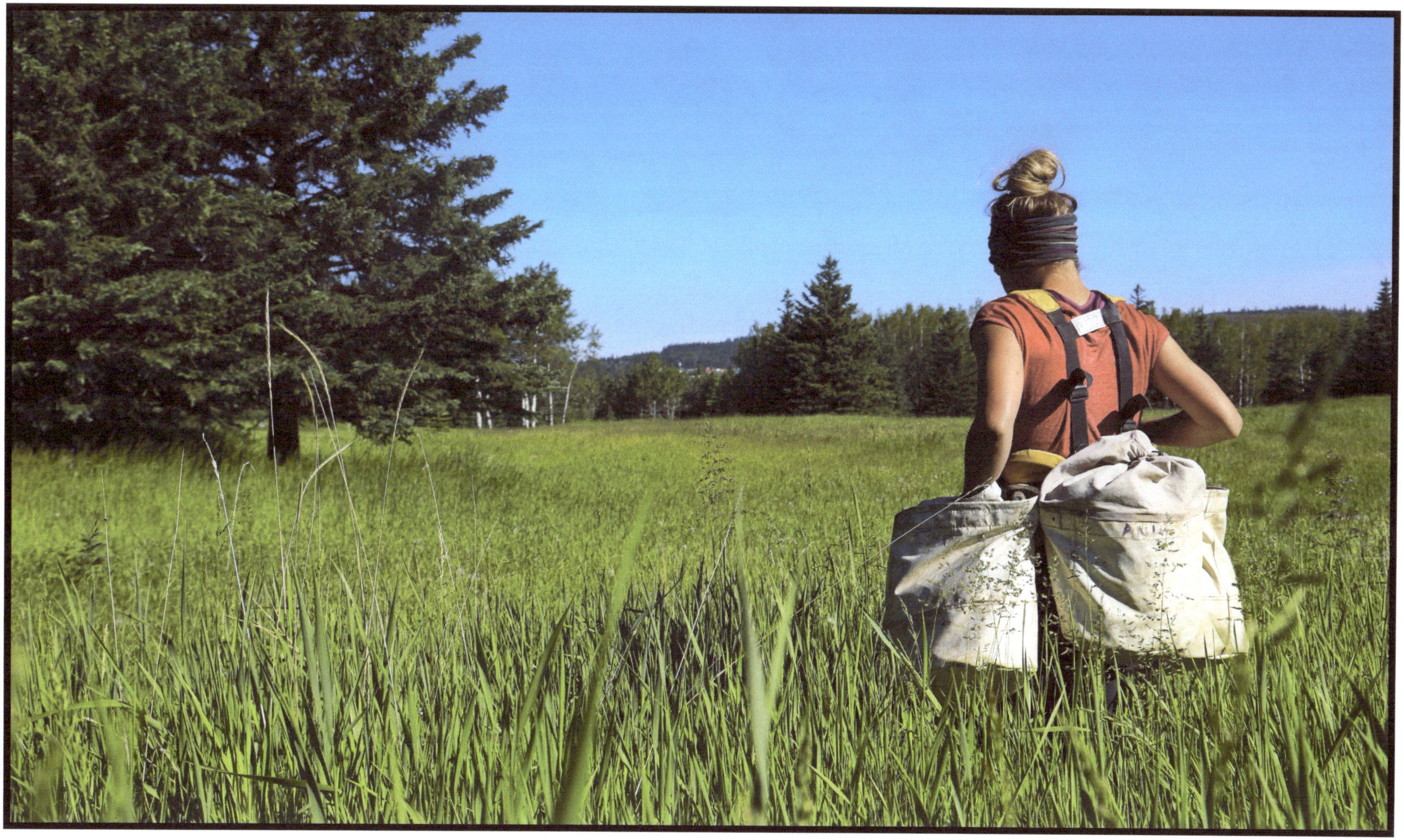

This was not a normal cut-block. This was a "special mission" that involved planting a thousand trees on the private property of someone who did a large favor for the mill that we were working for. The trees probably had a really hard time getting established with all of that grass competition.

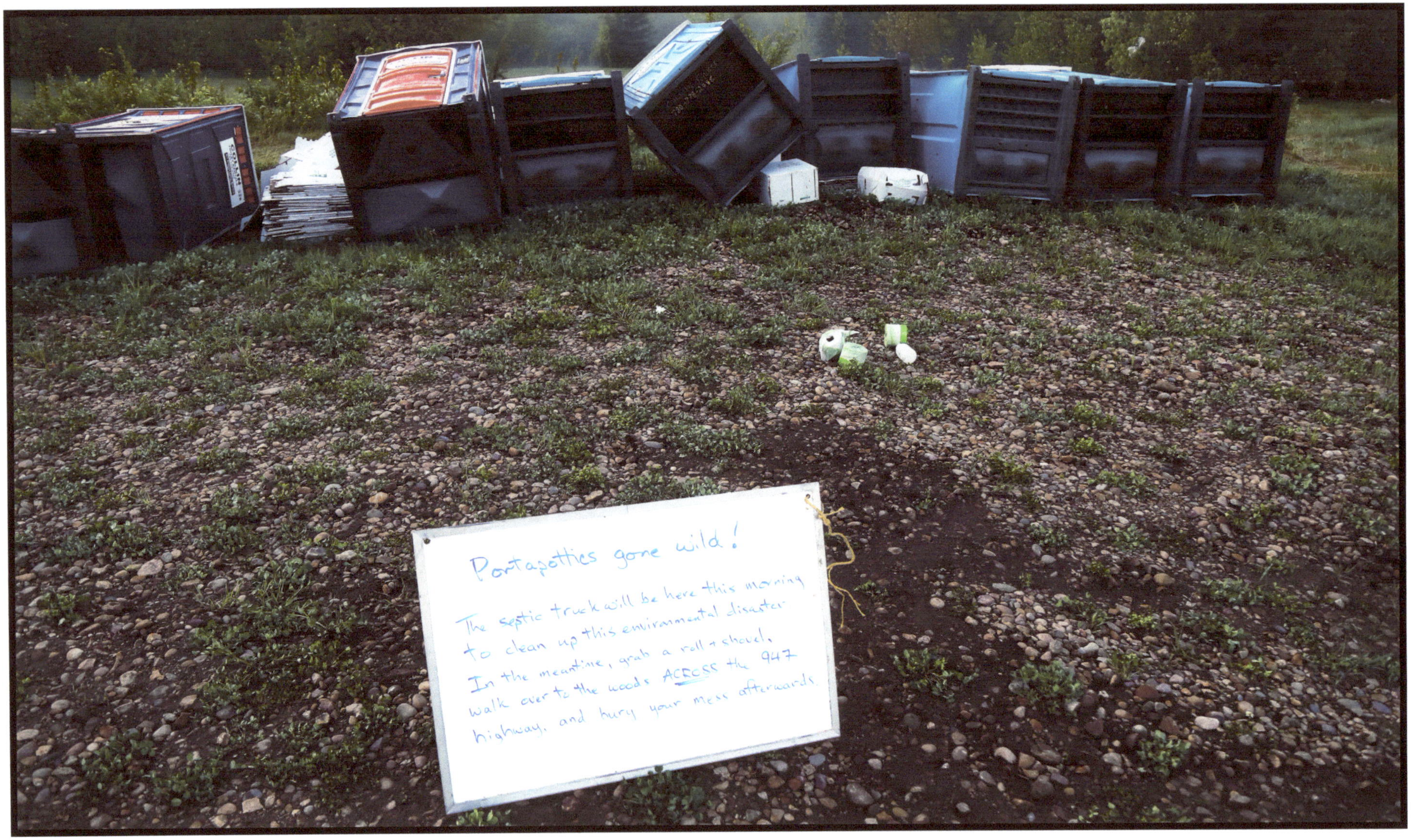

Expect the unexpected. One morning we woke up to discover that a massive windstorm had knocked over all of the camp port-a-potties. Incidentally, this is not a typical problem. That particular windstorm was such a fierce outlier that it flattened hundreds of hectares of mature forest throughout the region.

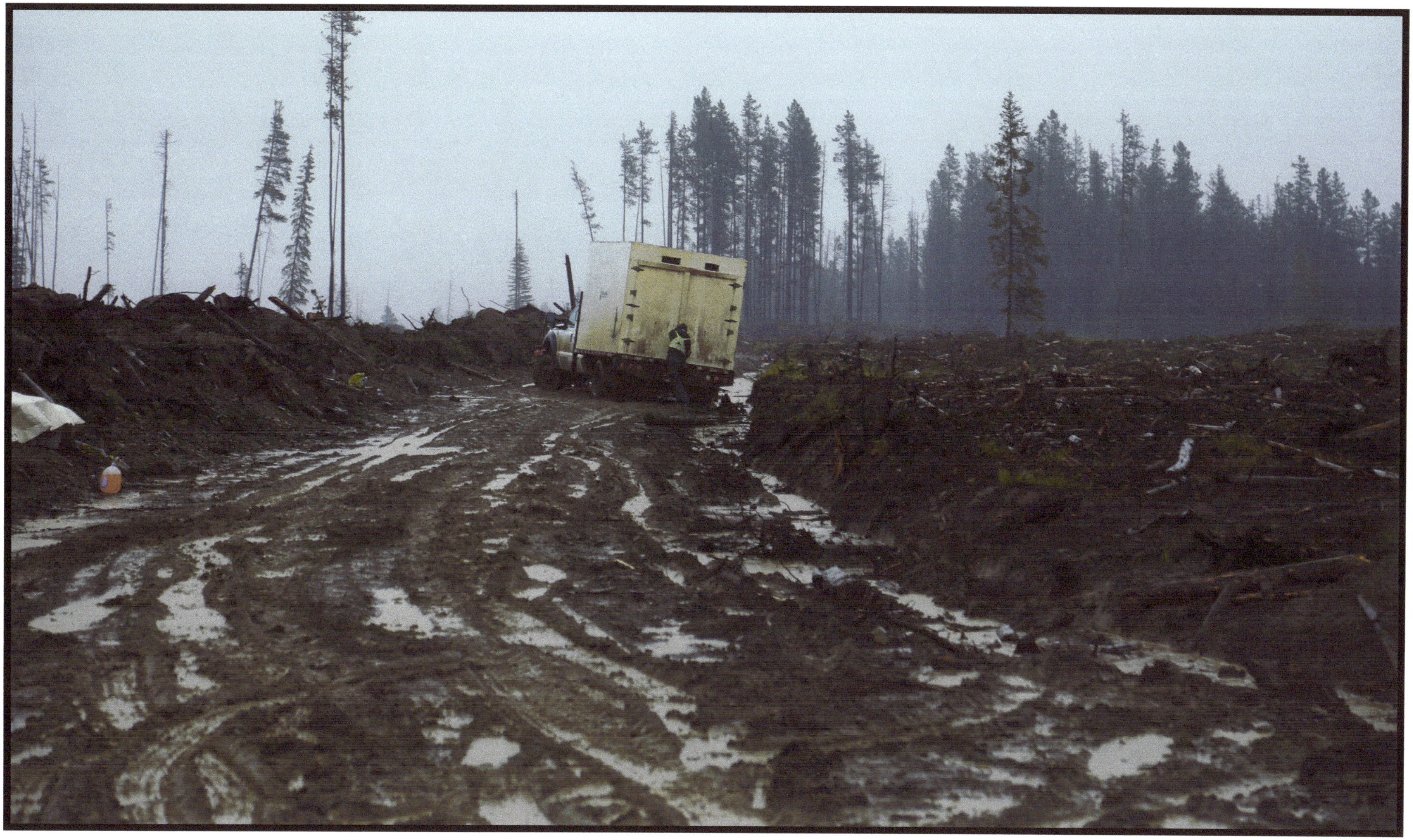

Be prepared for a little bit of misery when the rains come. As I mentioned earlier, we live in a world of mud.

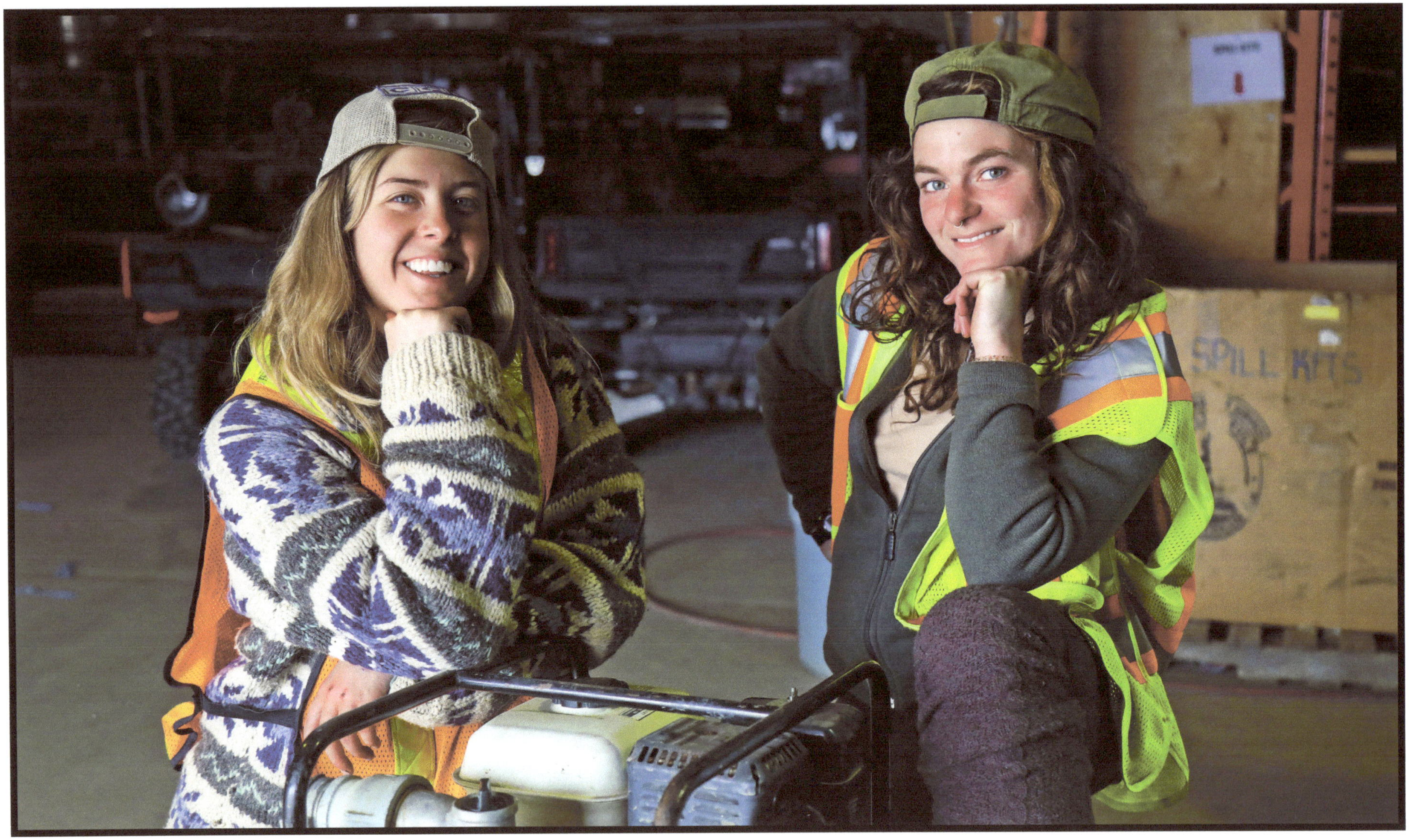

Laura and Laura, planting partners. They're also two of the best planters in my camp.

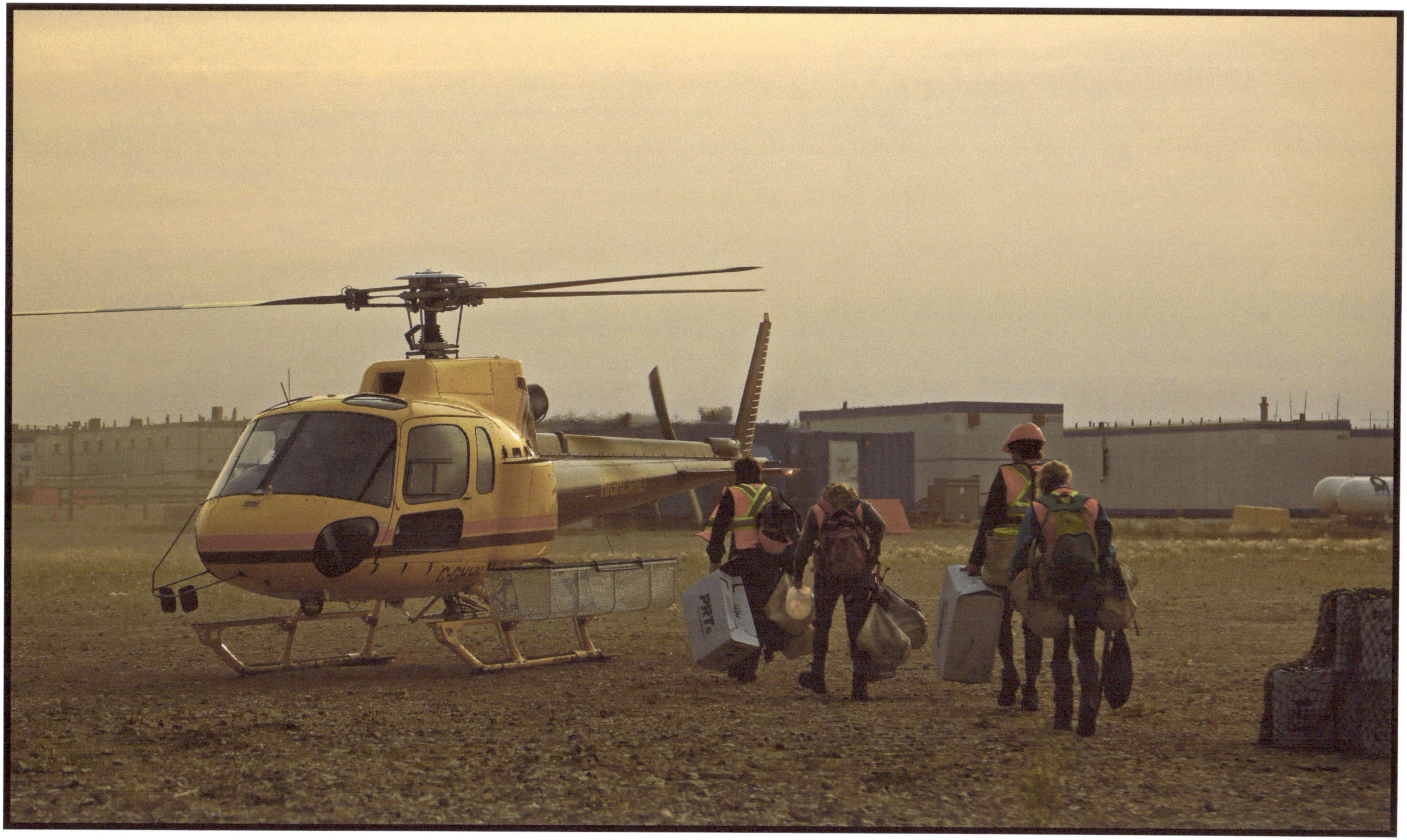

Although we frequently replant wildfire areas, large fires sometimes lead to short-term challenges for planters. This should have been a clear summer day, but the amount of smoke in the air was almost enough to shut down our operations because the helicopters barely had enough visibility for flight operations.

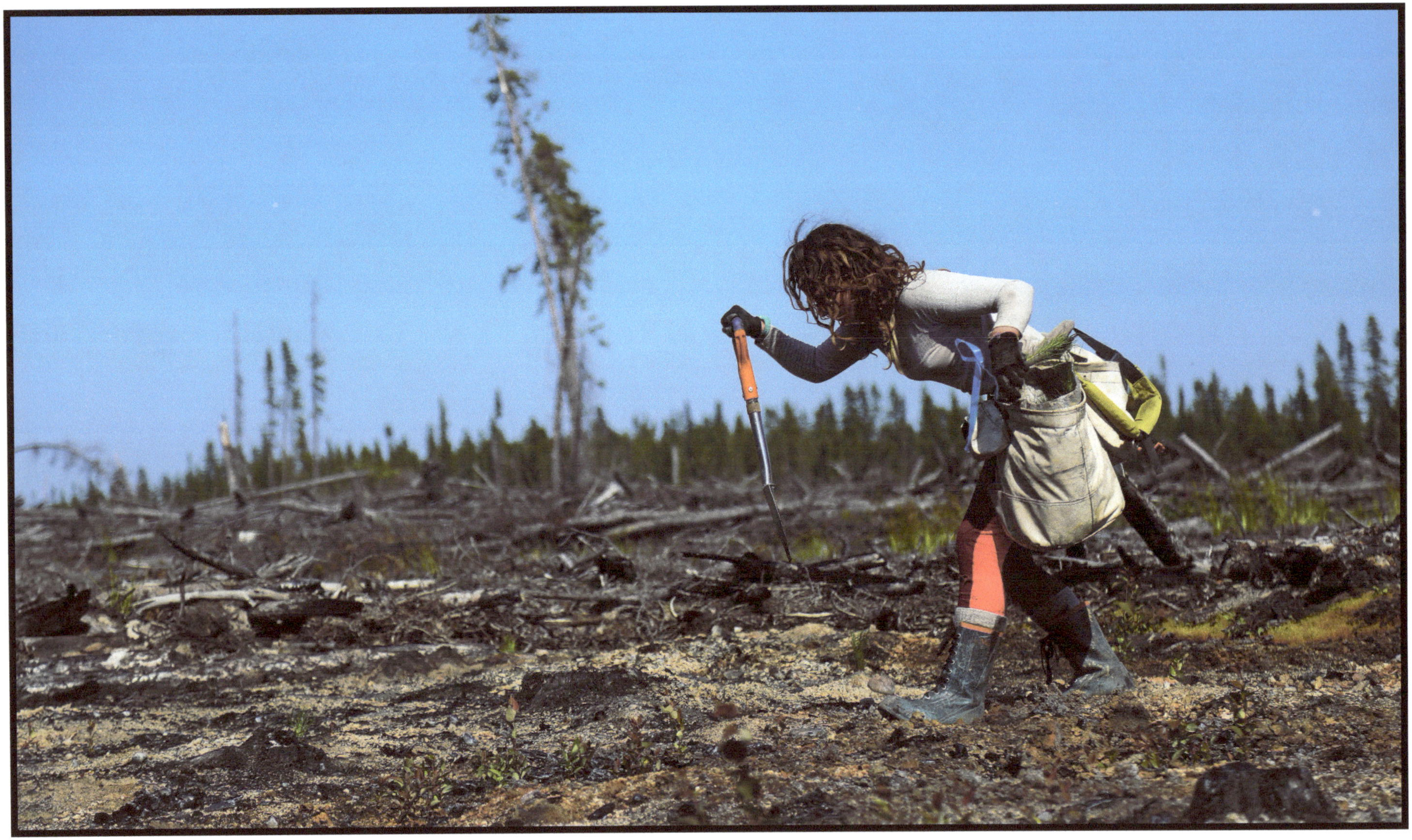

This is more like it. No smoke in the air for Laura.

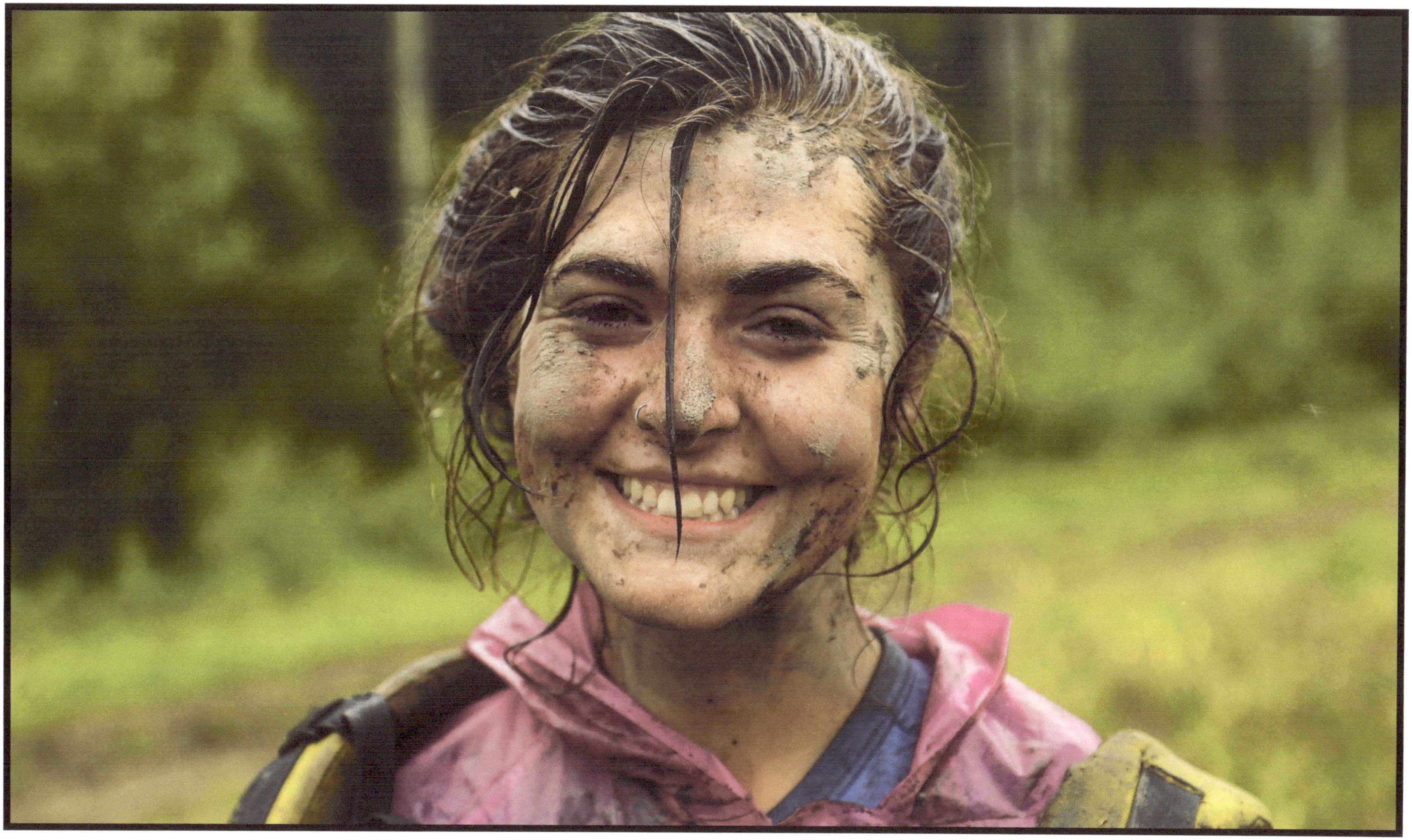

Here's a completely genuine smile at the end of a wet and muddy day. Some people don't let poor weather get them down. Marlie studies theatre at Queens, and hopes to pursue a professional career in acting after she finishes her studies at university.

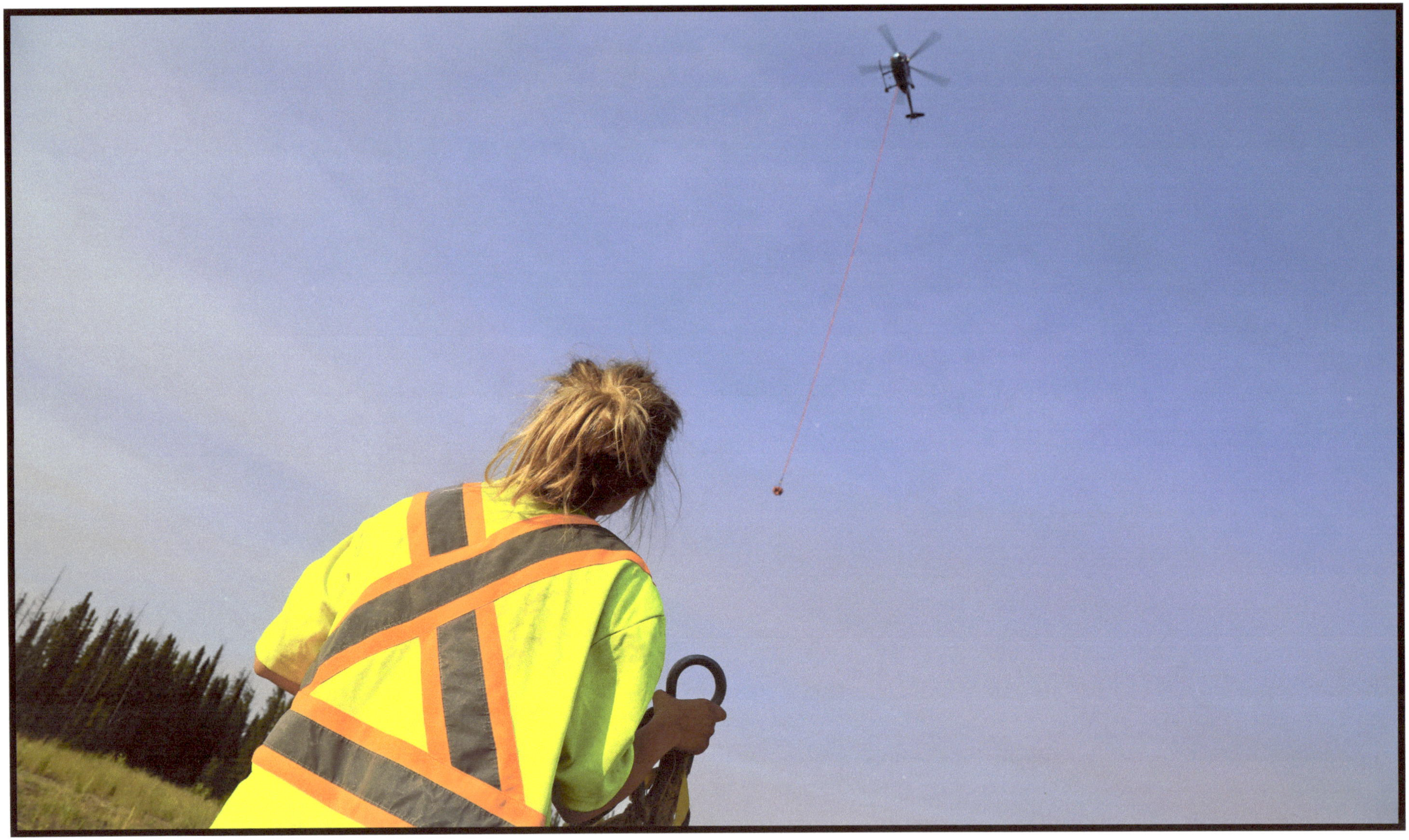

When a quad isn't able to safely move trees into a cut-block, a helicopter may be necessary. This helicopter has a "long line" and a hook attached underneath. It's coming in to pick up a sling full of tree boxes. The sling will be set down into the middle of the cut-block a few minutes later. Sometimes the planters will walk into these blocks, if the trucks can get close to the blocks and if there's no significant water in the way. Otherwise, the planters will fly.

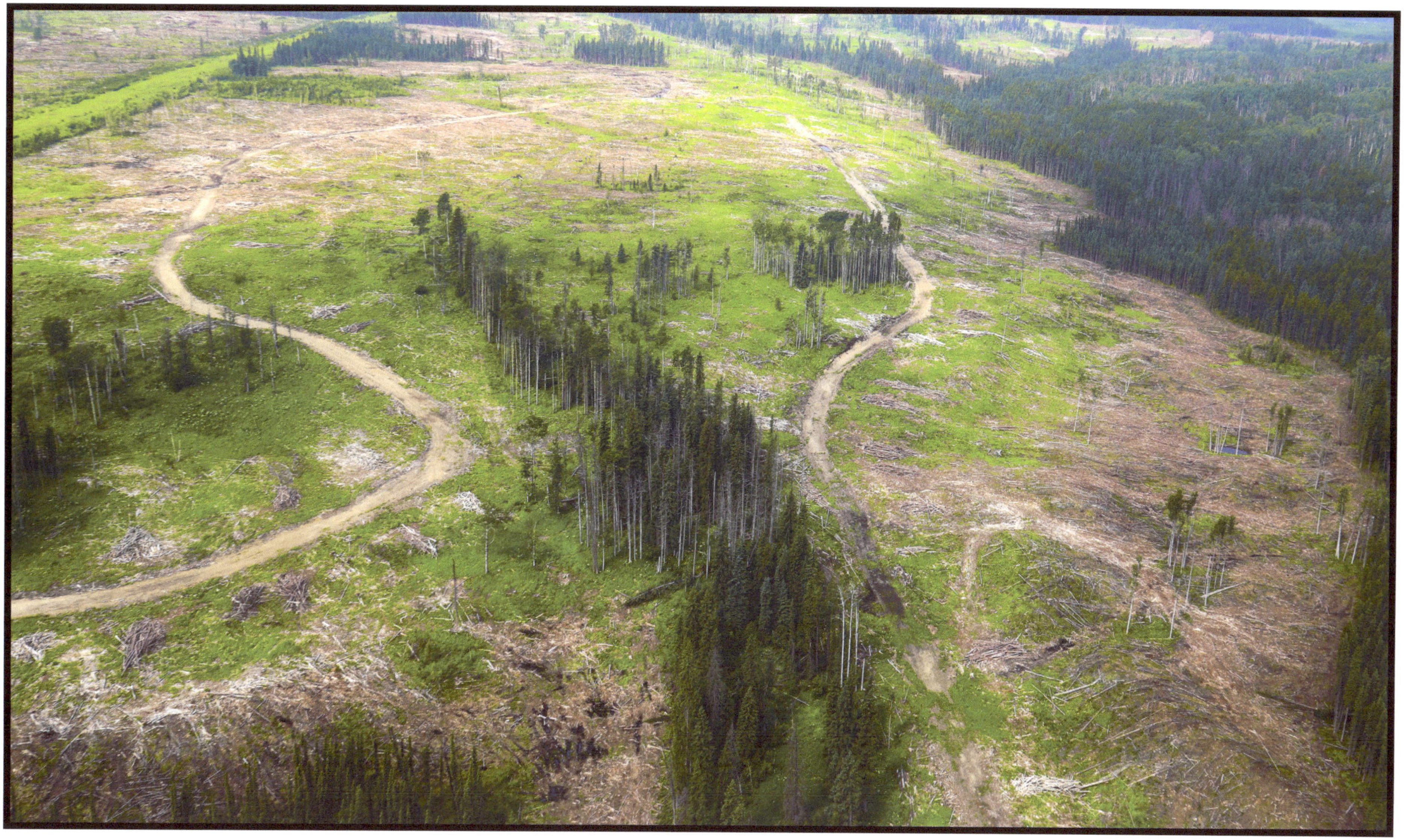

Here's an aerial view of a cut-block in Alberta, showing what the helicopter pilot sees on the way into a typical block. This block is starting to get a lot of grassy patches in the middle.

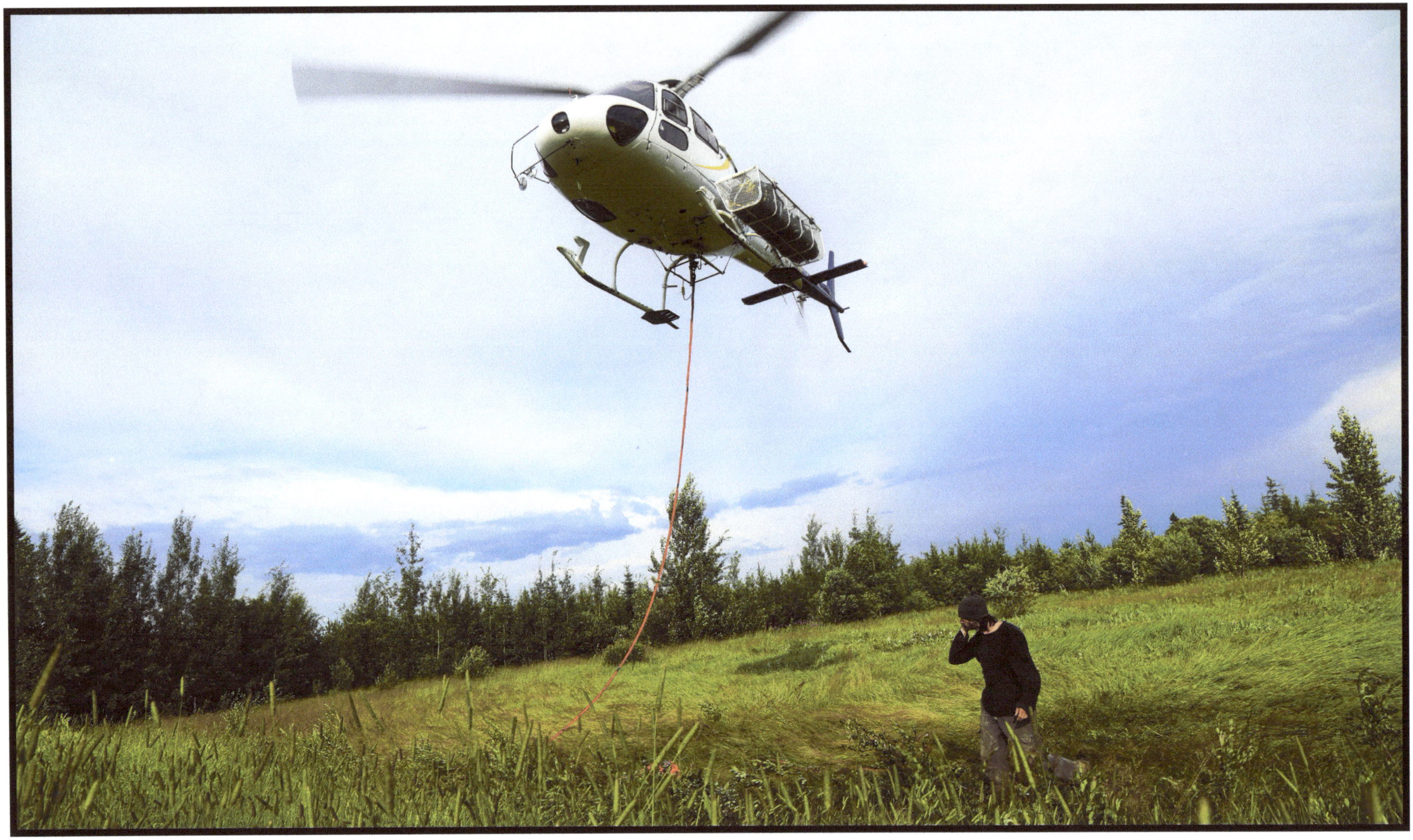

This helicopter is getting ready to go pick up some slings, after I've attached the long-line to the belly.

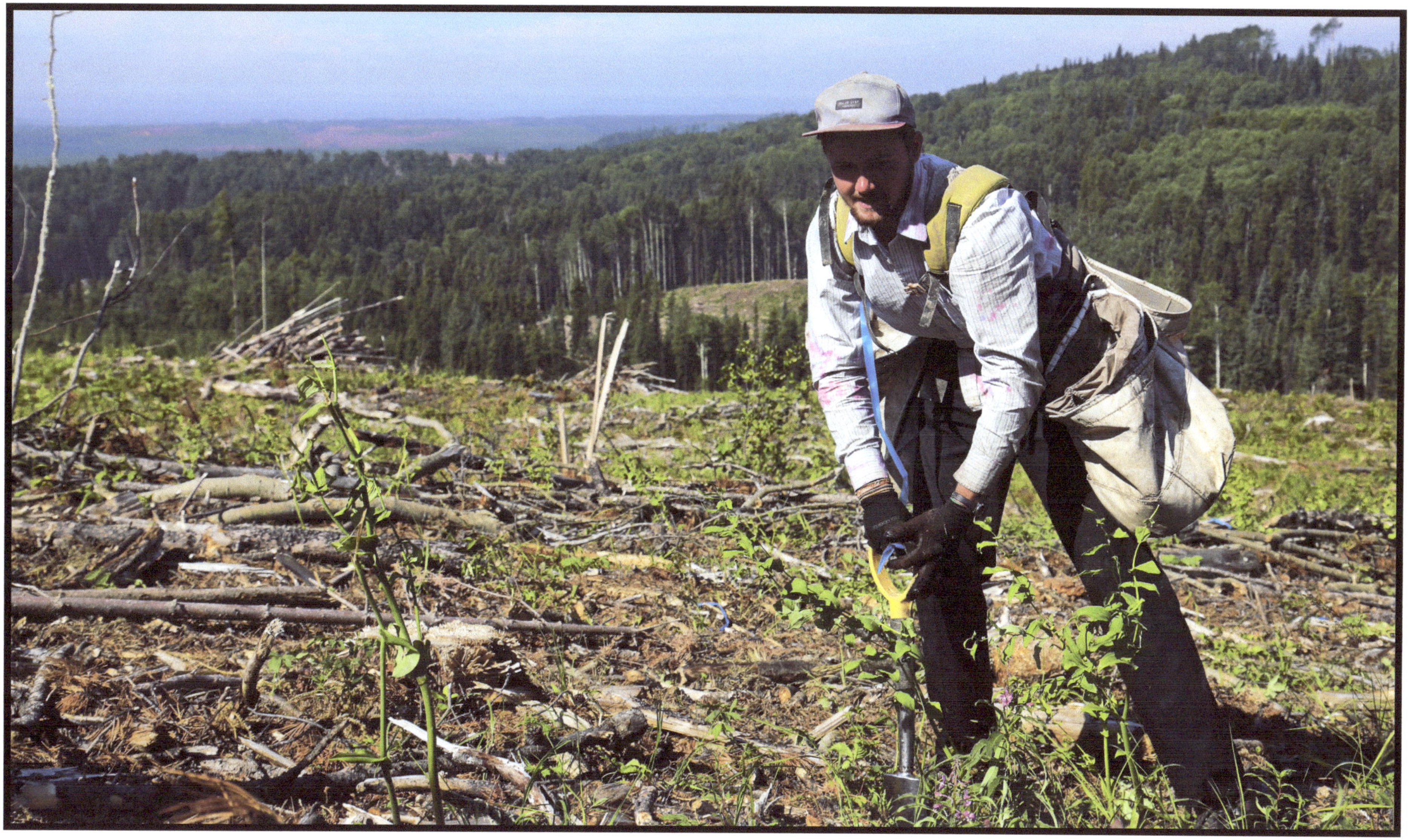

Luke, one of our most experienced planters, went to Thompson Rivers University in Kamloops to obtain his Adventure Guide diploma. In the off-season, Luke spends most of his free time skiing or rock climbing.

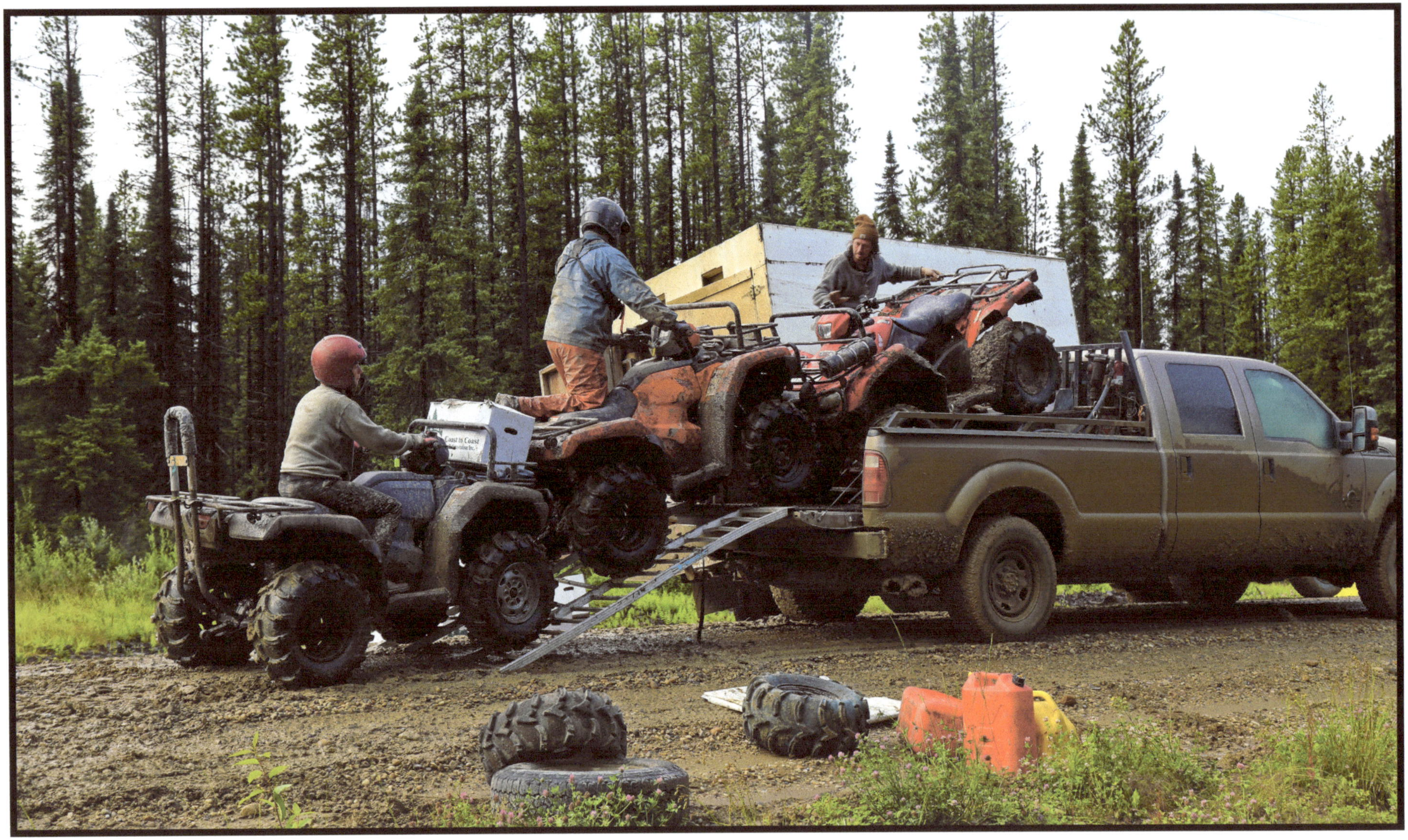

On the day that this photo was taken, the red quad died. We used the blue quad and the orange quad to push it up into the back of the truck, so I could take it to town for repairs.

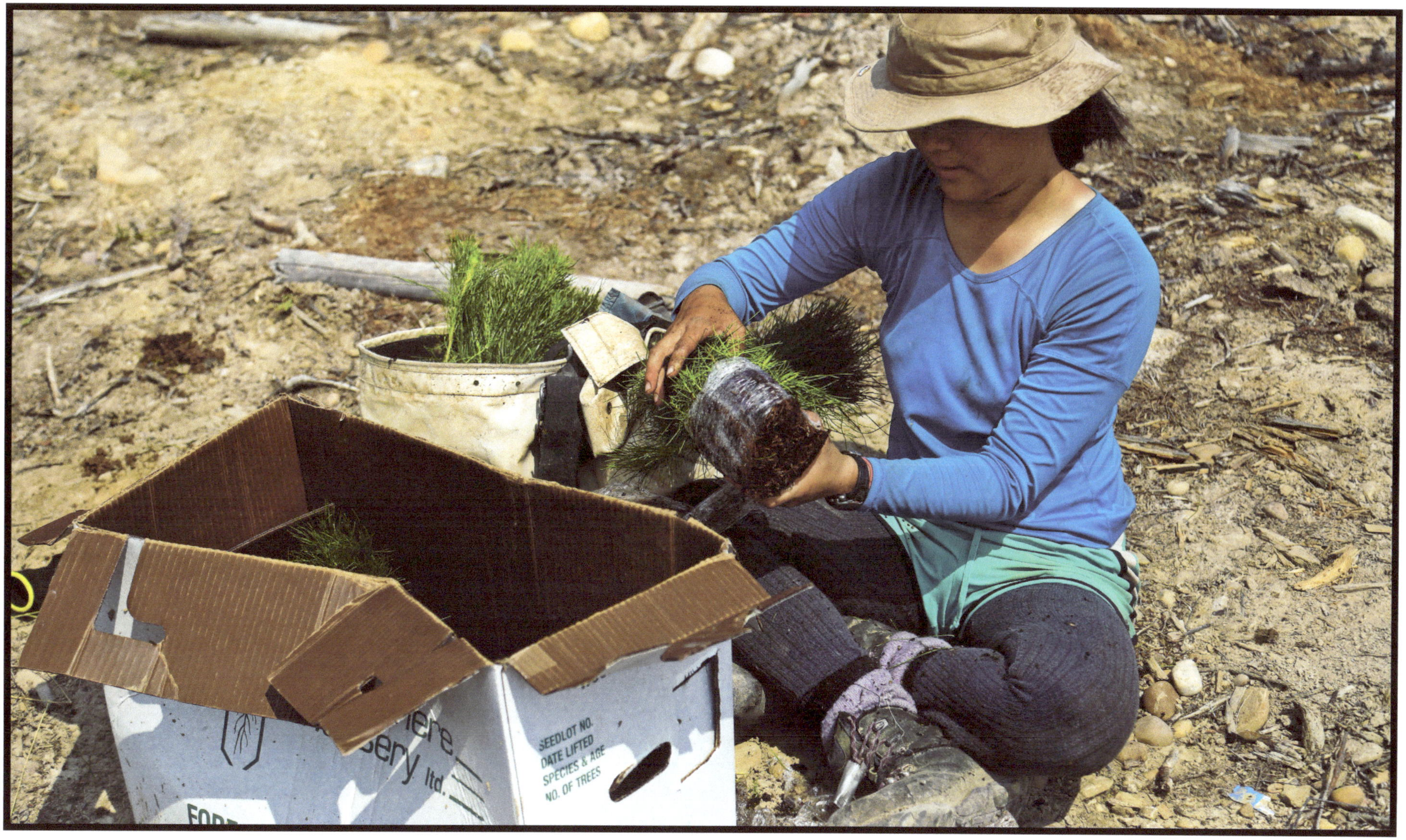

Rachel is loading a box of trees into her planting bags, called "bagging up." Rachel graduated from UBC, and currently works as the Operations Manager of the Espirit Orchestra. Rachel says that, "Although classical music and tree planting seem like very different worlds, I have found that they complement each other and require many of the same skills, including but not limited to discipline, organization, attention to detail, self-motivation, and having a strong work ethic."

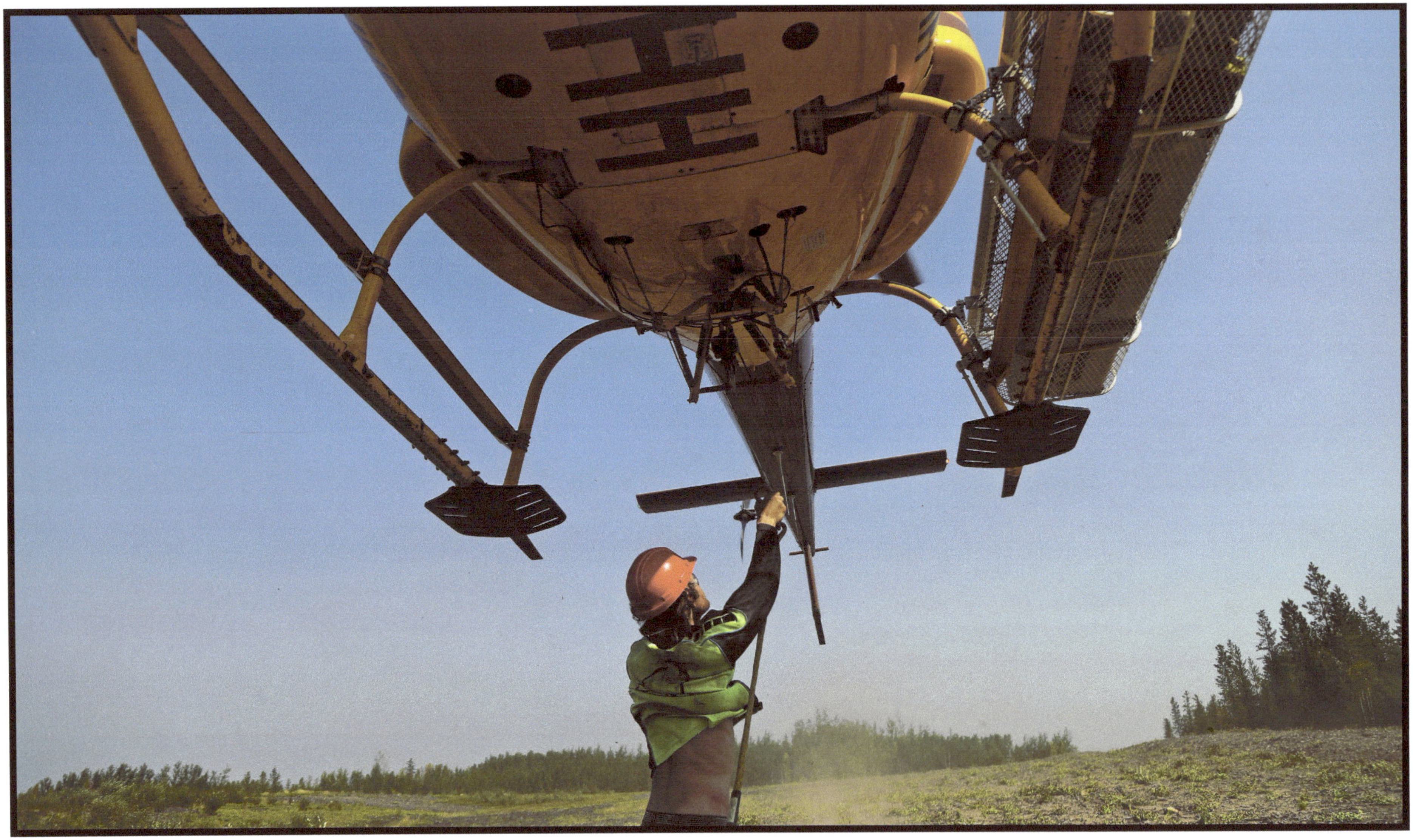

Here we see Sebastian hooking a sling cable directly to the bottom of an approaching helicopter, without the use of an intermediary long-line. This approach is often faster than using a long-line, although it's only suitable for very clean blocks with no brush, where the helicopter is able to safely hover very close to ground level.

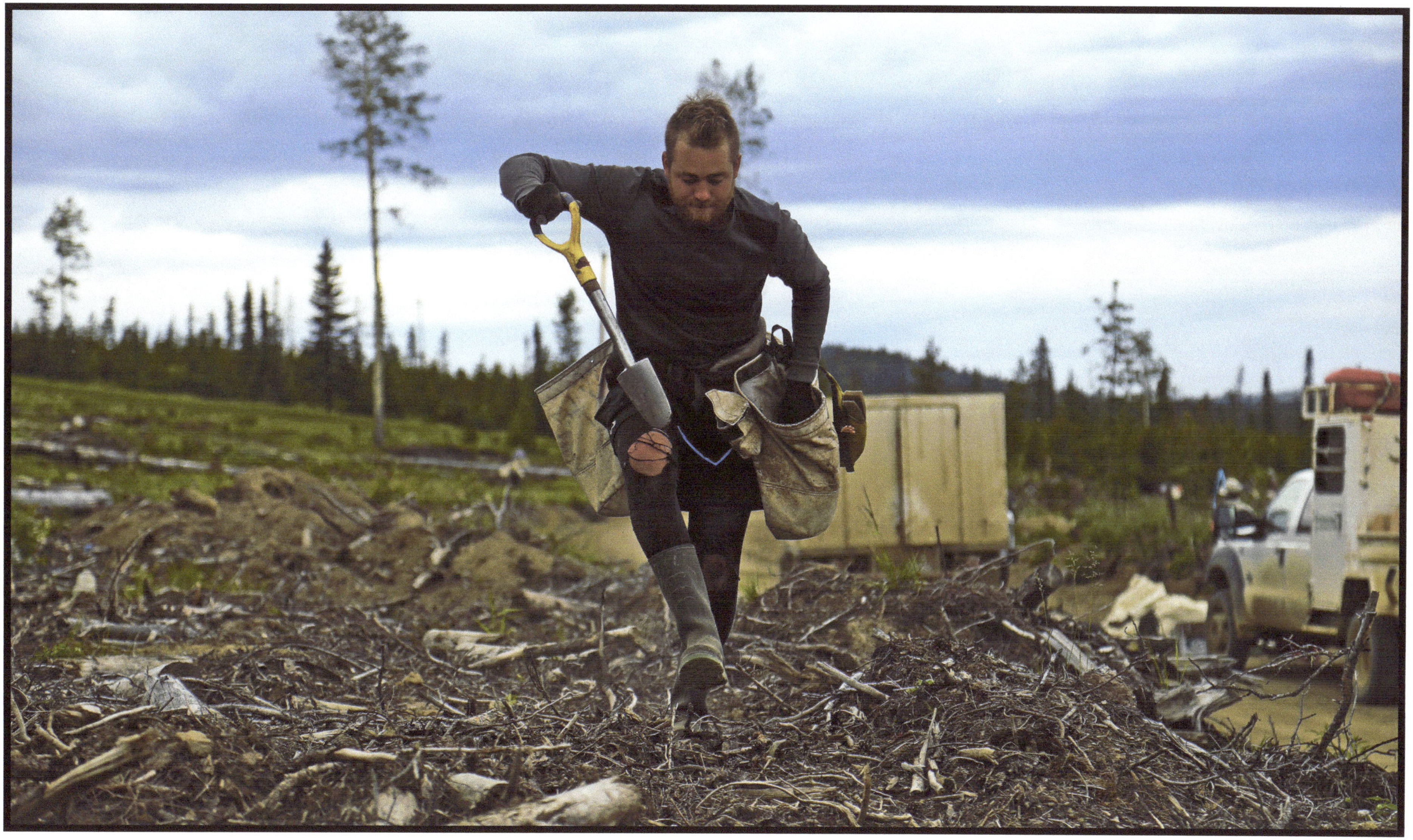

Zach is from Pembroke. He studied at Mount Allison University in New Brunswick, where he obtained a Bachelor of Science majoring in Aviation. He currently works as a flight instructor in New Brunswick, and his other hobbies and interests including camping and skiing.

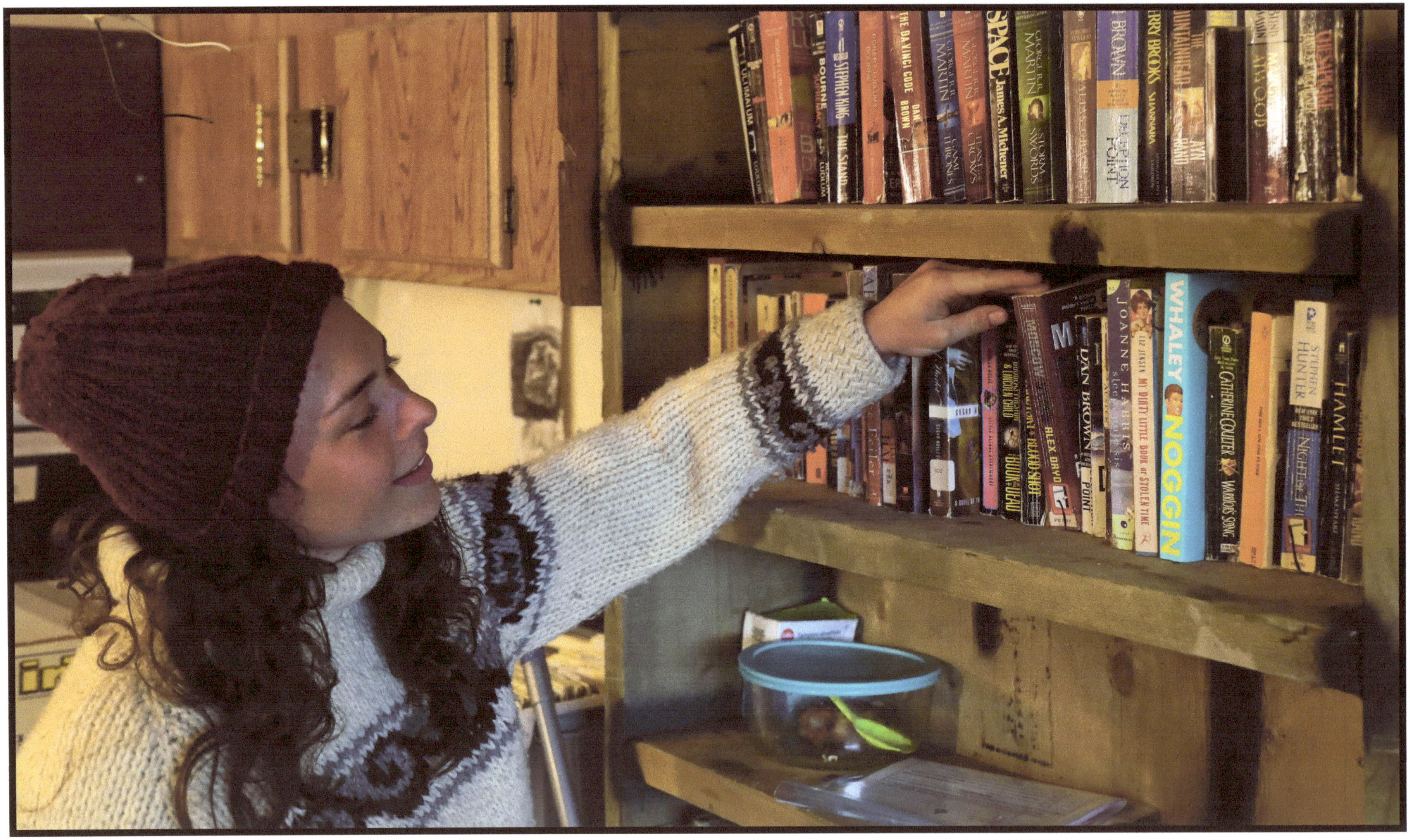

Although most planters are exhausted and fall asleep soon after dinner, a planting camp is a great place to catch up on some recreational reading. If a planting camp has a library with a few hundred books, it's usually quite popular.

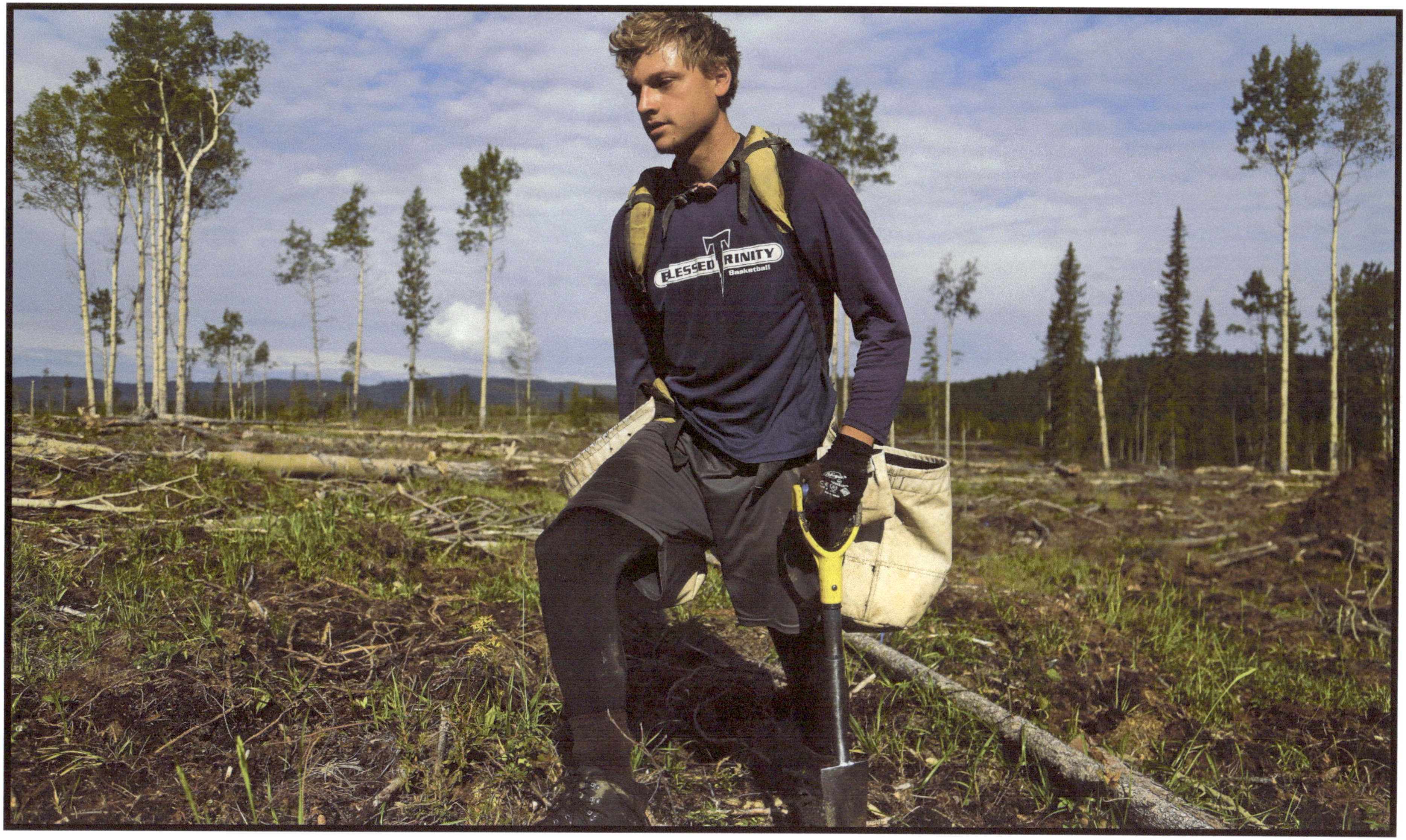

Lucas is from Hamilton. He went to the University of Waterloo, where he studied Environmental Science. He planted for a few seasons in Ontario before he joined the industry in western Canada.

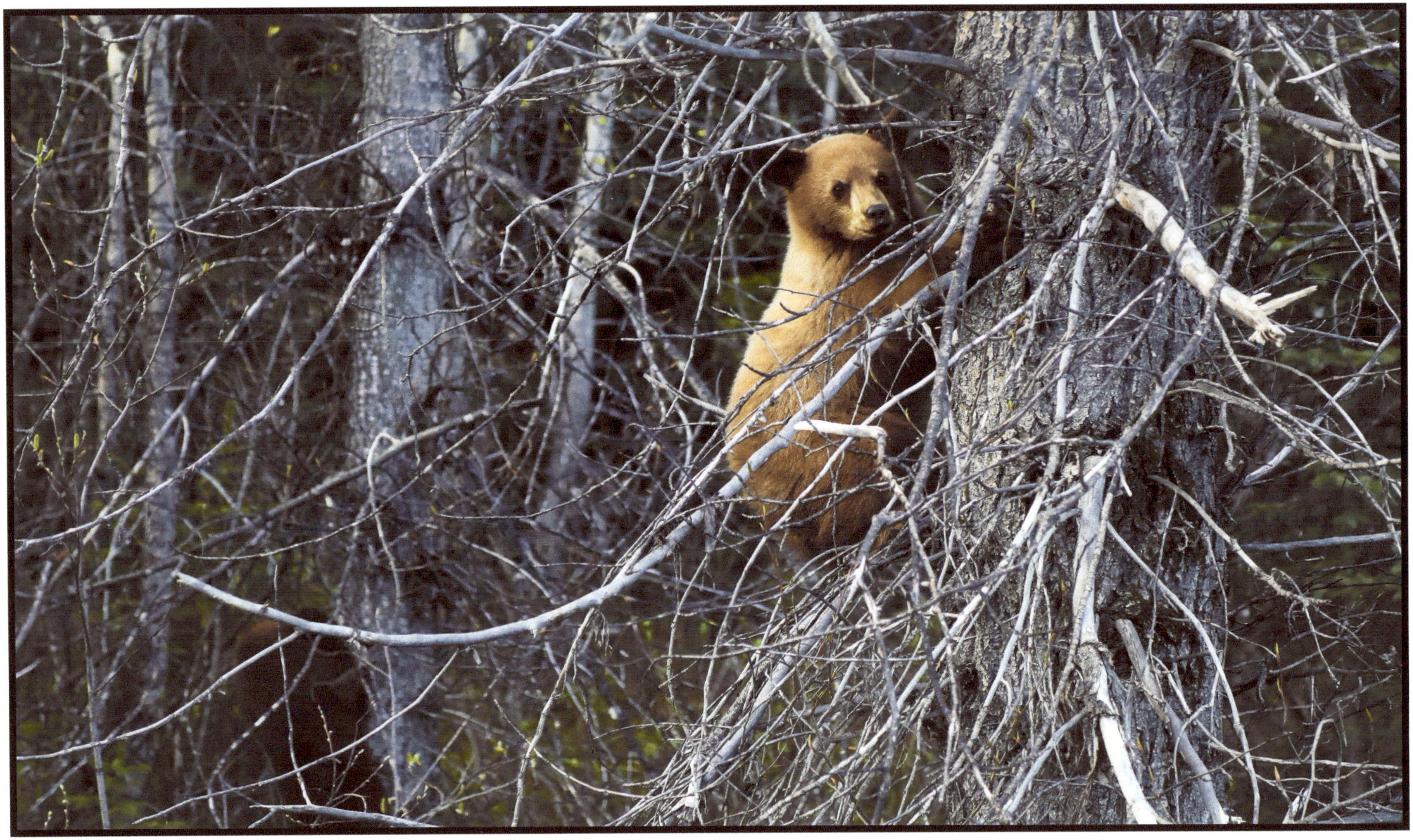

One of the most interesting things about this video is not that it shows a bear in a tree, but that there's a second bear in the tree behind it. I didn't notice the second bear when I originally took this photo.

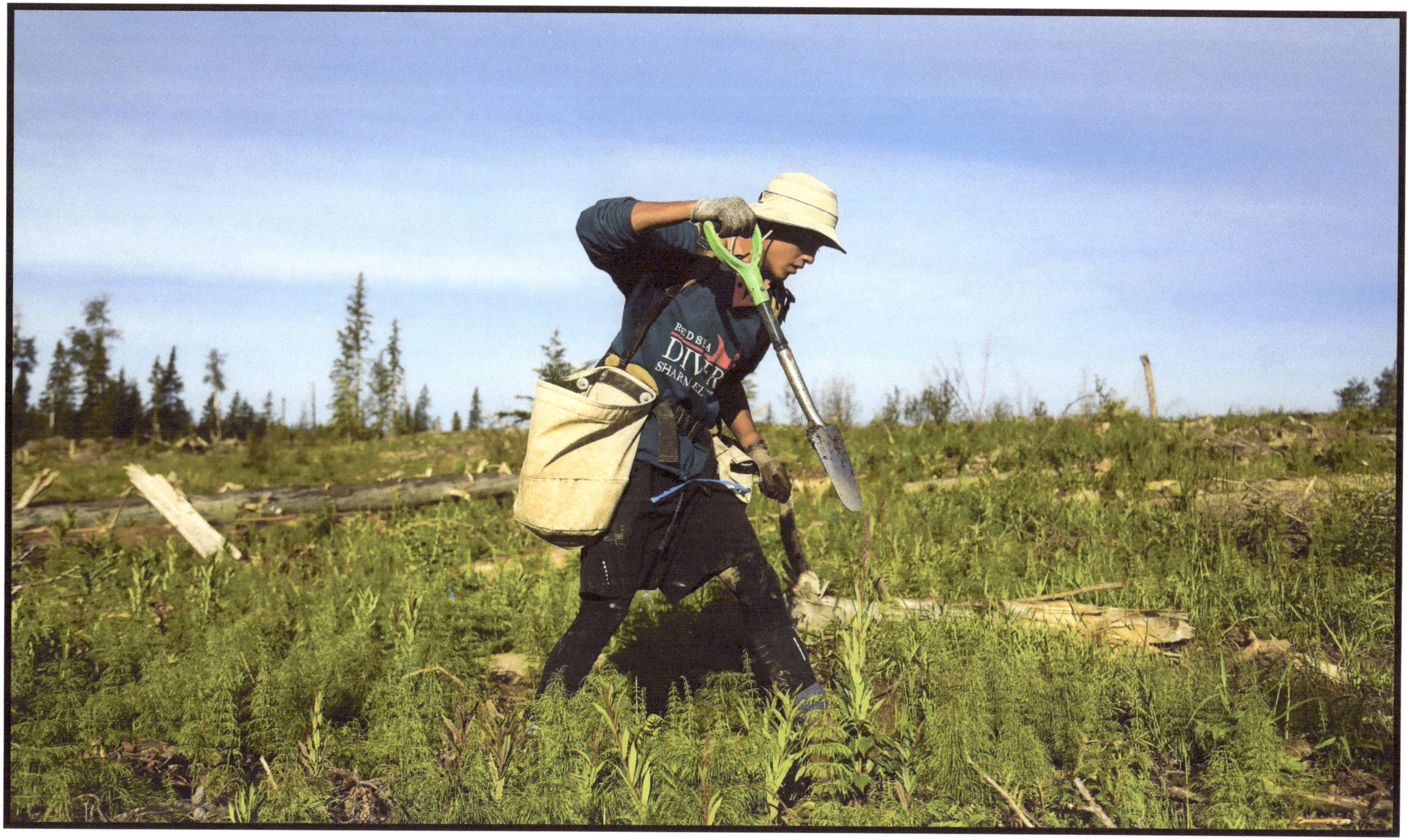

Connor is from Mississauga. He studies Computer Engineering at the University of Western Ontario in London, and loves playing rugby.

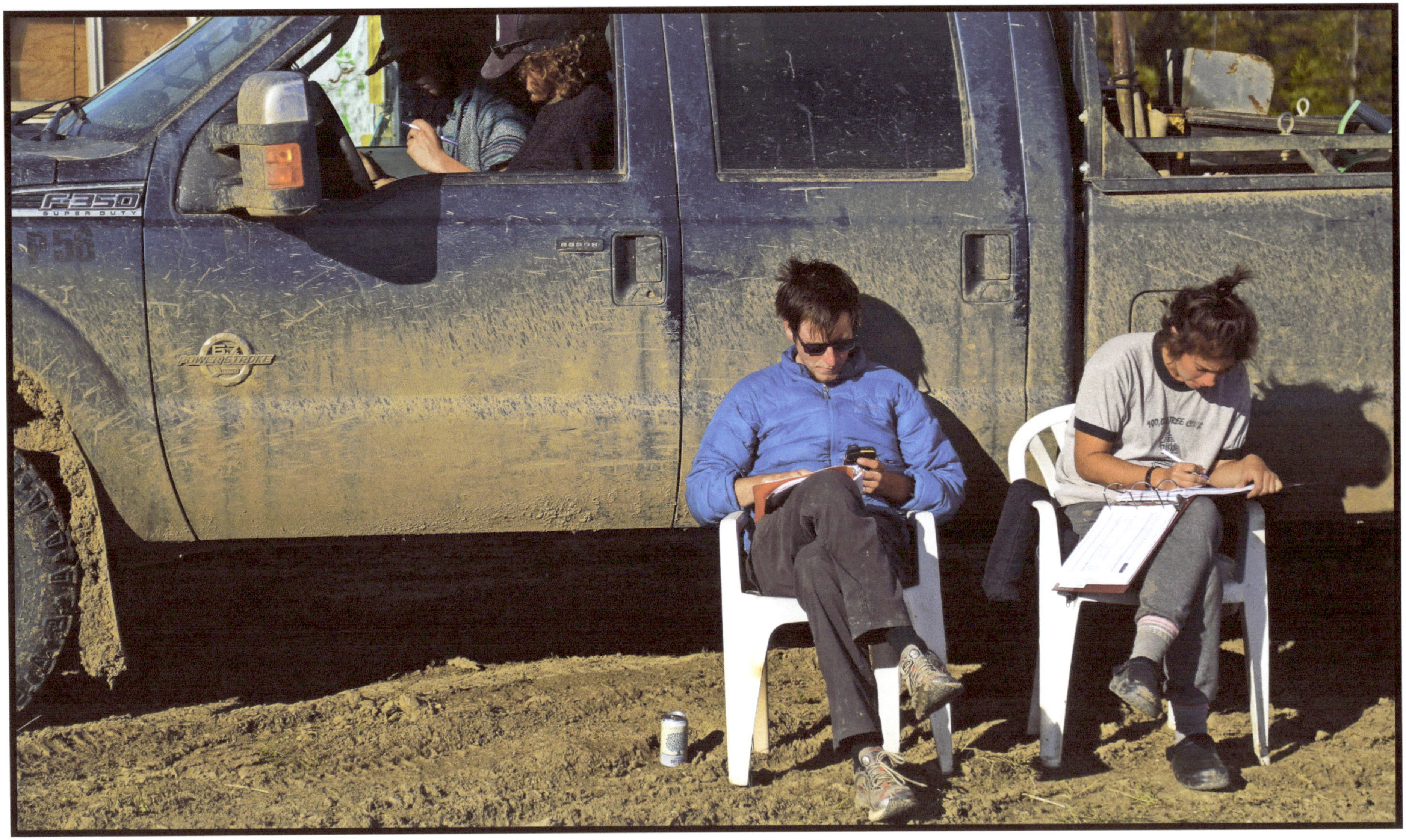

For the management personnel in camp, filling out paperwork seems to be a never-ending responsibility.

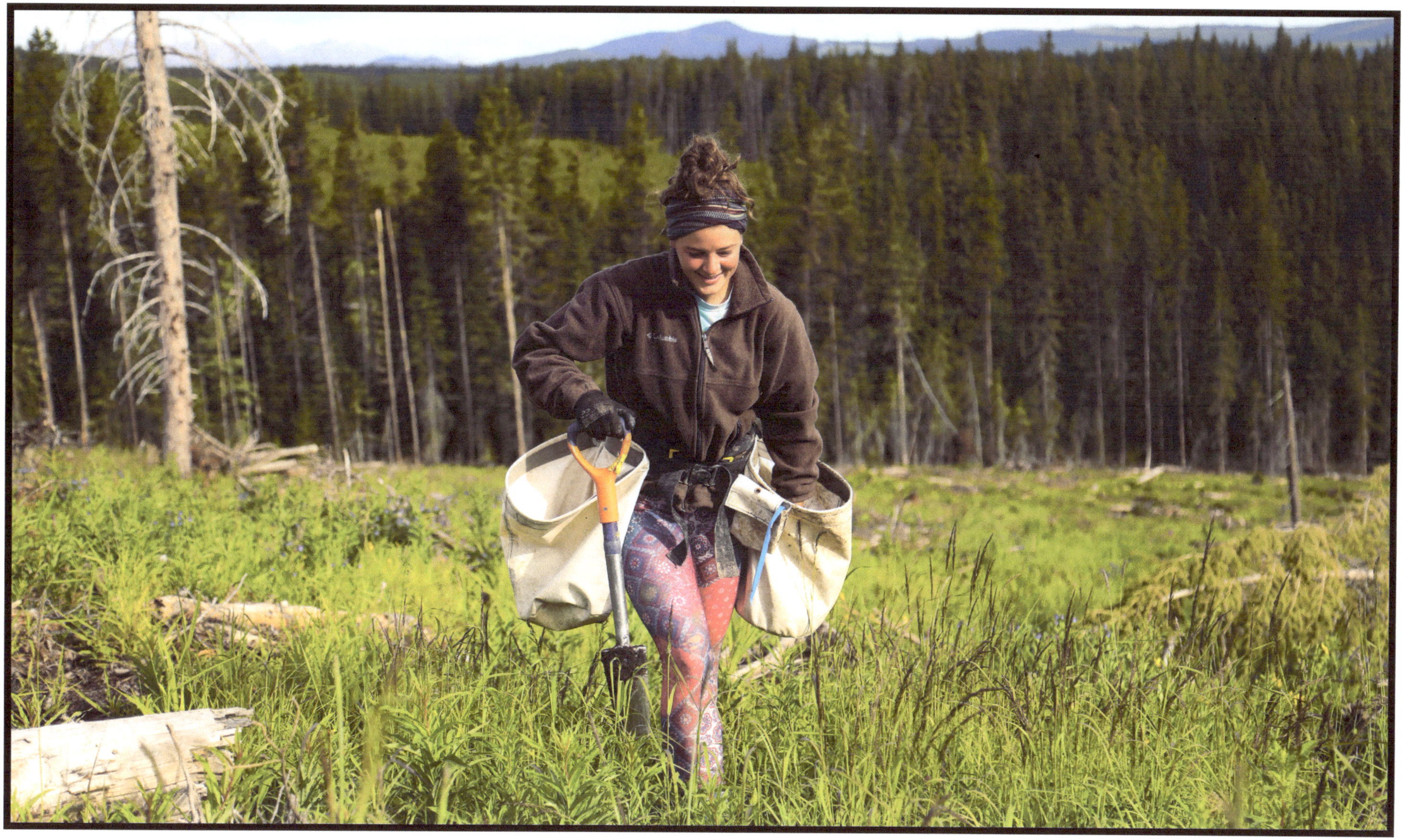

Katie attends York University, where she is finishing a Bachelor of Fine Arts degree majoring in Visual Arts with a minor in Disaster and Emergency Management. Her eventual goal is to work in a humanitarian organization, which would allow her to work on relief projects around the world. In terms of her art, her main focuses are black & white film photography and screen printing.

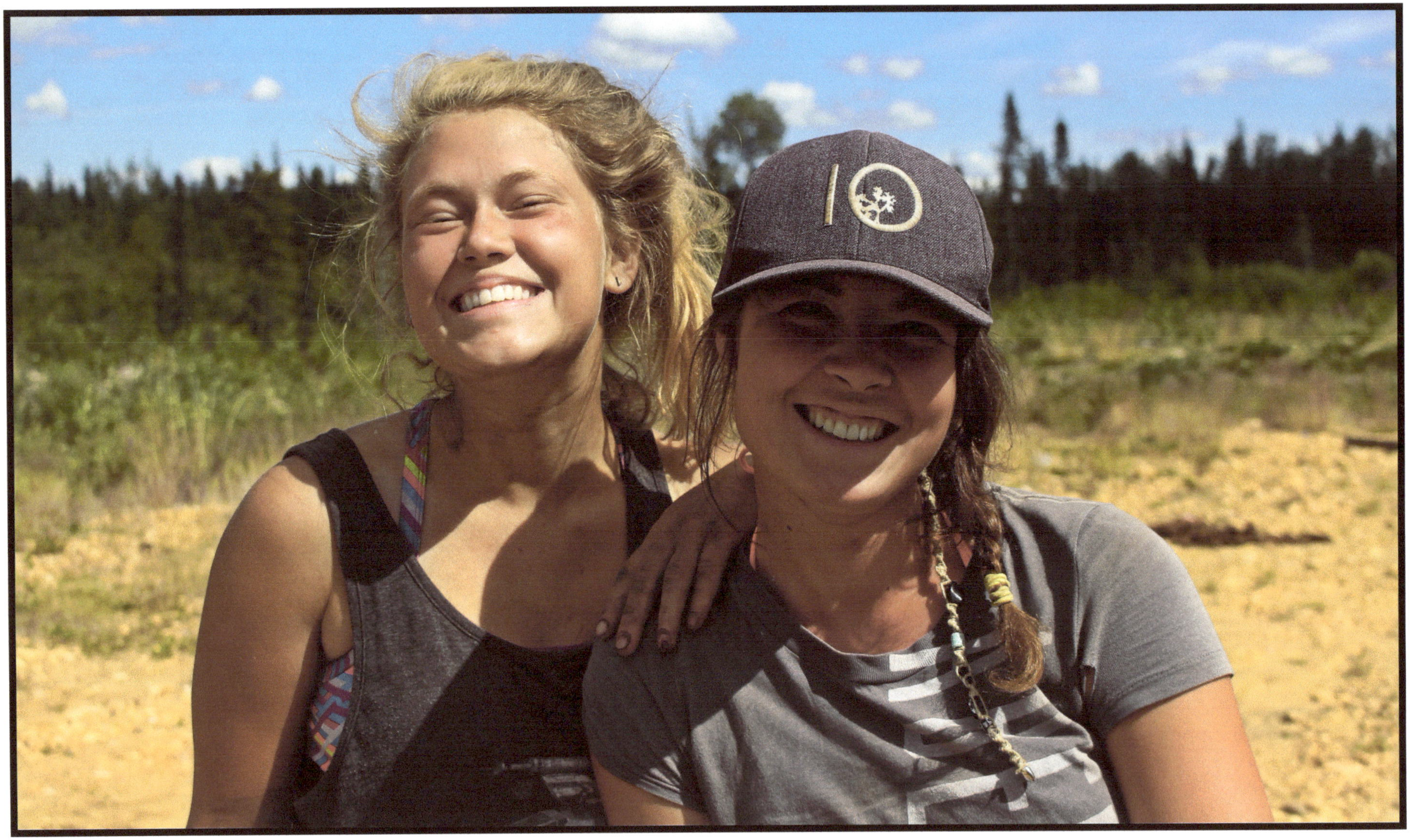

Based on these smiles, it must be a day off.

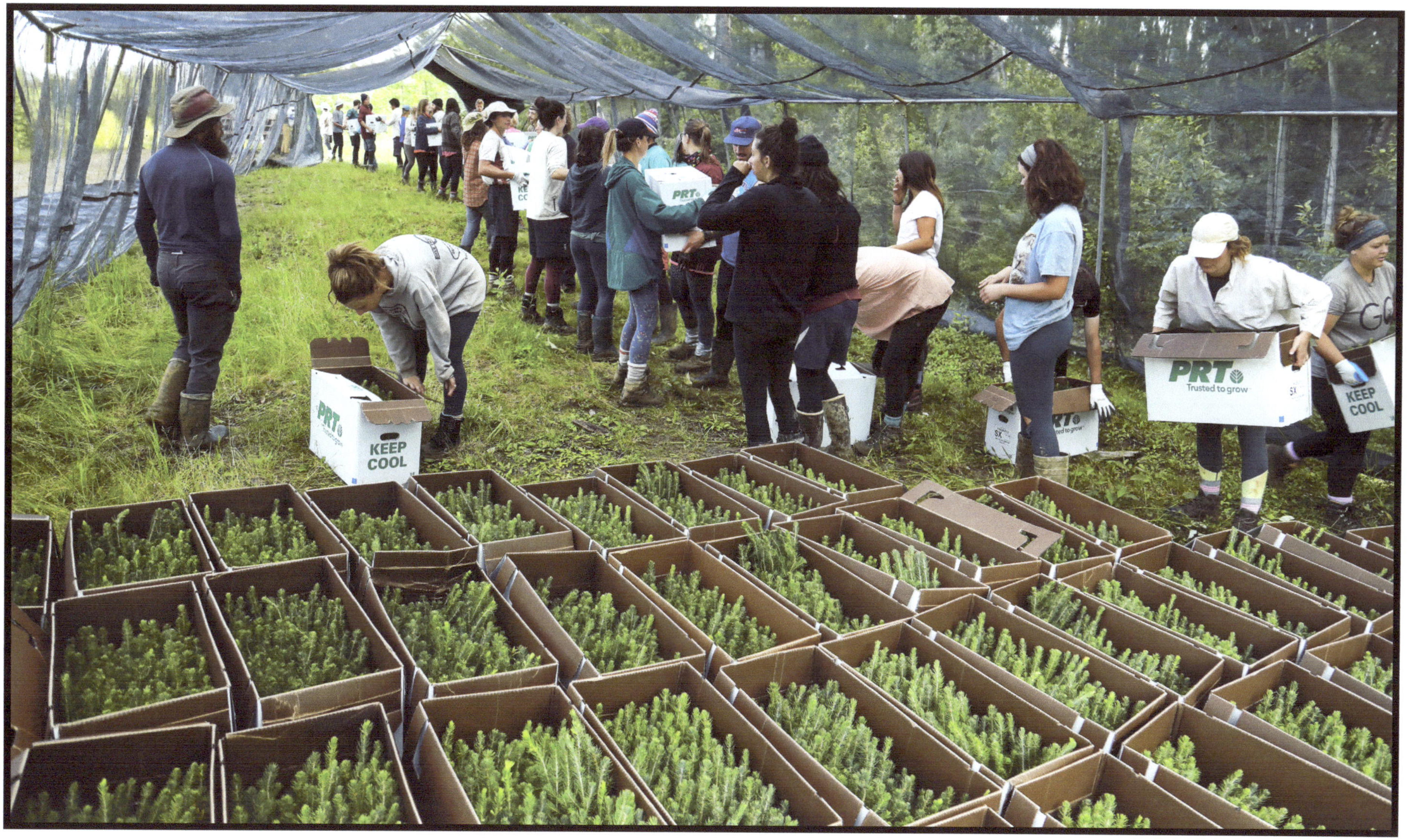

A typical tree shipment contains up to a thousand or more boxes of trees. Usually, it takes a dozen people slightly over an hour to unload a full tractor trailer load of trees, but in this case, we used more people to get the job done faster.

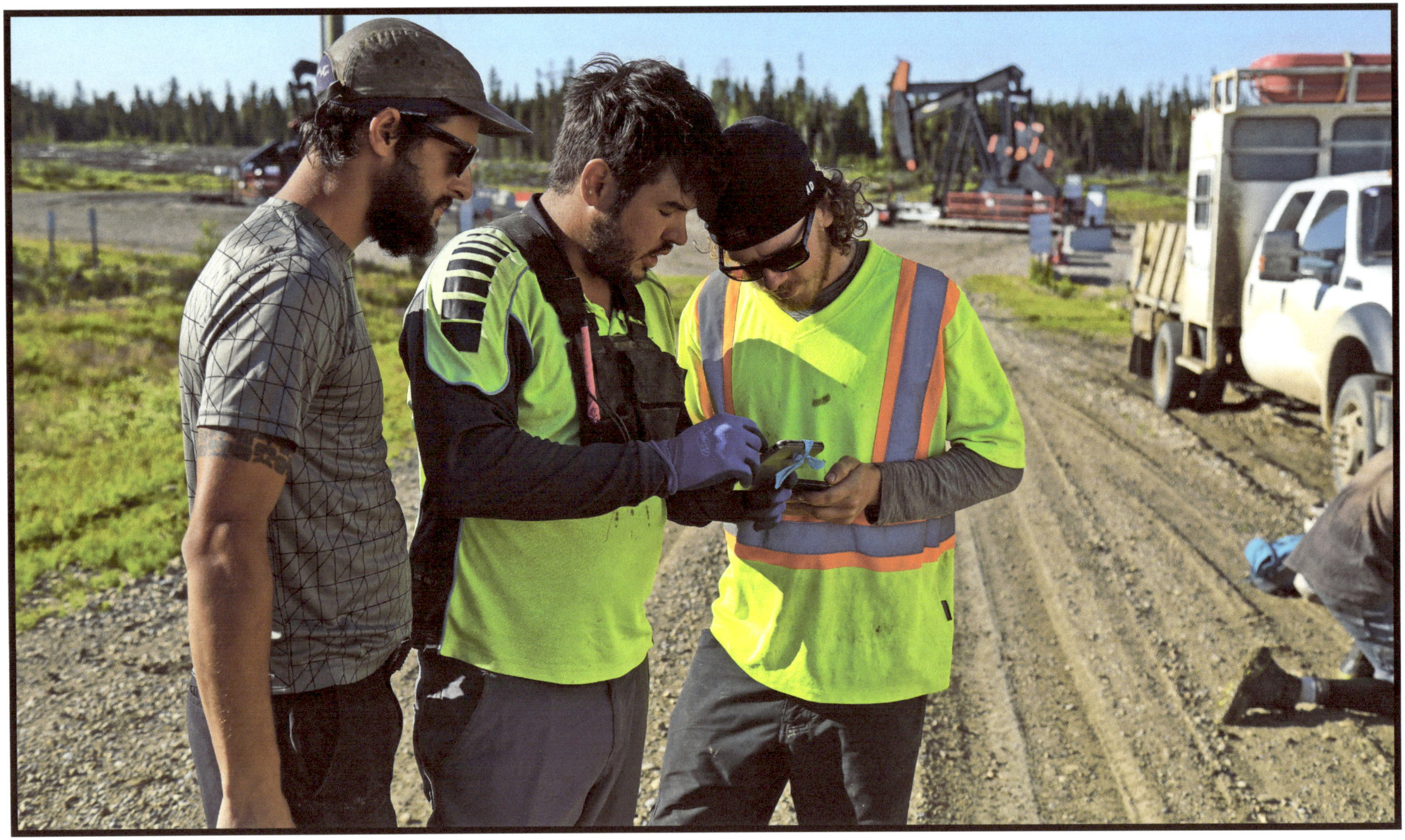

Here we see some of the crew leaders collaborating with the use of digital maps on their phones, to ensure that planters are spread out efficiently across the block.

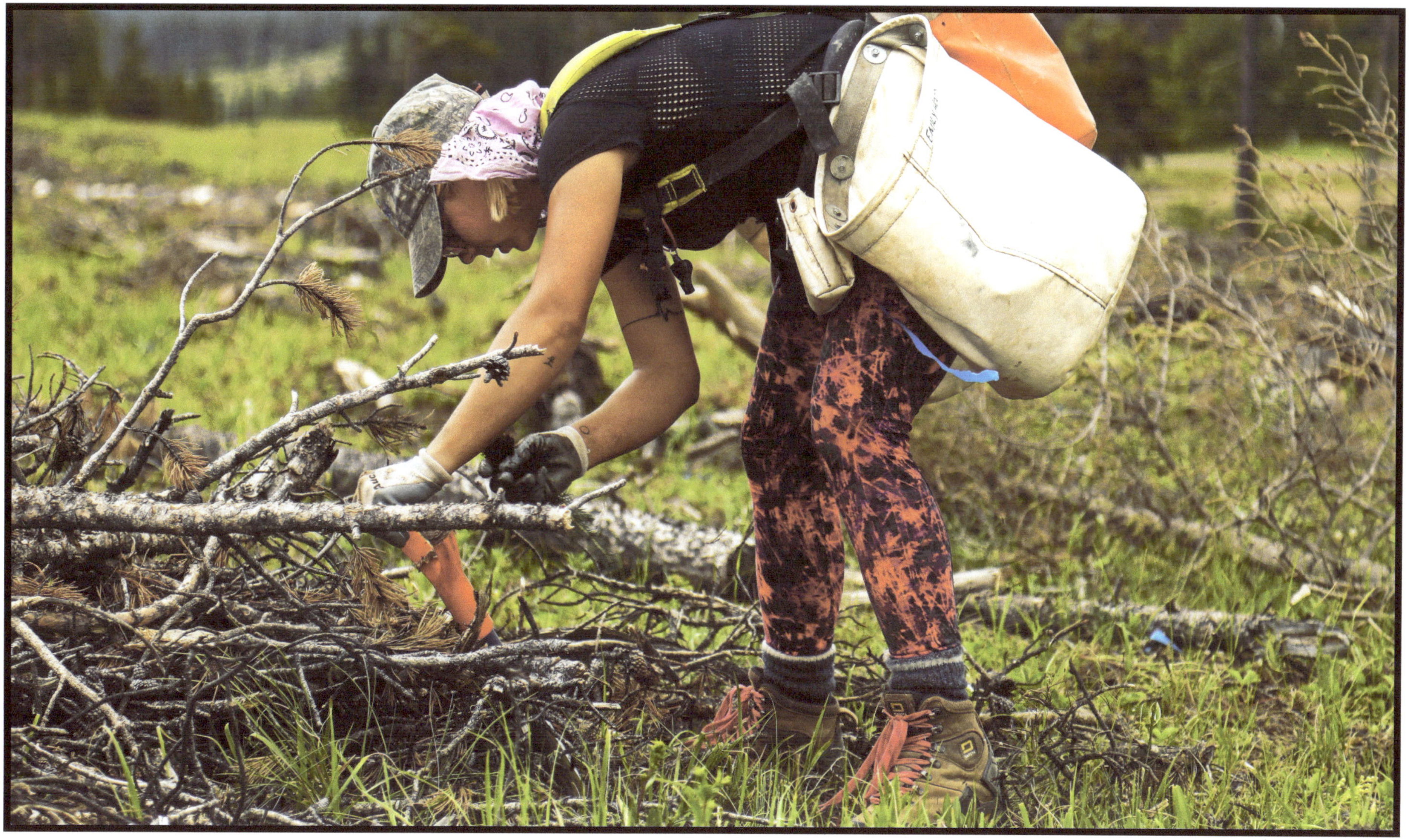

Emily knows that a bandana on the back of the head can keep sunlight or rain off the back of your neck. Emily is from Guelph, but she did an Environmental Sciences degree at Western. When she's not knitting socks, she loves to roller derby.

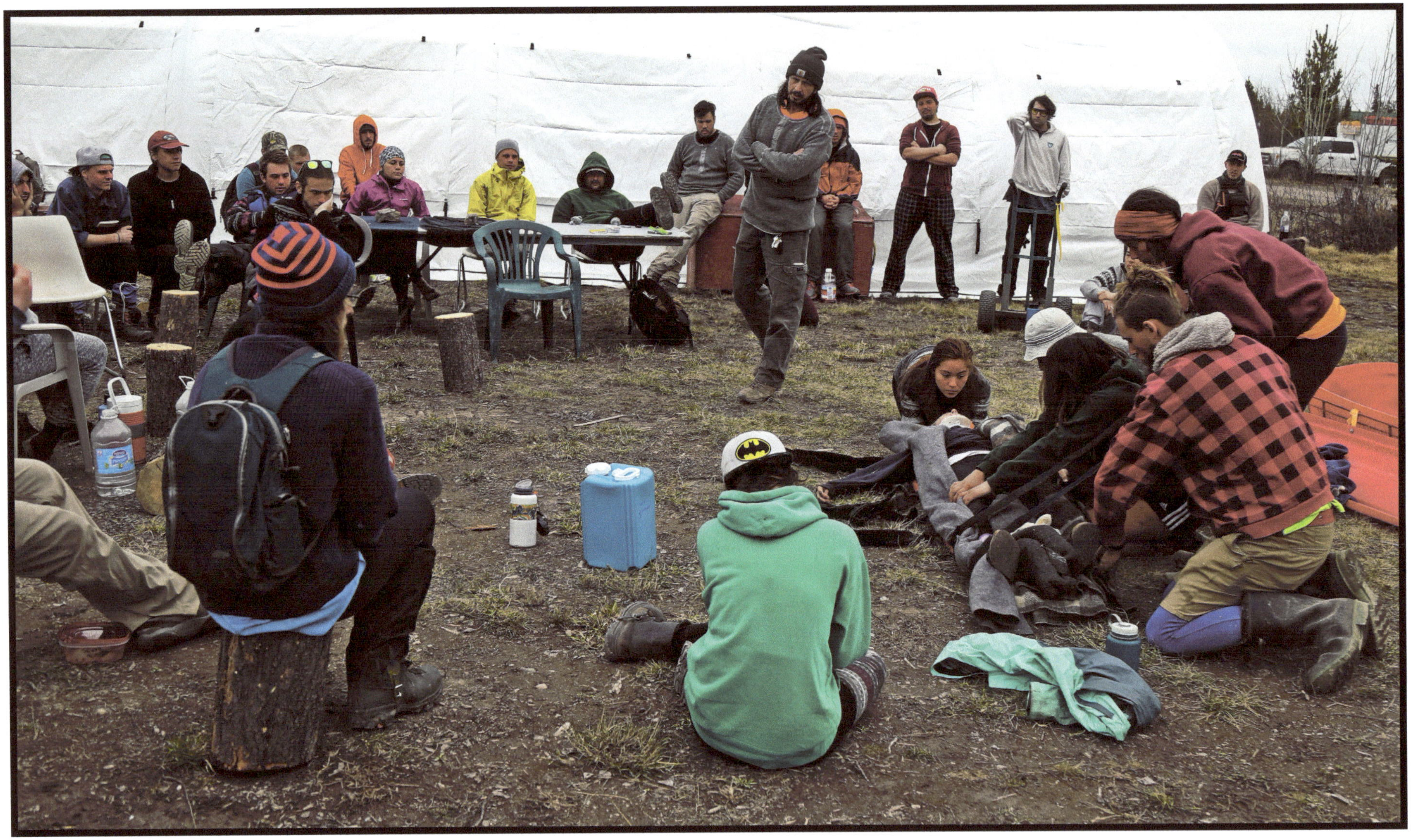

At the start of the season, it's a wise idea to do a first-aid & emergency evacuation drill, so everyone understands what would need to be done if there was a serious emergency.

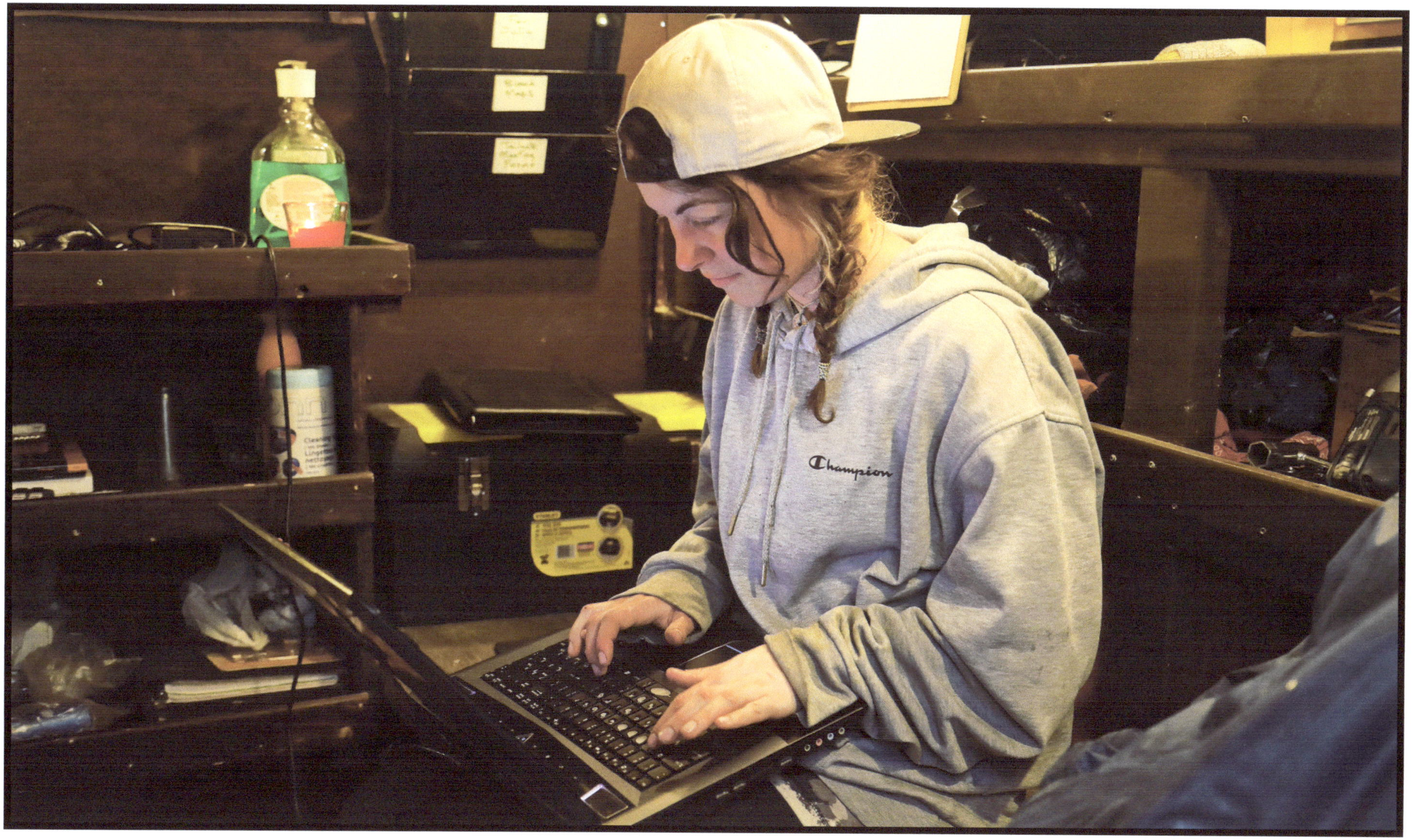

A summary of the events that occurred during the emergency evacuation drill should be shared with the local forester, to demonstrate due diligence for liability management.

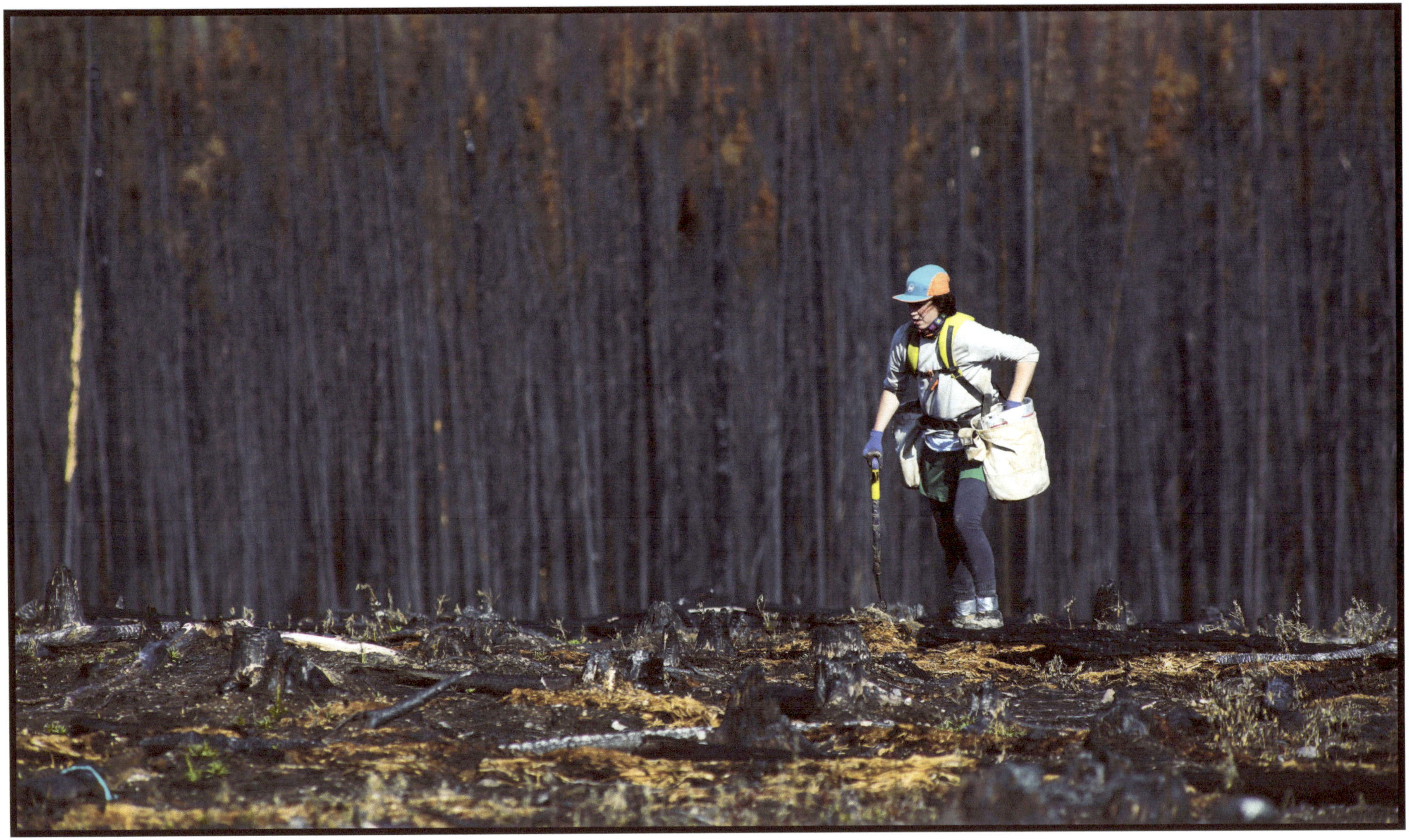

It can be depressing to plant on a cut-block that has been burned by a wildfire, when you think about the damage that has been done to the environment, but it's also good to know that we're helping to repair that damage.

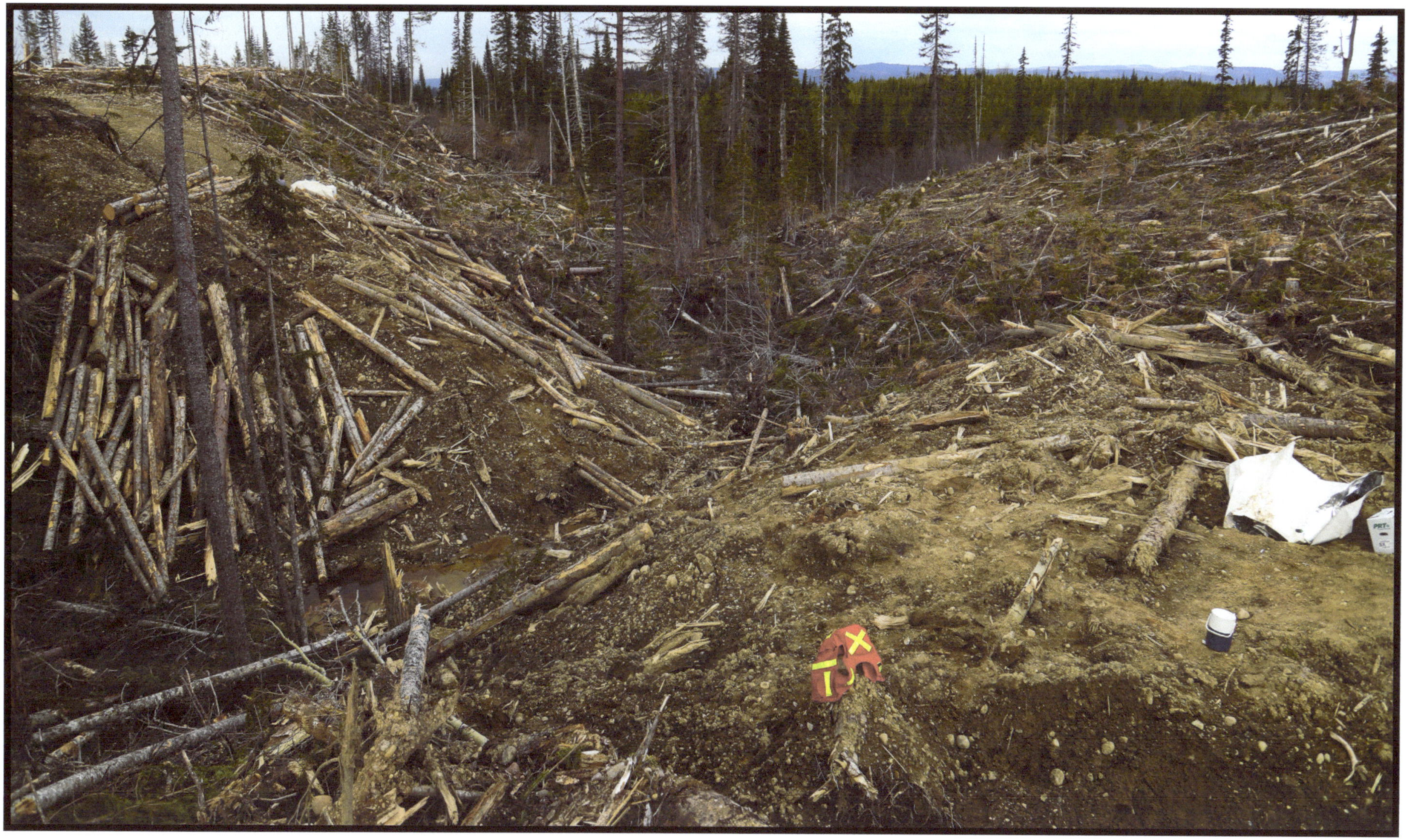

This was a particularly annoying logistical challenge that we had to deal with. On this block, the road was completely removed by a reclamation crew before we started planting, so all of the sixty-plus boxes needed for the back half of the block had to be carried by hand through this large gully.

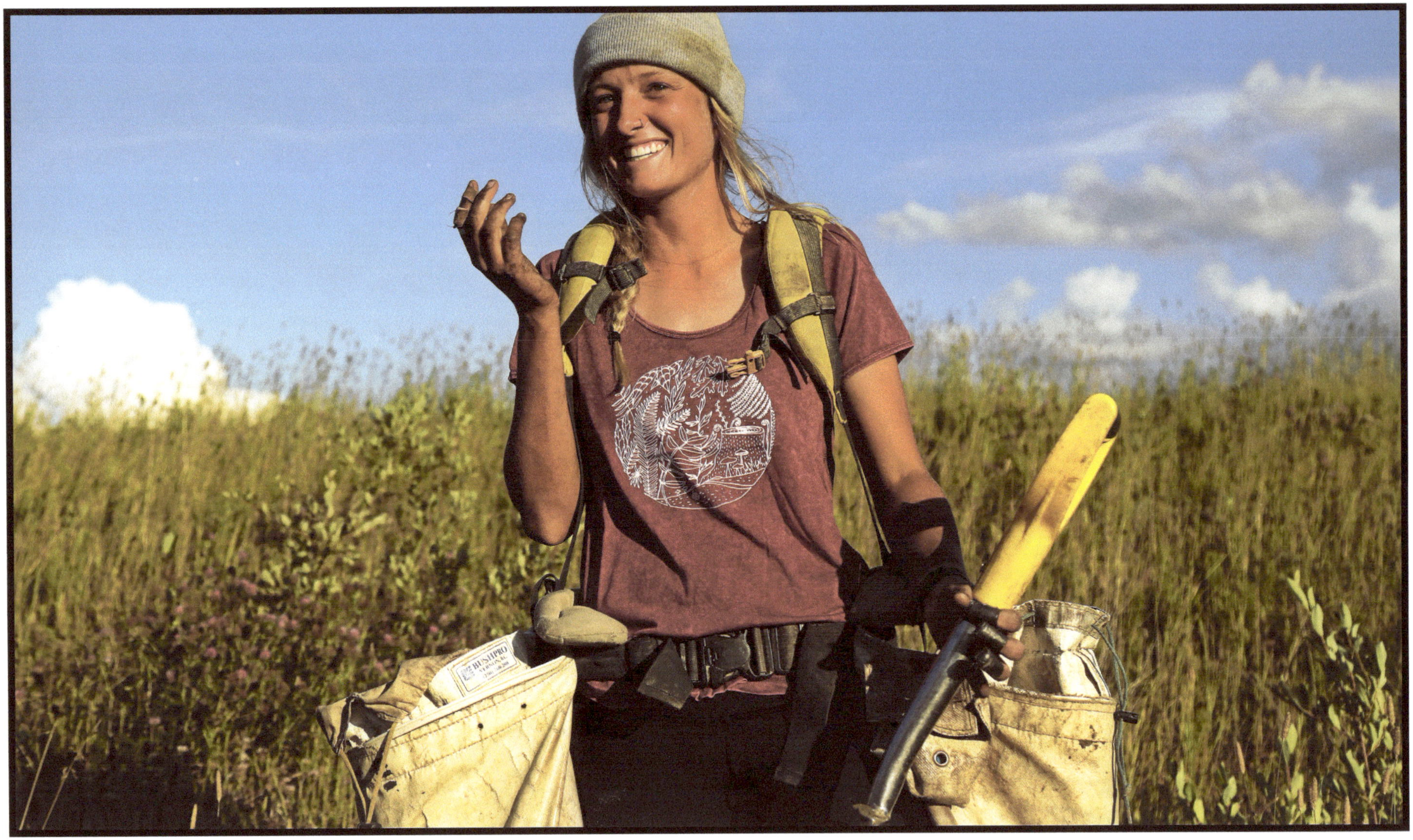

Becca graduated from Ryerson with a Sociology degree, and she eventually wants to become a police officer. In addition to being a very strong planter, Becca is a stellar soccer player.

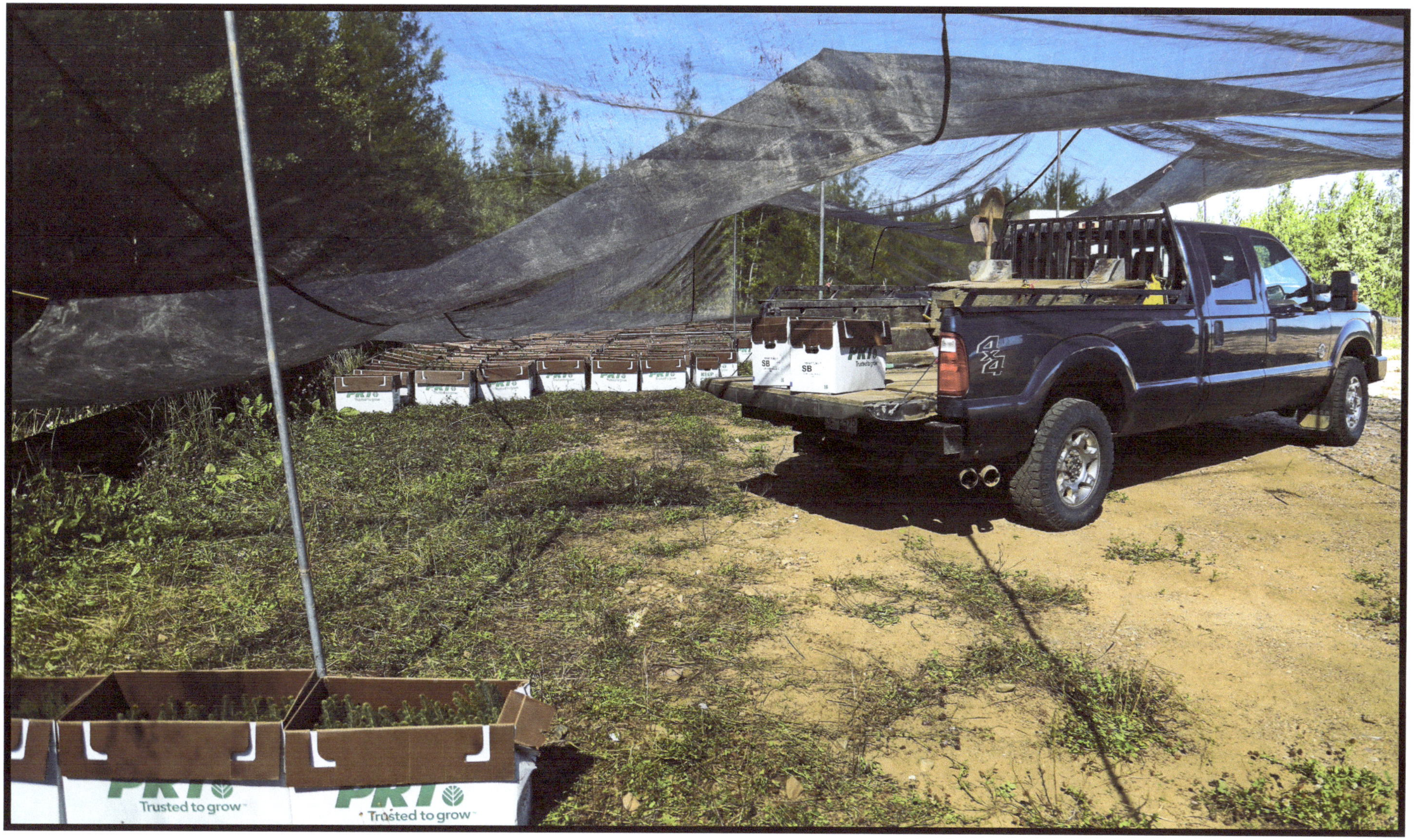

Trees that we plant in the spring were lifted and boxed the previous fall, then frozen in a dormant state over the winter. In the latter half of the summer planting season, most of the trees that we plant are "hot-lifted" trees that come directly from the nursery. These trees are wide awake and growing quickly, and are usually stored out in the open with a black mesh "shade cloth" overhead. This allows the trees to get enough sunlight to aid in photosynthesis and respiration.

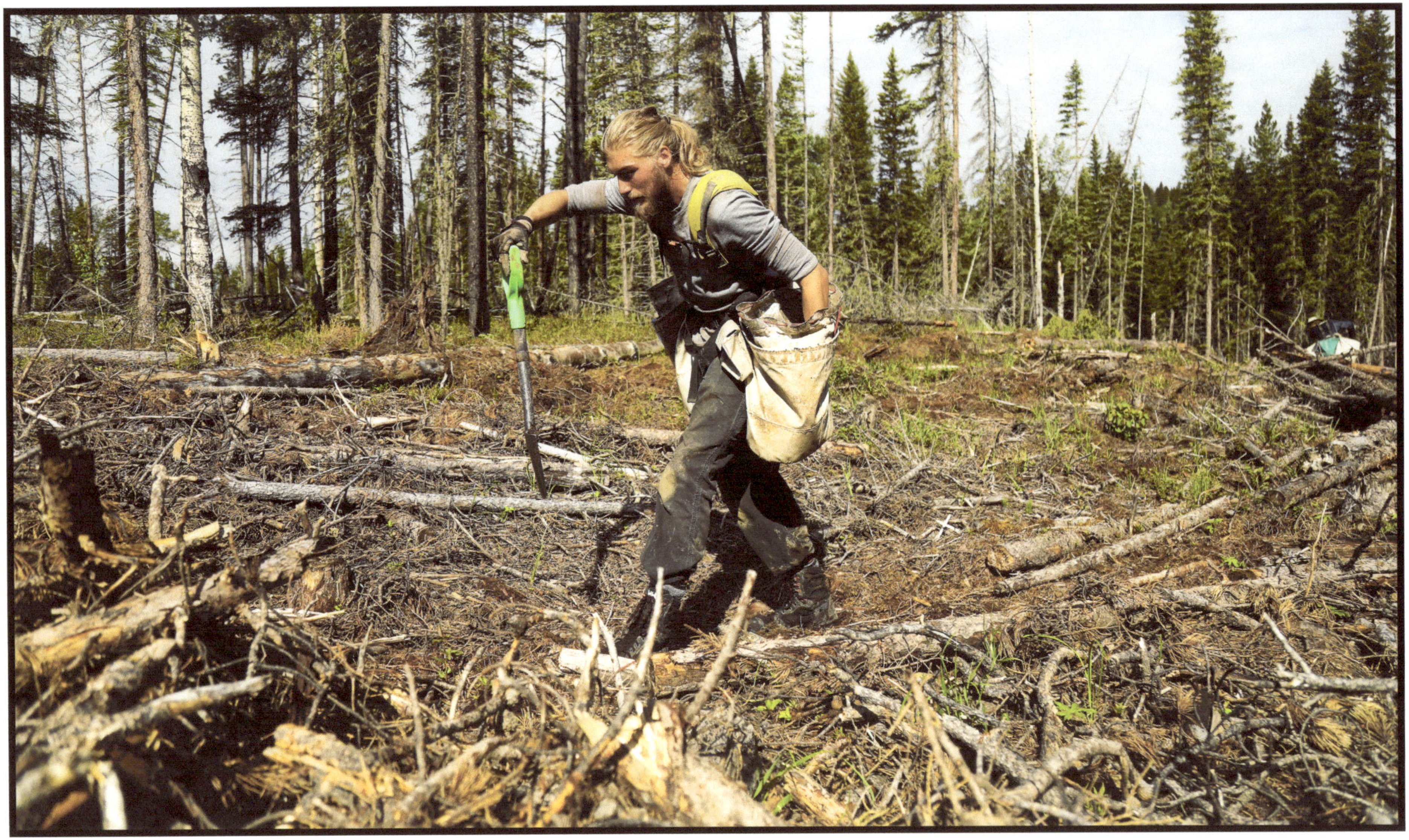

Jackson, who grew up in Niagara Falls, studies History at Brock with a minor in German language. Jackson describes himself as a serious workaholic who restricts himself to a plant-based diet. He plans to spend time travelling and working in several additional countries in the next few years, including New Zealand, Iceland, Germany, and China.

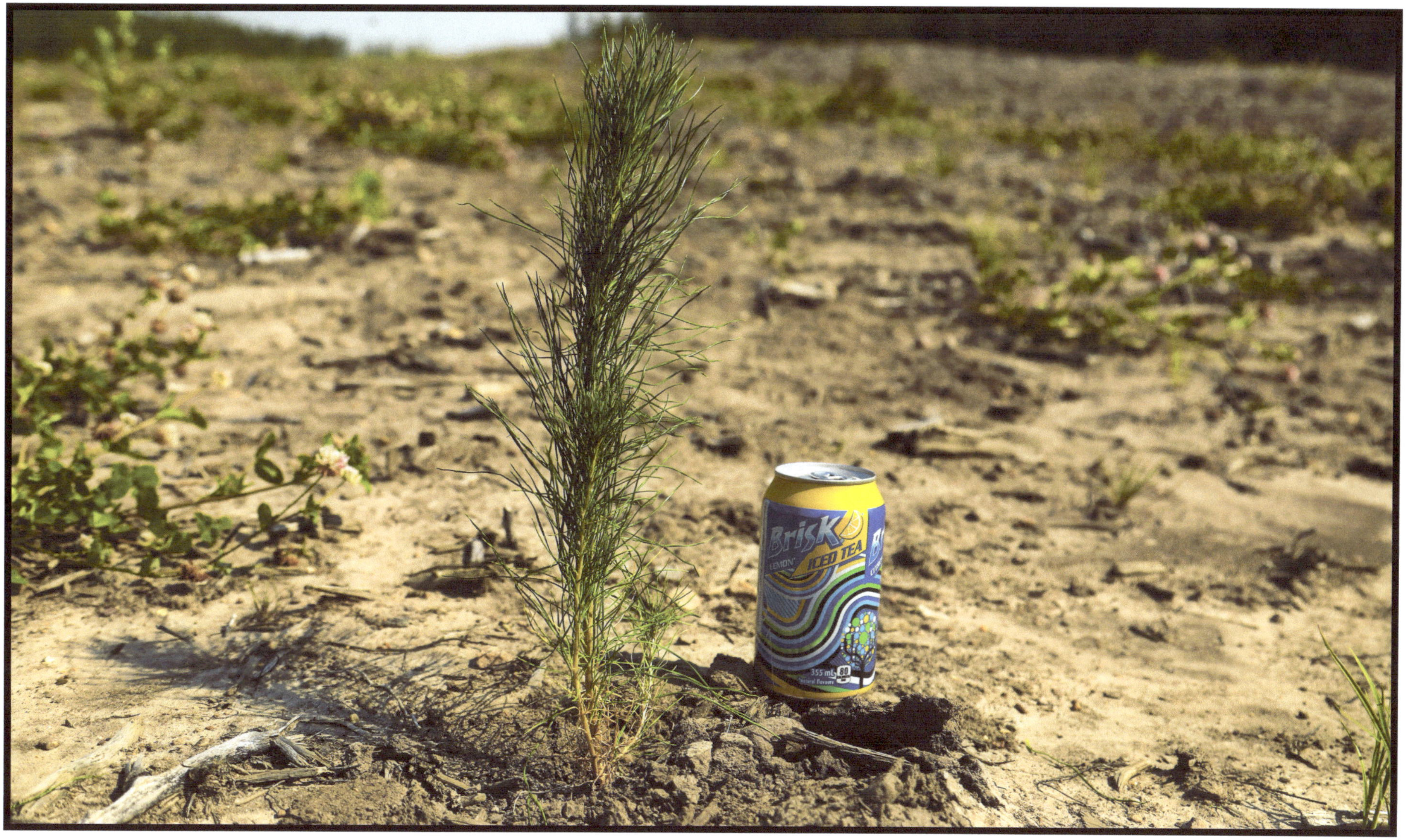

The Jack Pine seedlings that we were planting on this particular project were much larger than the seedlings that we typically plant. The iced tea can is simply there for scale.

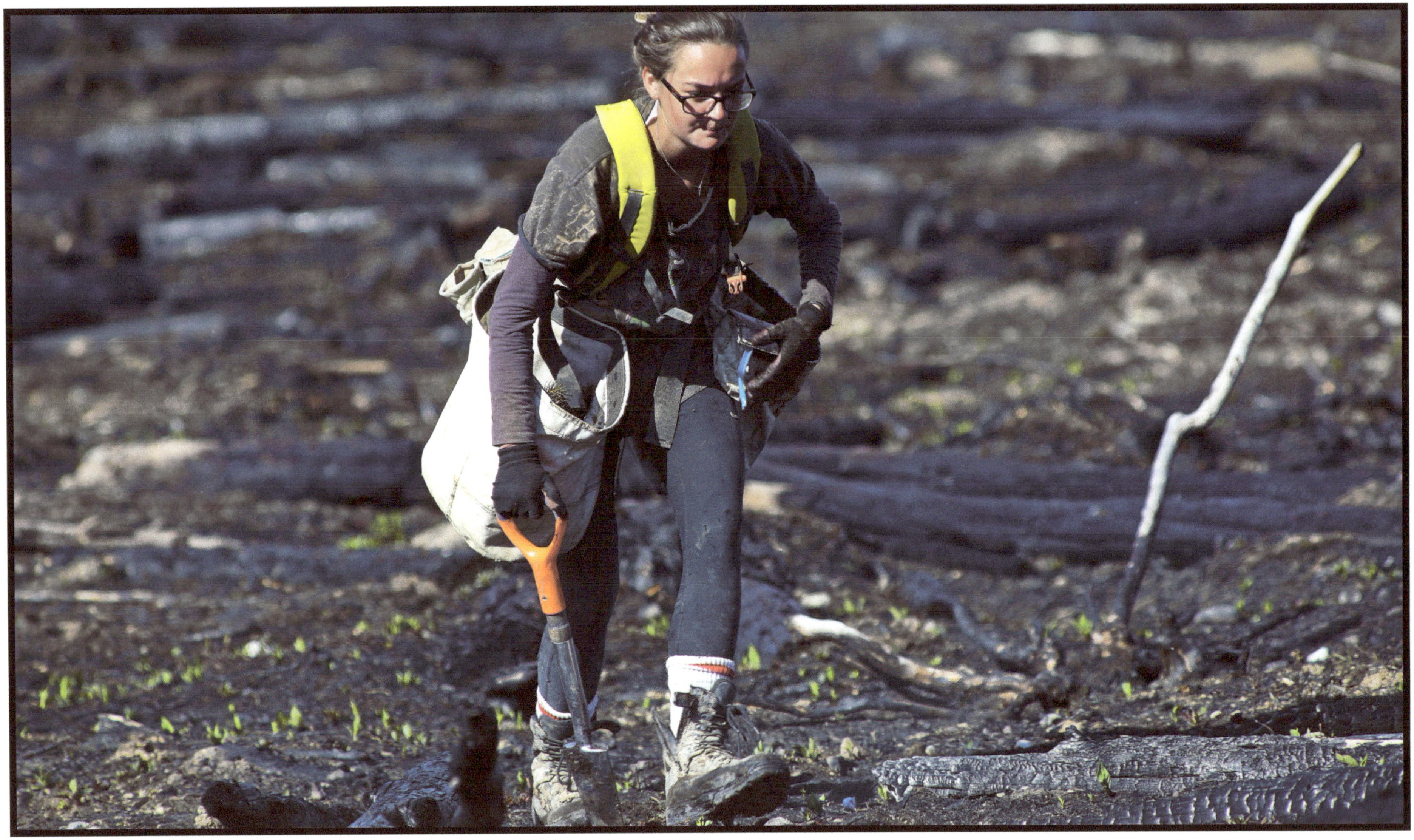

Here's the face of a determined planter.

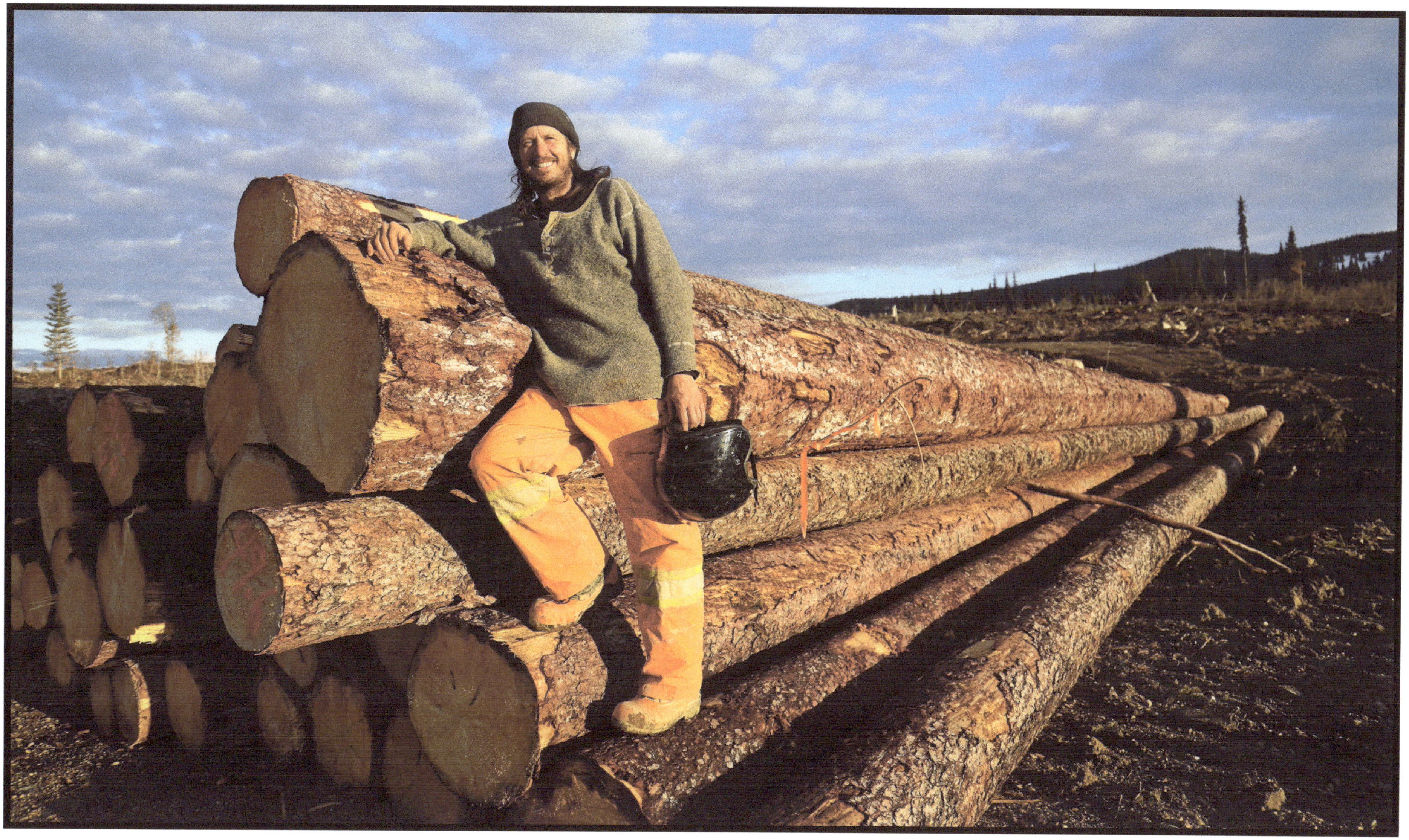

I took a few minutes here to pose for a photo on a log deck while delivering trees to a block after dinner.

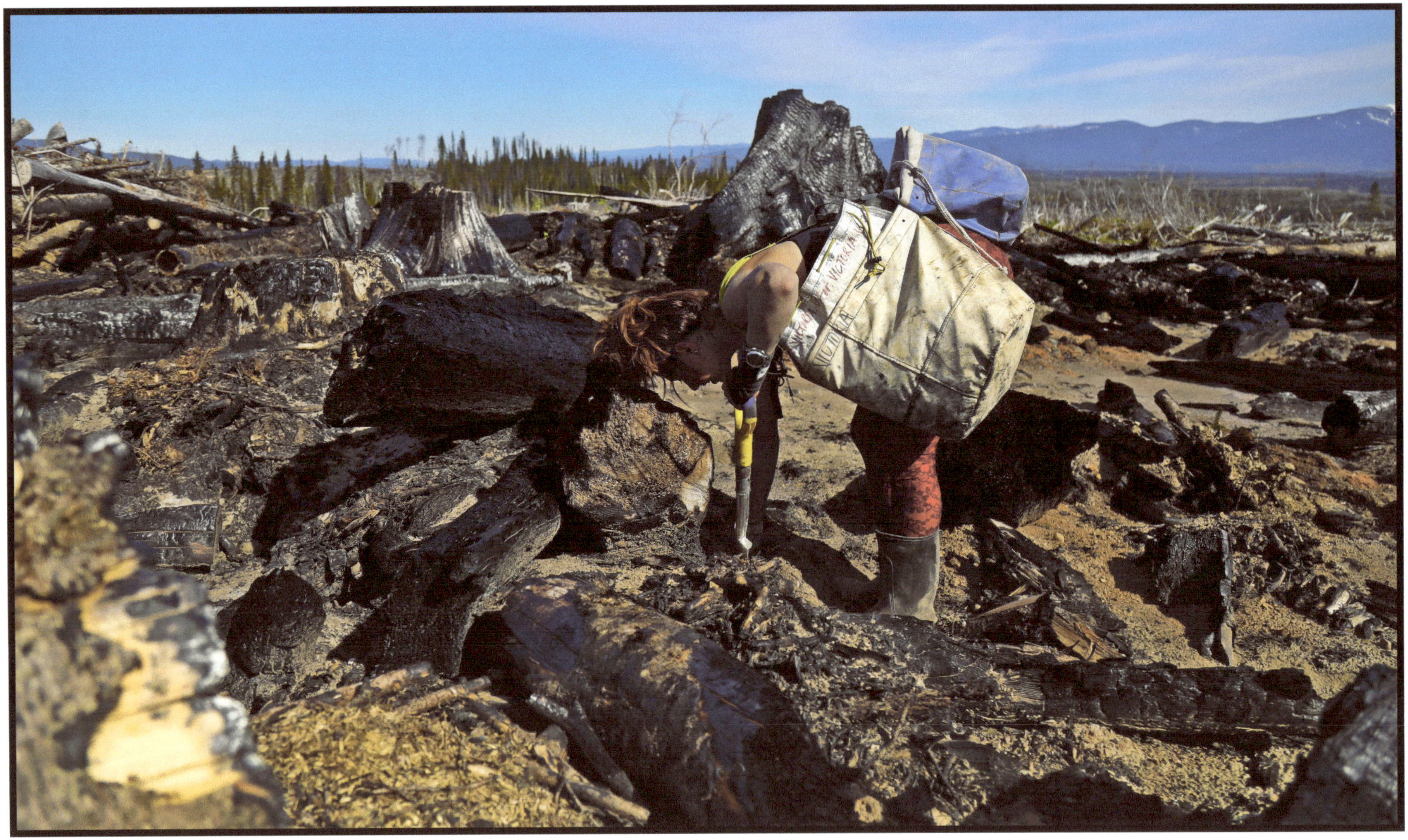

When there is a lot of slash on a block, the logging company will sometimes clean it up into piles that are subsequently burned in the winter. This ensures that the slash doesn't take up otherwise productive ground. Planters usually enjoy finding burns in their pieces, because the planting in those small areas is so easy.

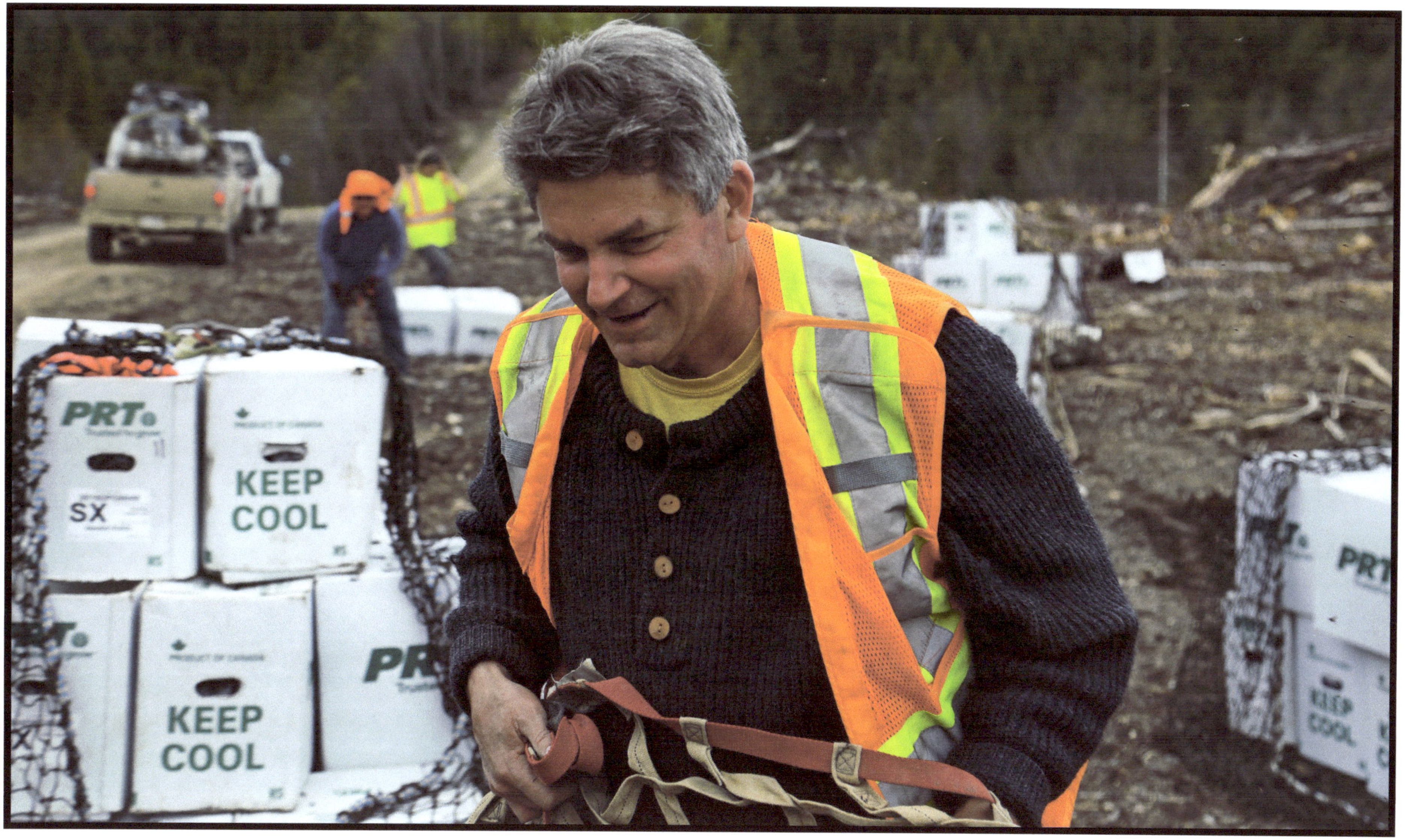

Jim has been a well-known figure in the industry since the early 1980's. He currently works with Folklore Contracting, a company based in Prince George, BC.
www.folklorecontracting.ca

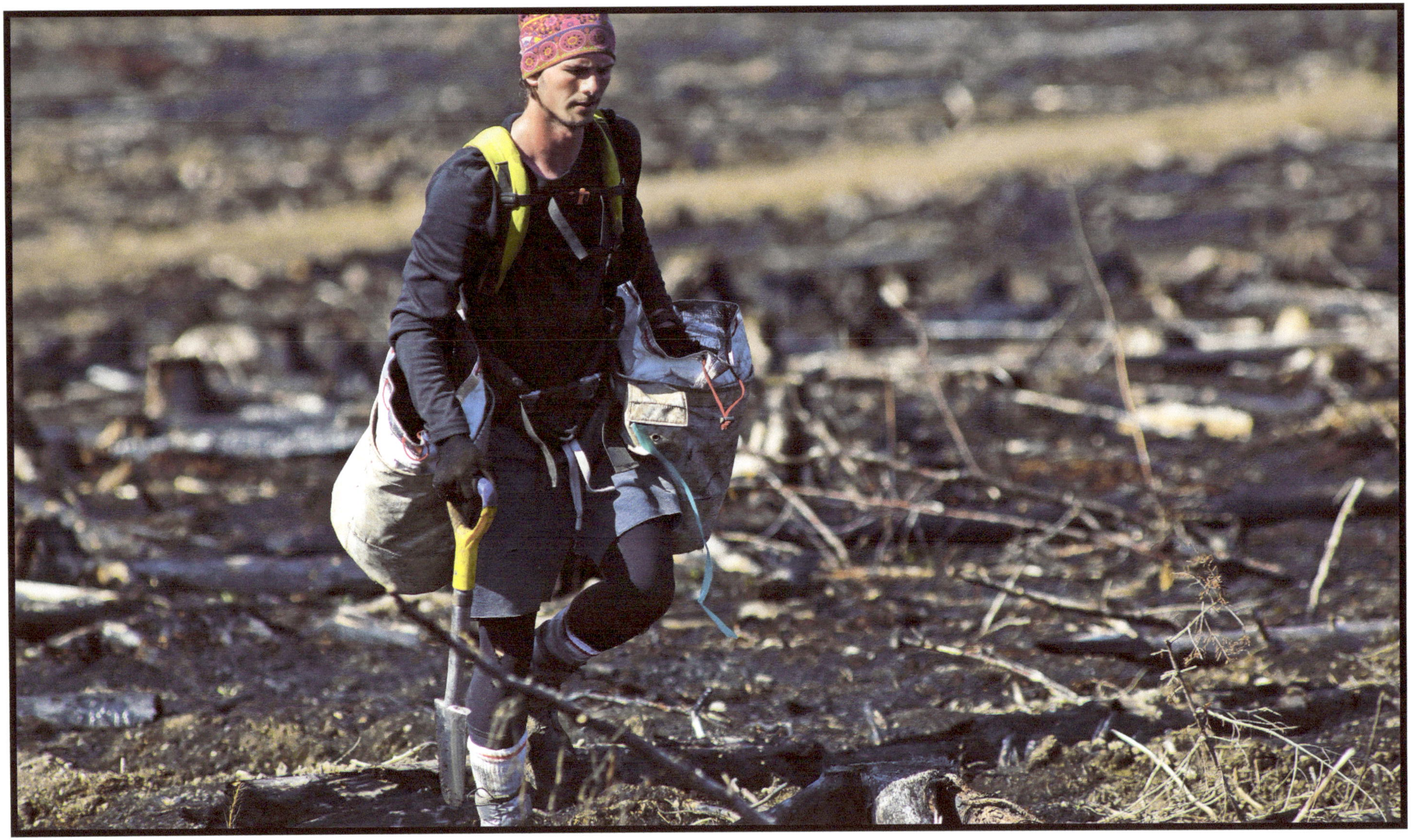

Jordan is from Ottawa, but he's currently studying at St. Francis Xavier University in Nova Scotia where he's enrolled in a Bachelor of Business Administration program with a major in finance.

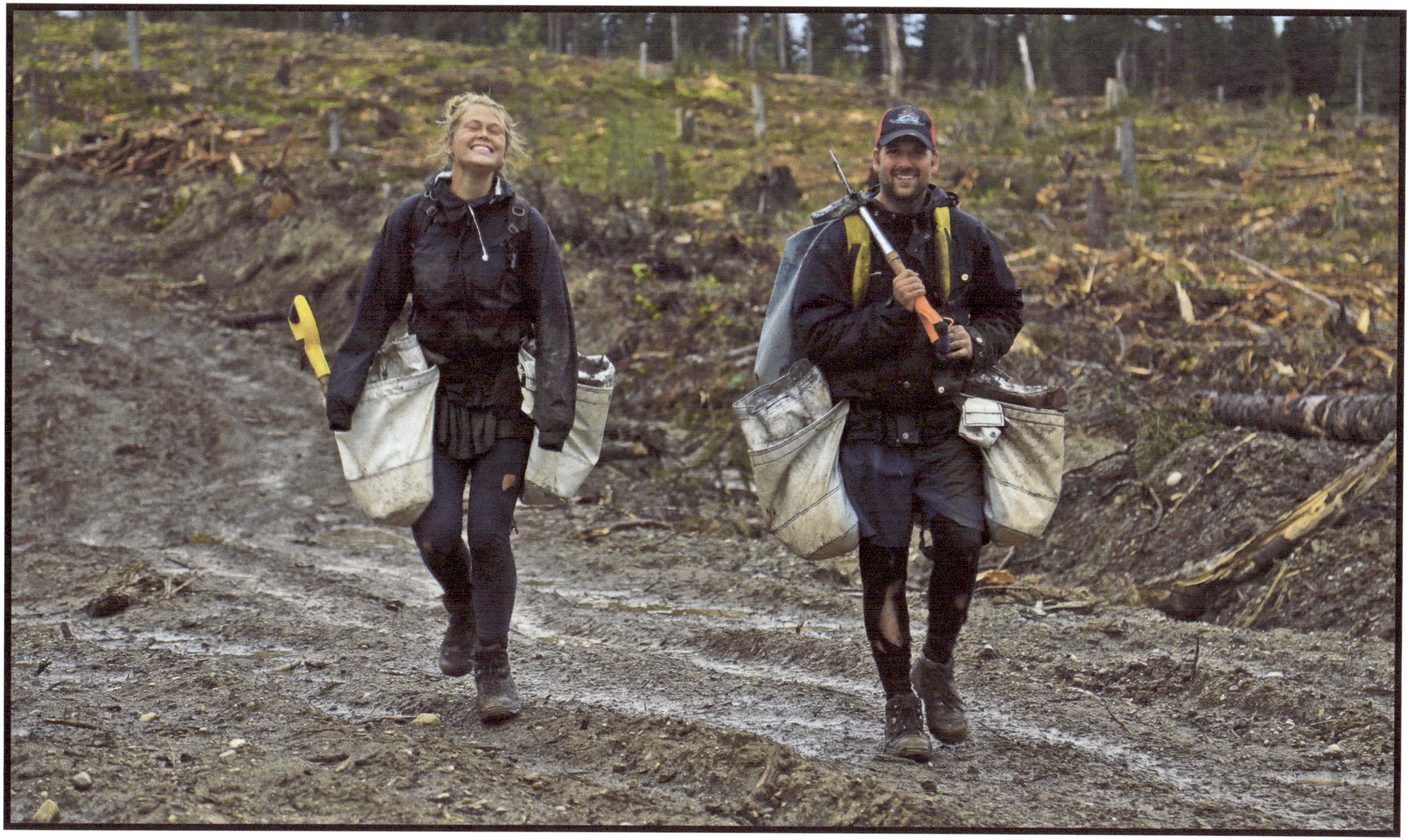

Shaylagh and Cam look pretty happy at the end of this wet day of planting. Shaylagh had been having a challenging day: the land was rocky, there was freezing rain, and the sole tore off one of her new boots. Cam was able to come to the rescue with a warm sweater and a jolly rancher.

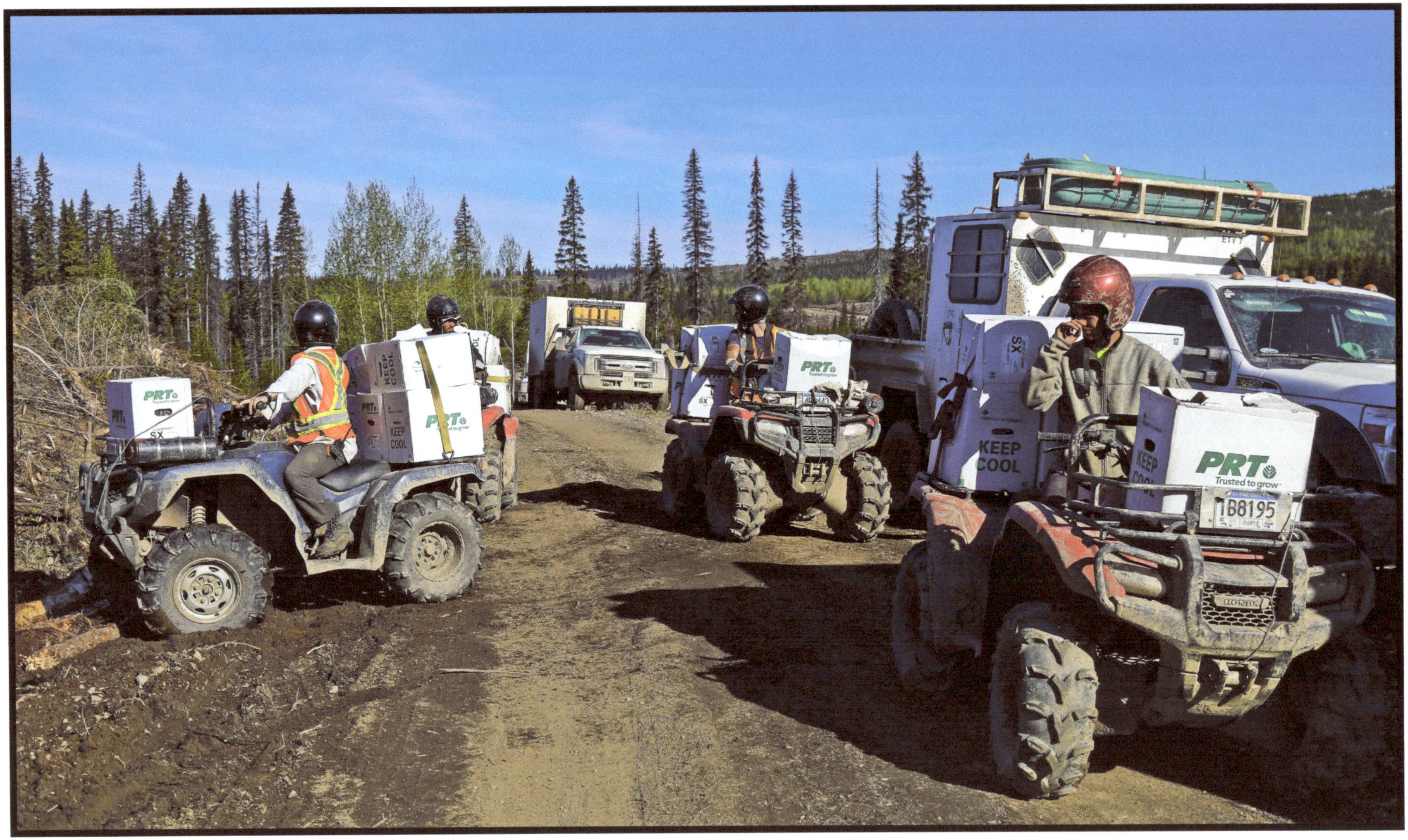

This photo was taken on the last day of a contract, when all the crews were working together to finish up a huge block. Here we see several of the foremen getting ready to deliver loads of boxes to the planters, immediately after a quick strategy meeting at the trucks.

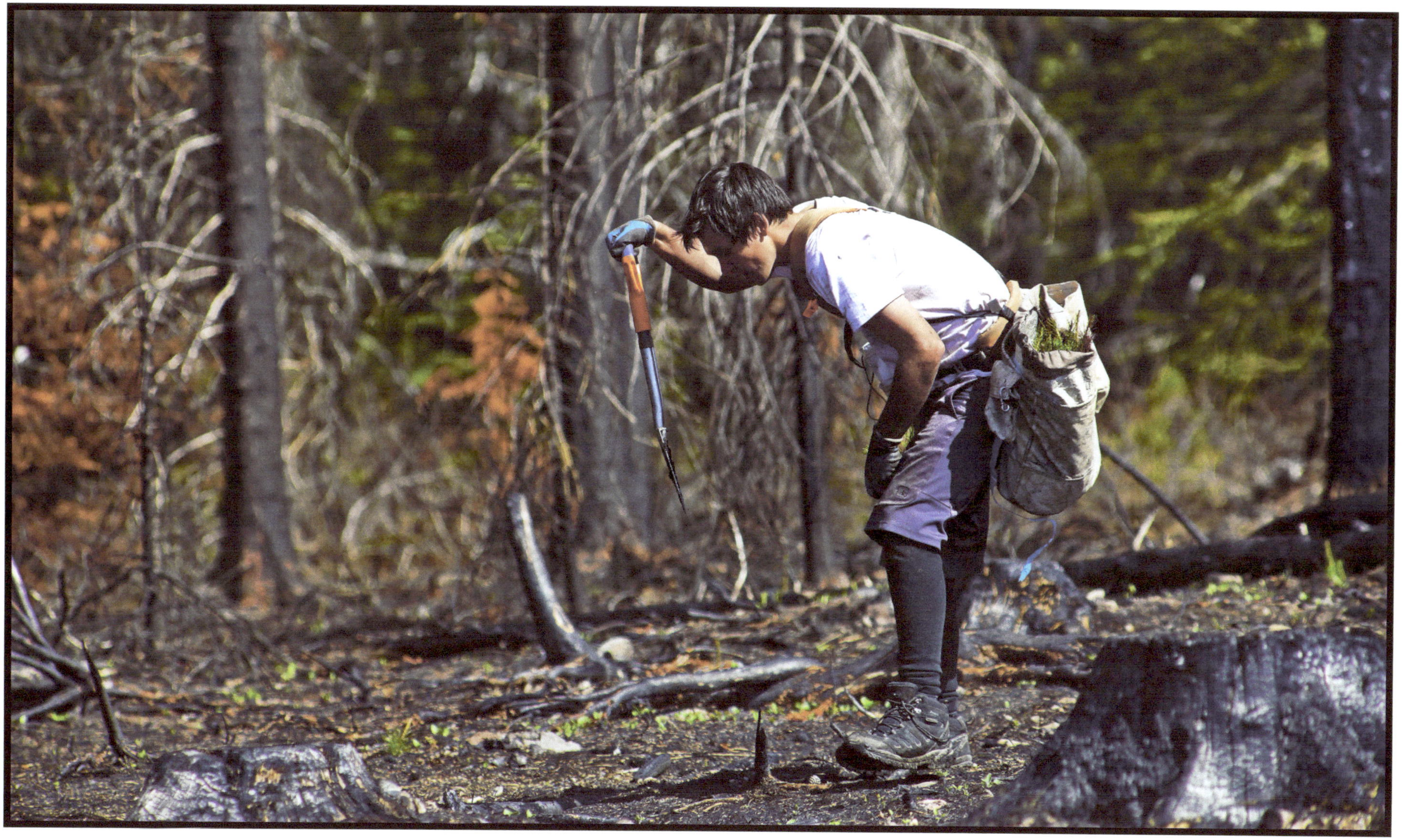

A planter uses a combination of arm strength and gravity to drive a shovel into the ground.

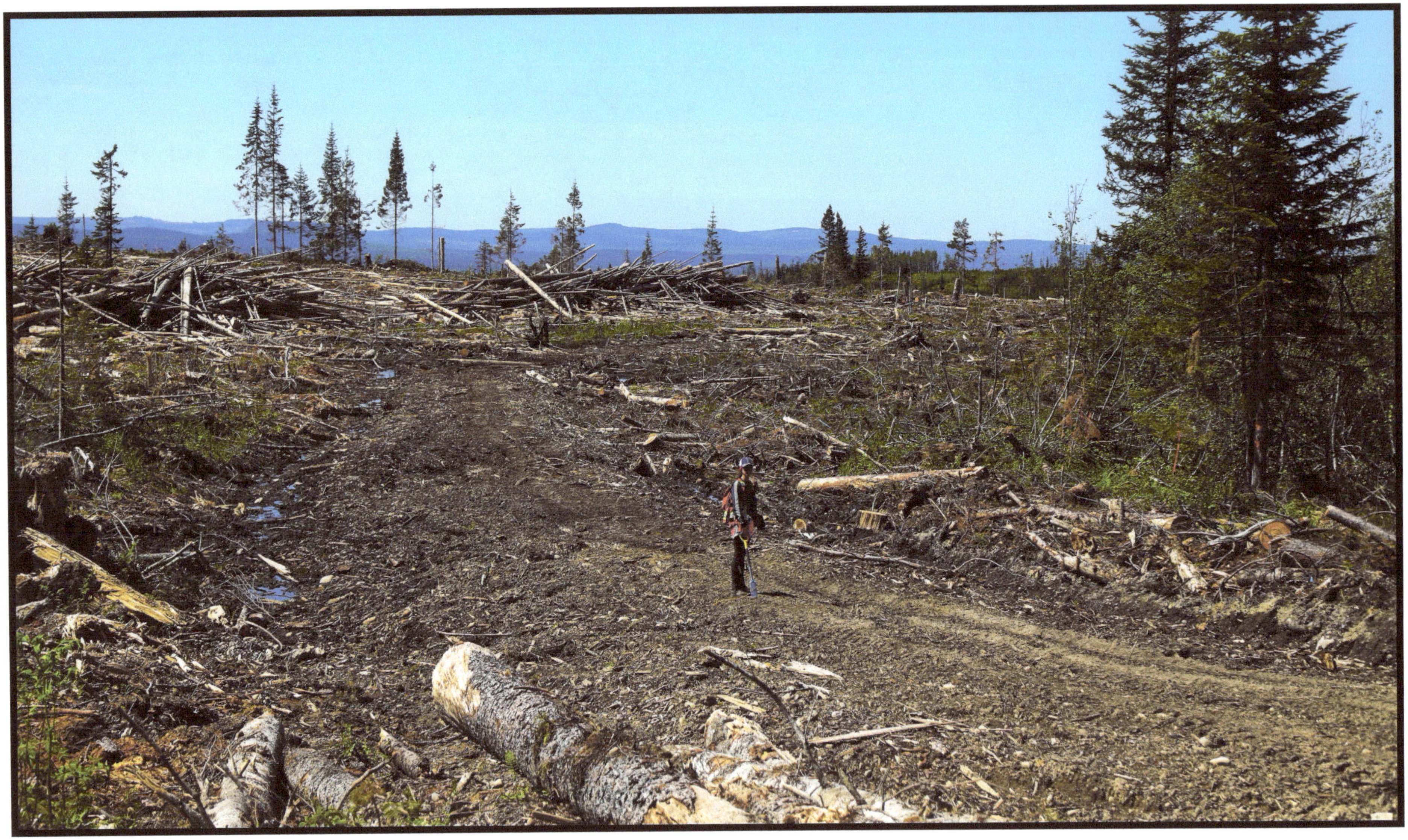

The job of the quality checker can be a lonely one.

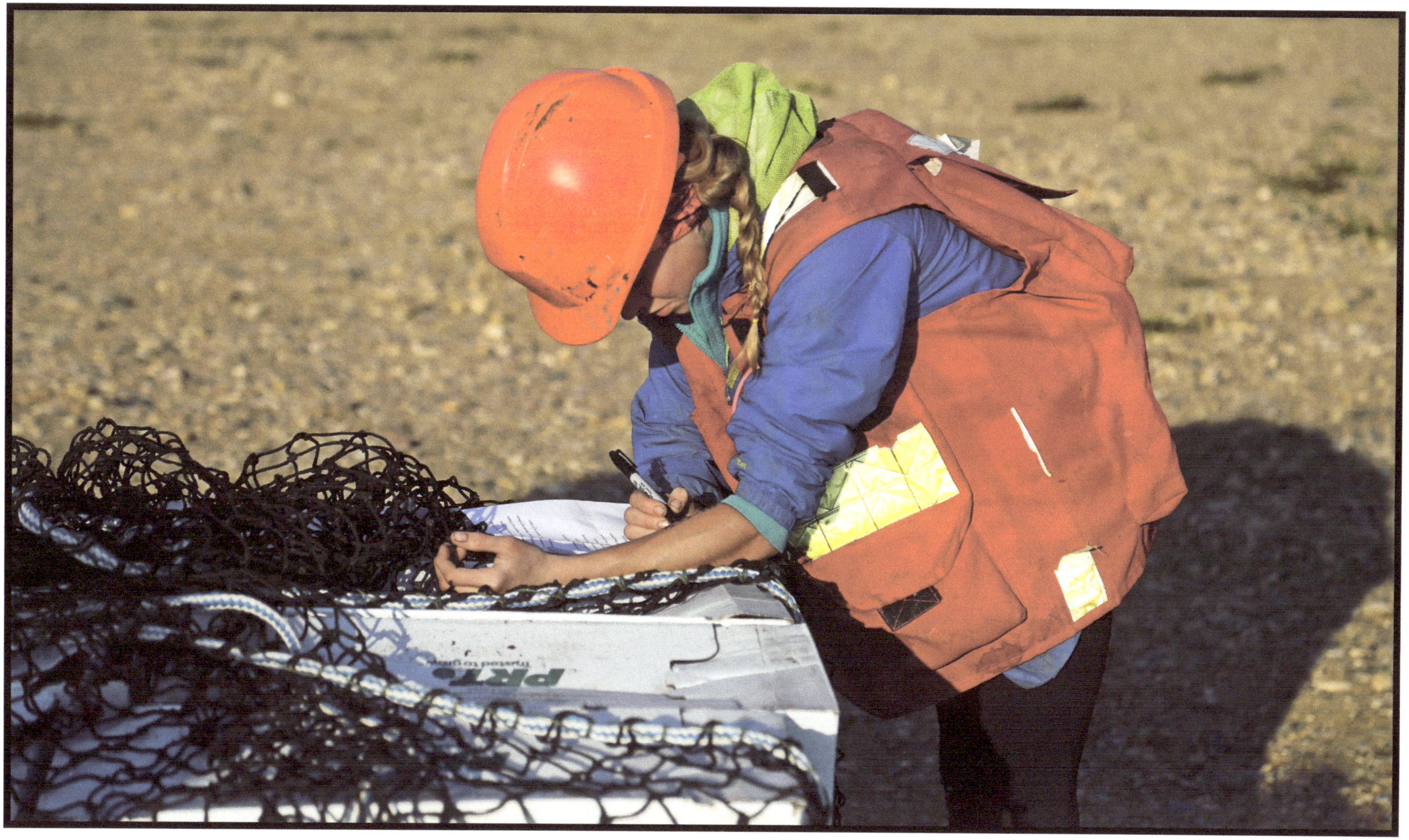

Here we see Hilary organizing the logistics for sending a number of slings out to a block.

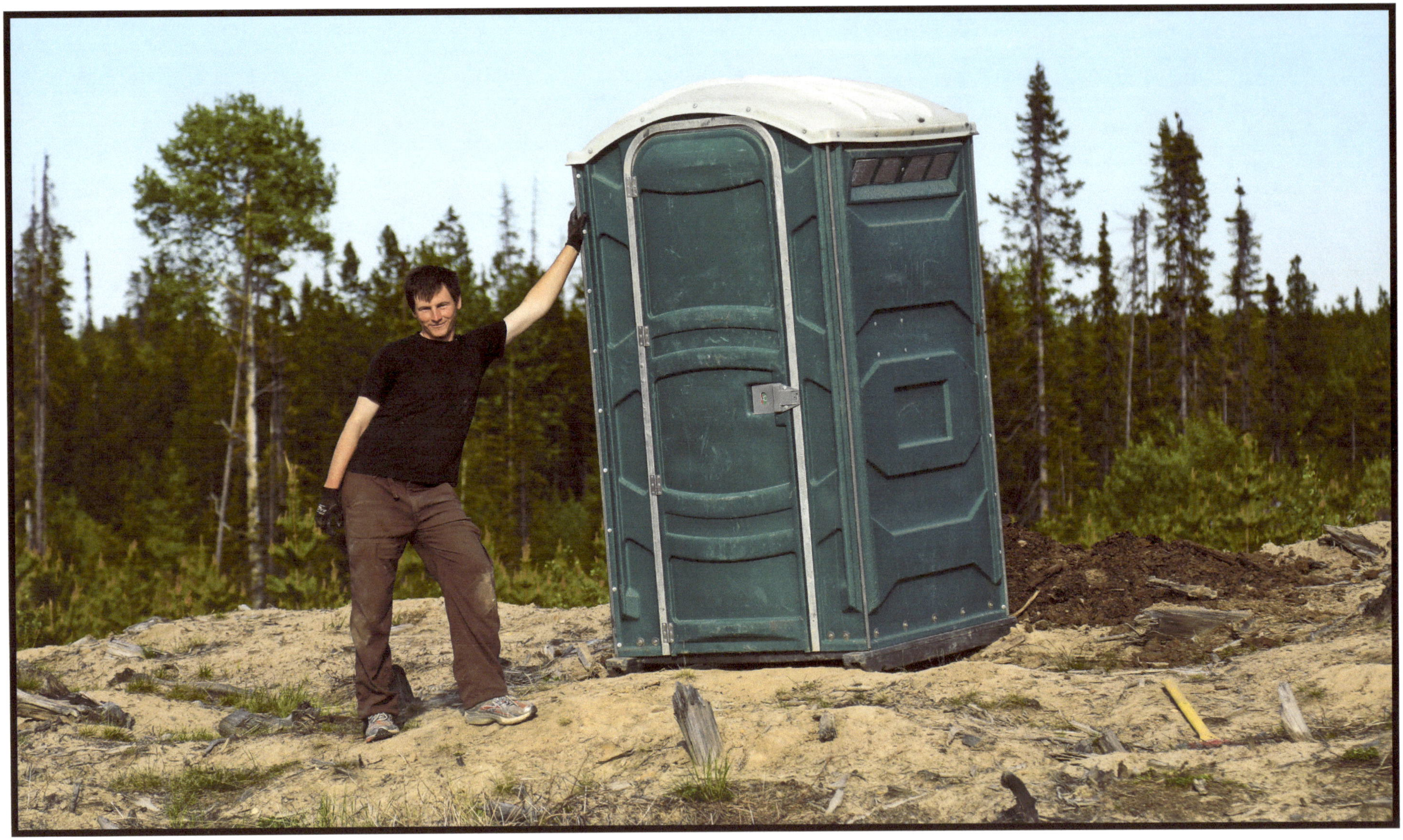

Sean appears to be in a good mood about setting up this port-a-potty.

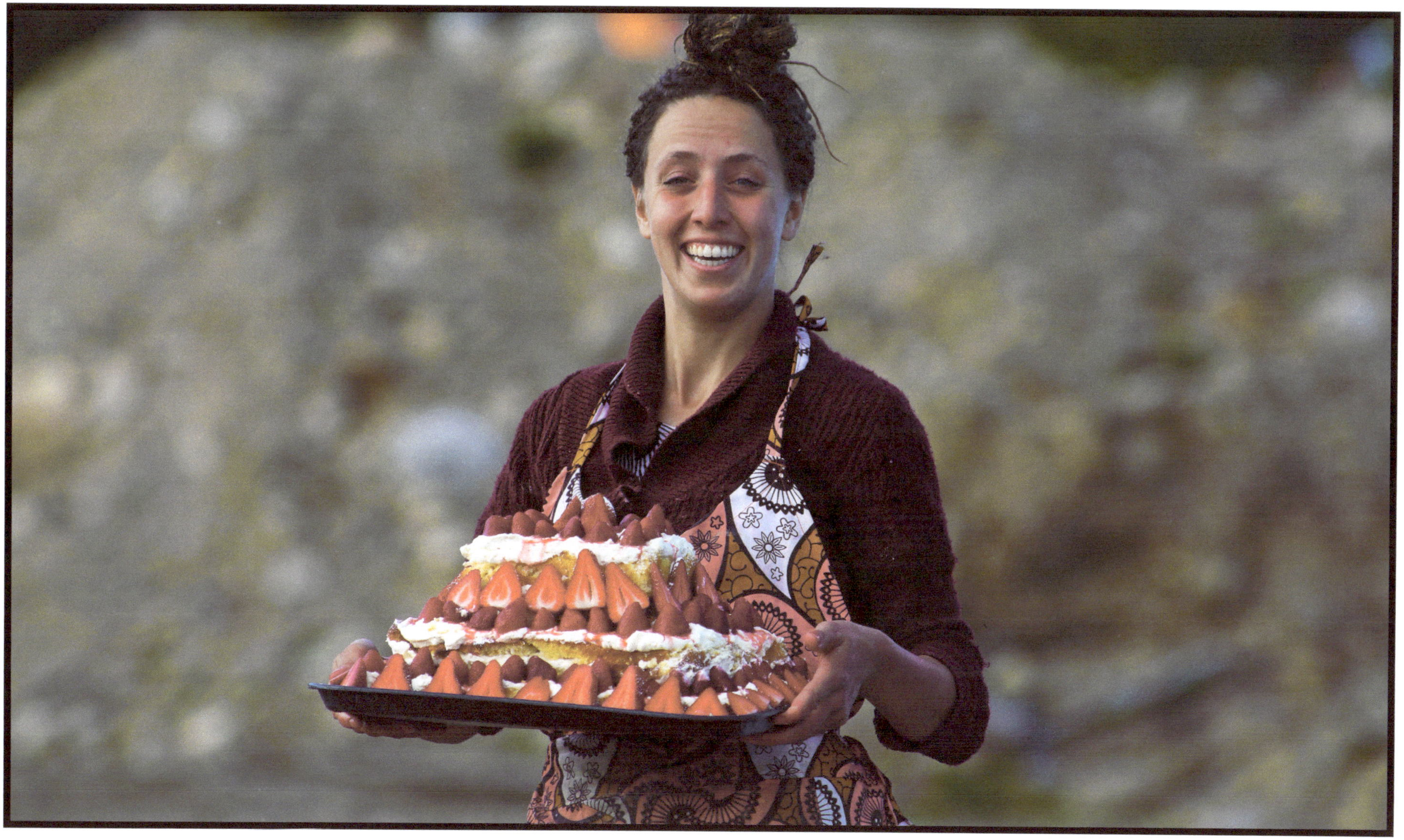

I'm often amazed by what our cooks can create in such a challenging working environment. It's especially difficult because tree planters eat about three times as much food as a normal person would consume.

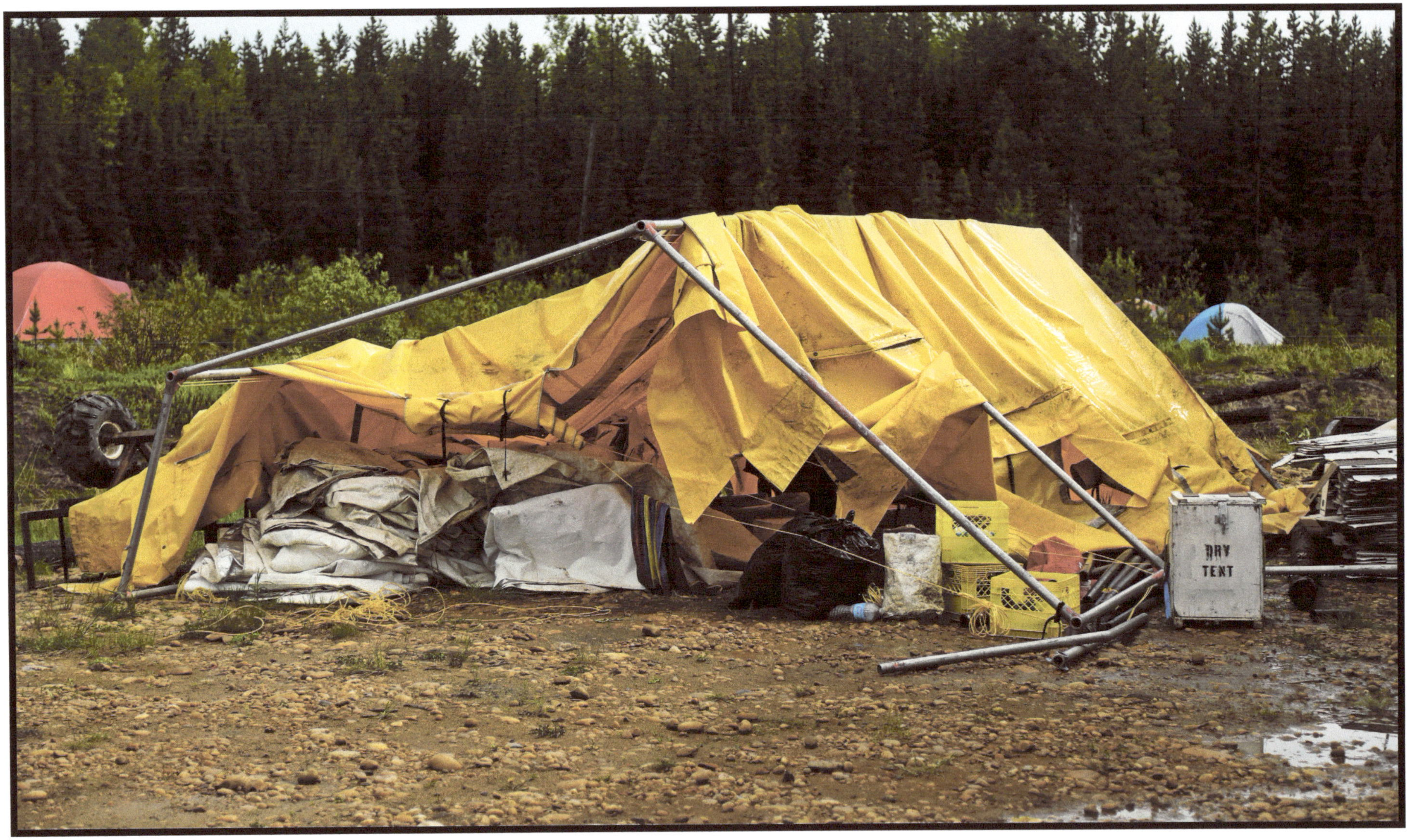

Here we see a dry tent in its natural state, collapsed after a strong wind storm. I'm not sure why, but our dry tent always seems to suffer the most damage after a storm. Perhaps this is because it has a pentagon-shaped profile, rather than the half-circle profile of our larger mess tent? Whatever the reason, it seems that we need to rebuild the dry tent after at least one storm every season.

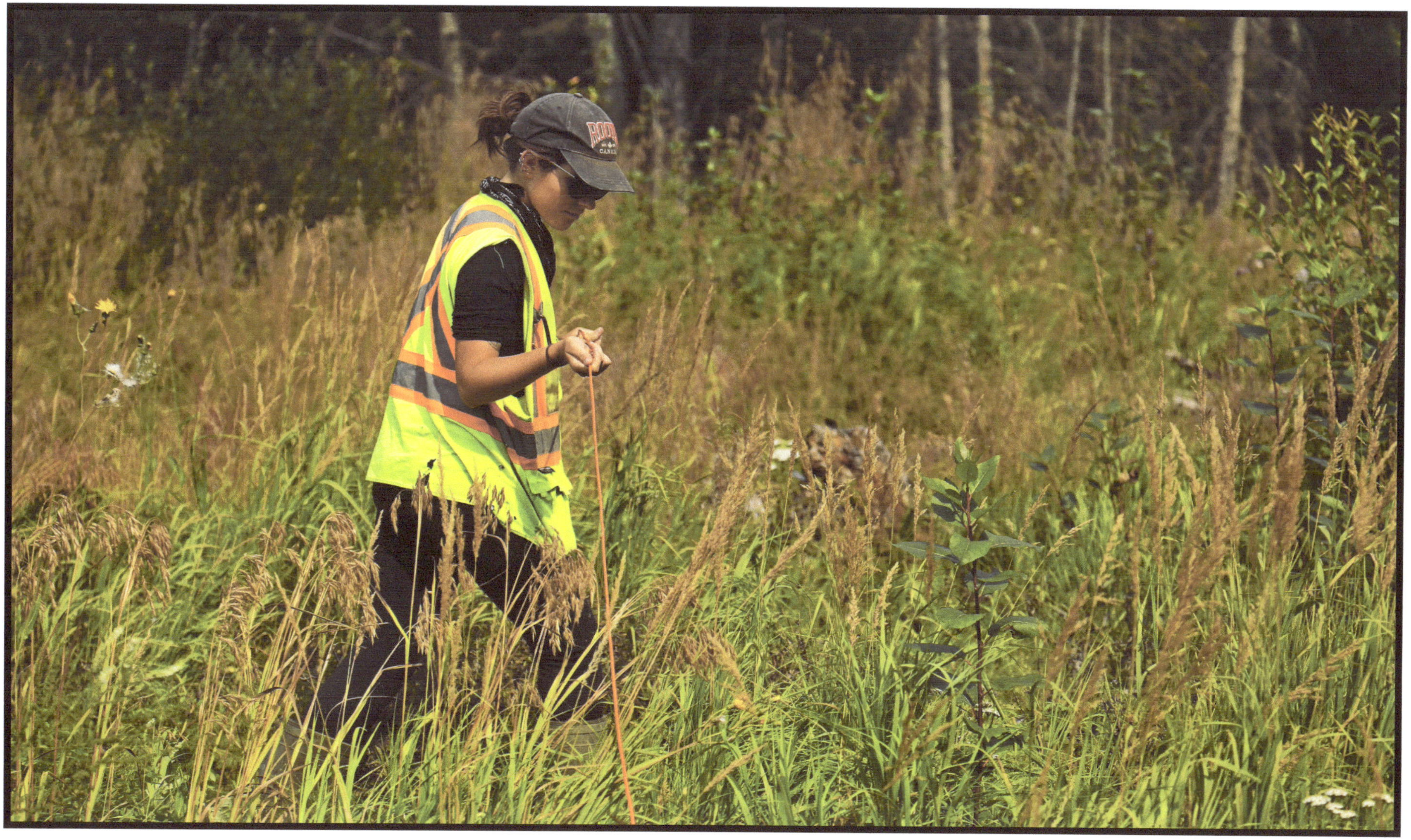

Here's a checker using a plot cord to "throw a plot," which allows her to make an estimate of the planting density in the block. She fastens one end of the 3.99m plot cord onto a shovel, then walks in a 360° circle around the plot center and counts how many seedlings were planted in that area. The number of seedlings within that circle give her a rough idea of the planting density in the area. Several dozen plots are typically thrown on each block.

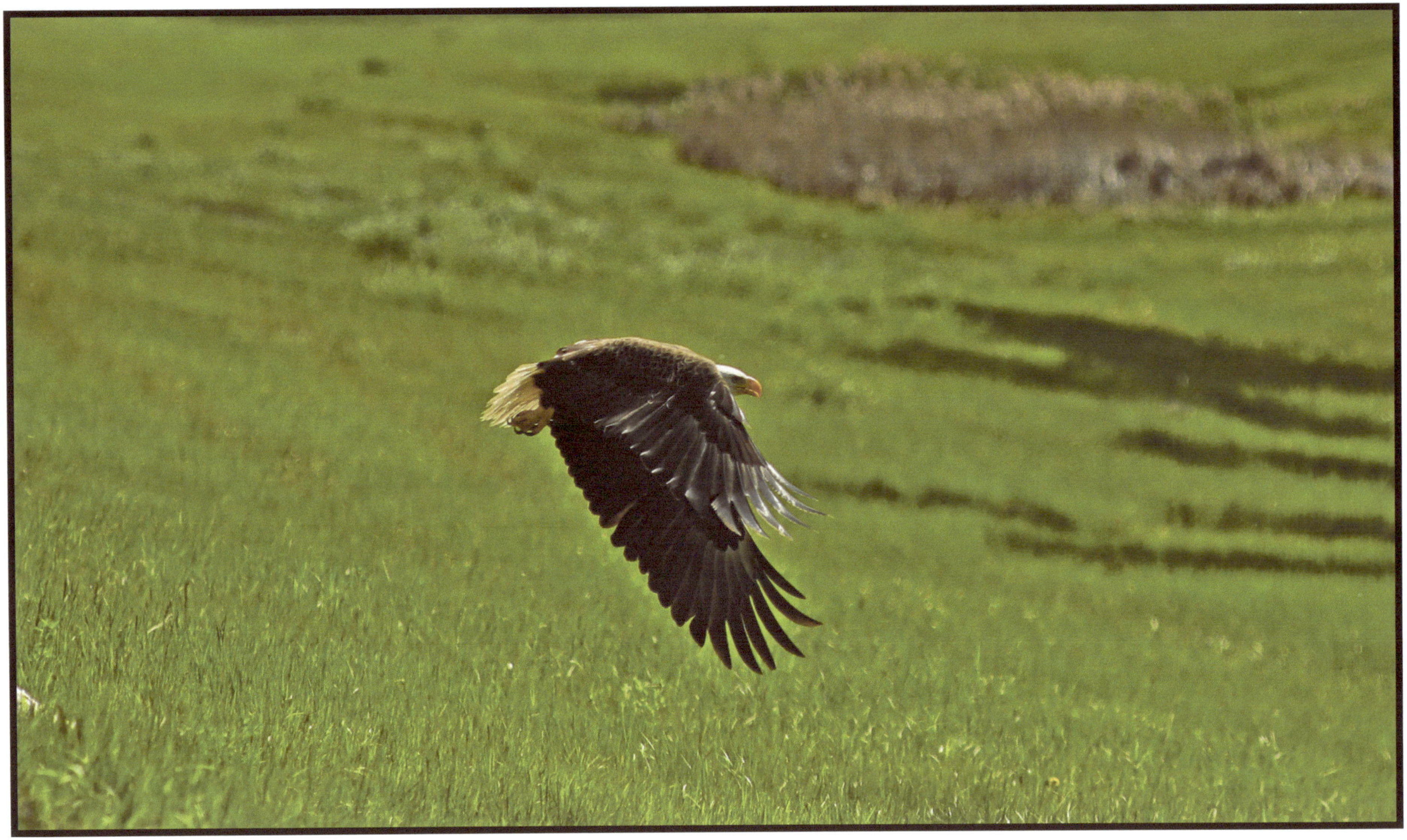

Tree planters get to see a lot more wildlife than most Canadians.

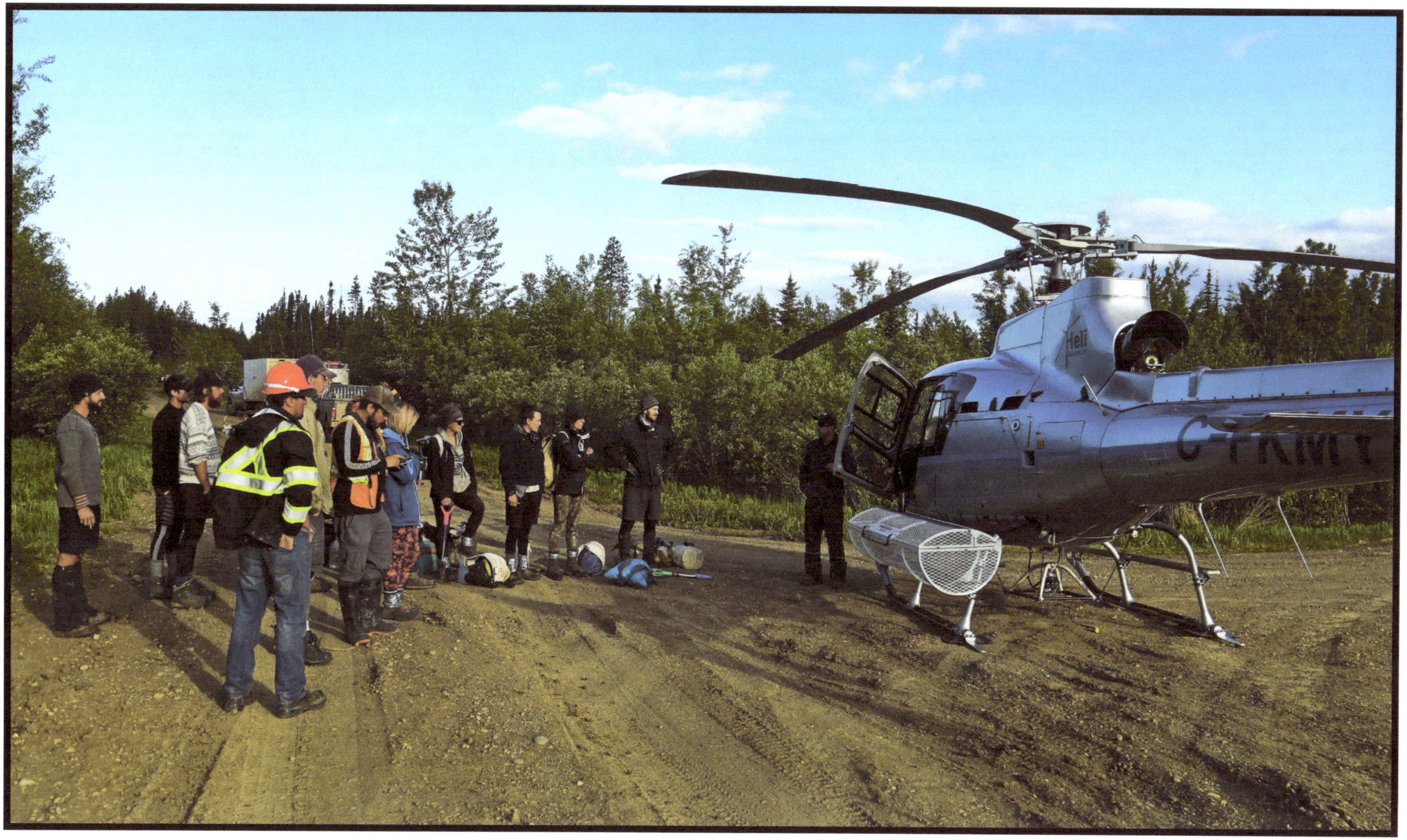

Here we see a helicopter pilot giving a safety pre-work talk to a crew before work starts for the day.

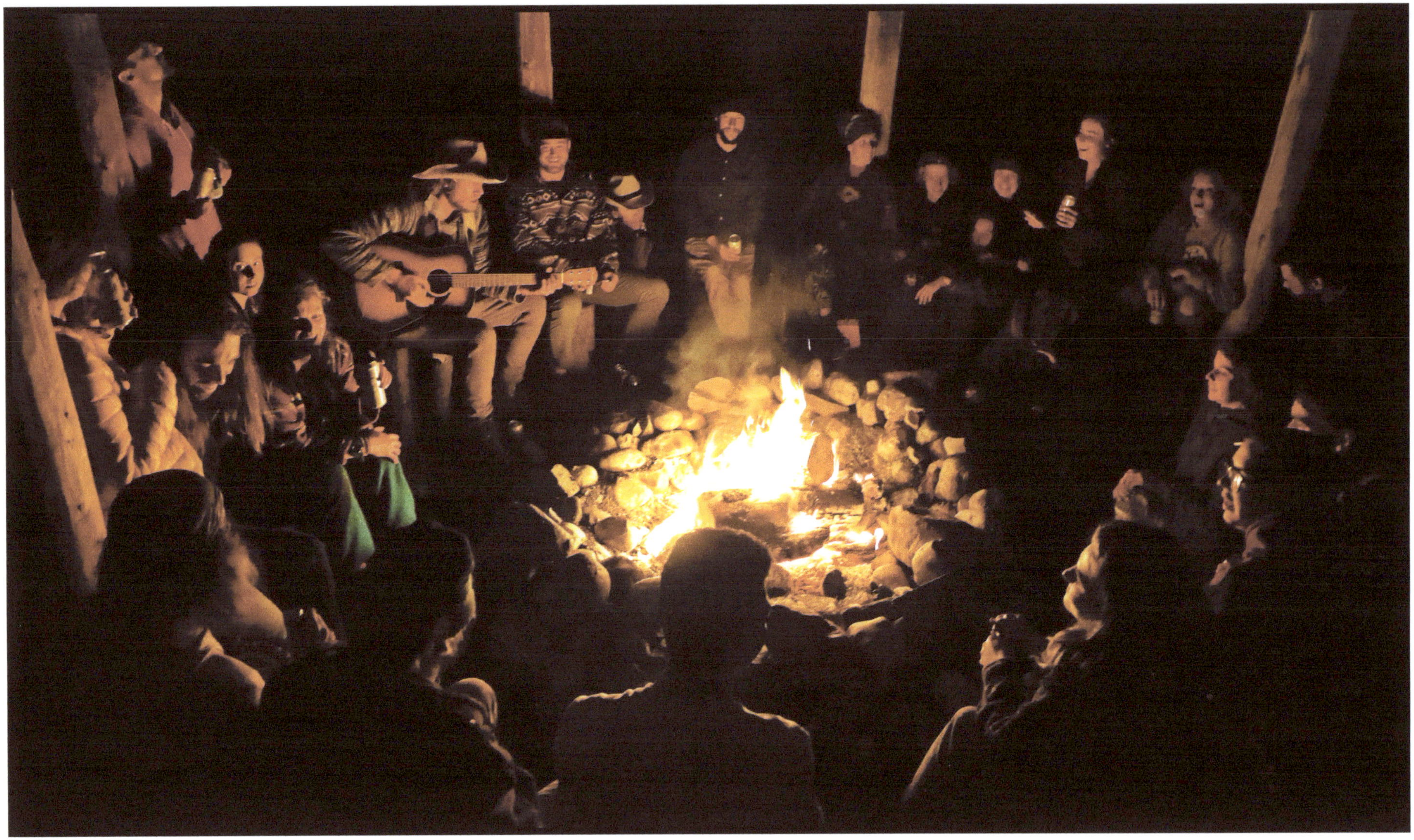

On a night off, if there are no fire restrictions due to the fire hazard index, a campfire can be a really enjoyable way for the crew to sit back and relax.

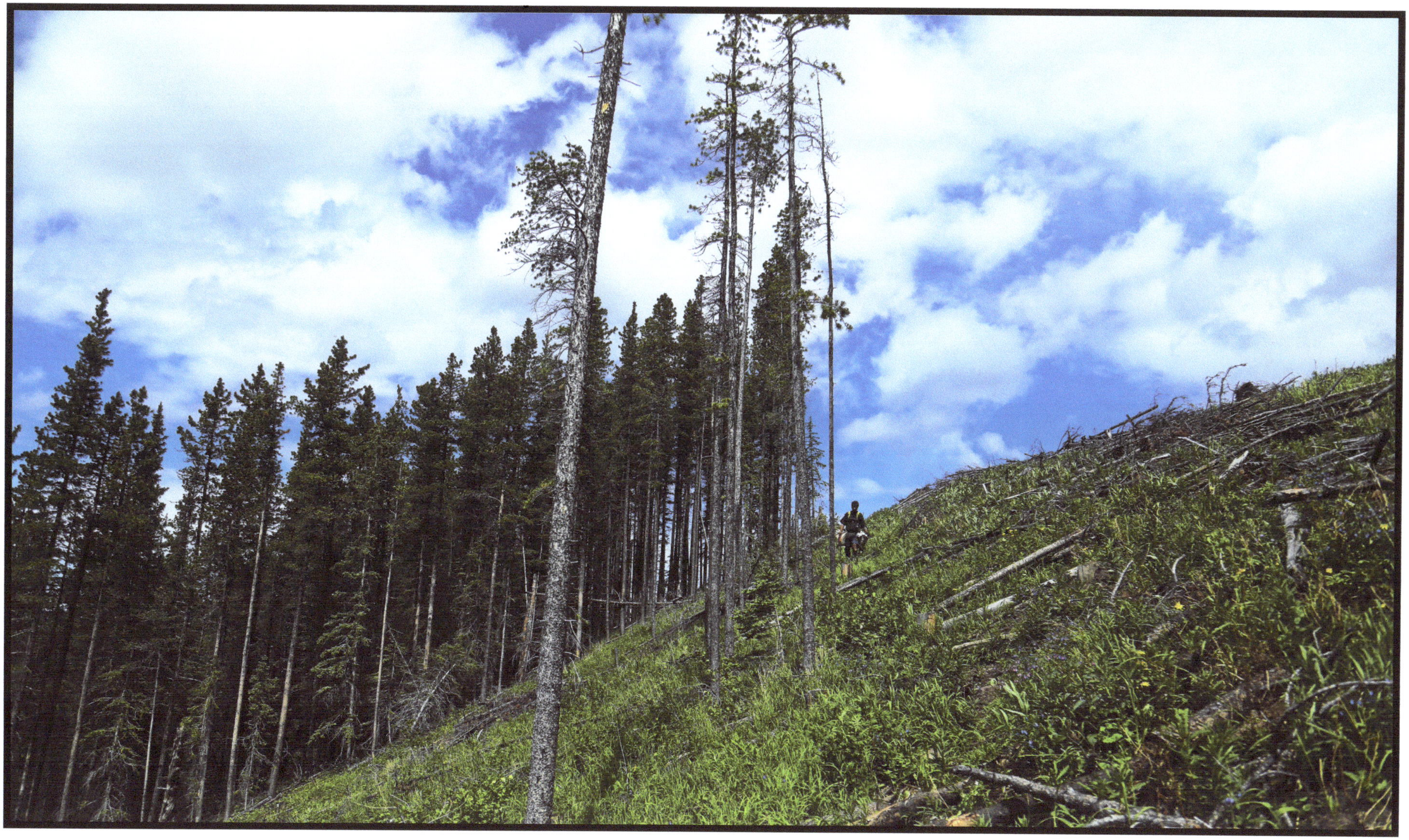

This block appears to be much steeper than most of the blocks in Alberta. Southwestern Alberta has some relatively steep ground.

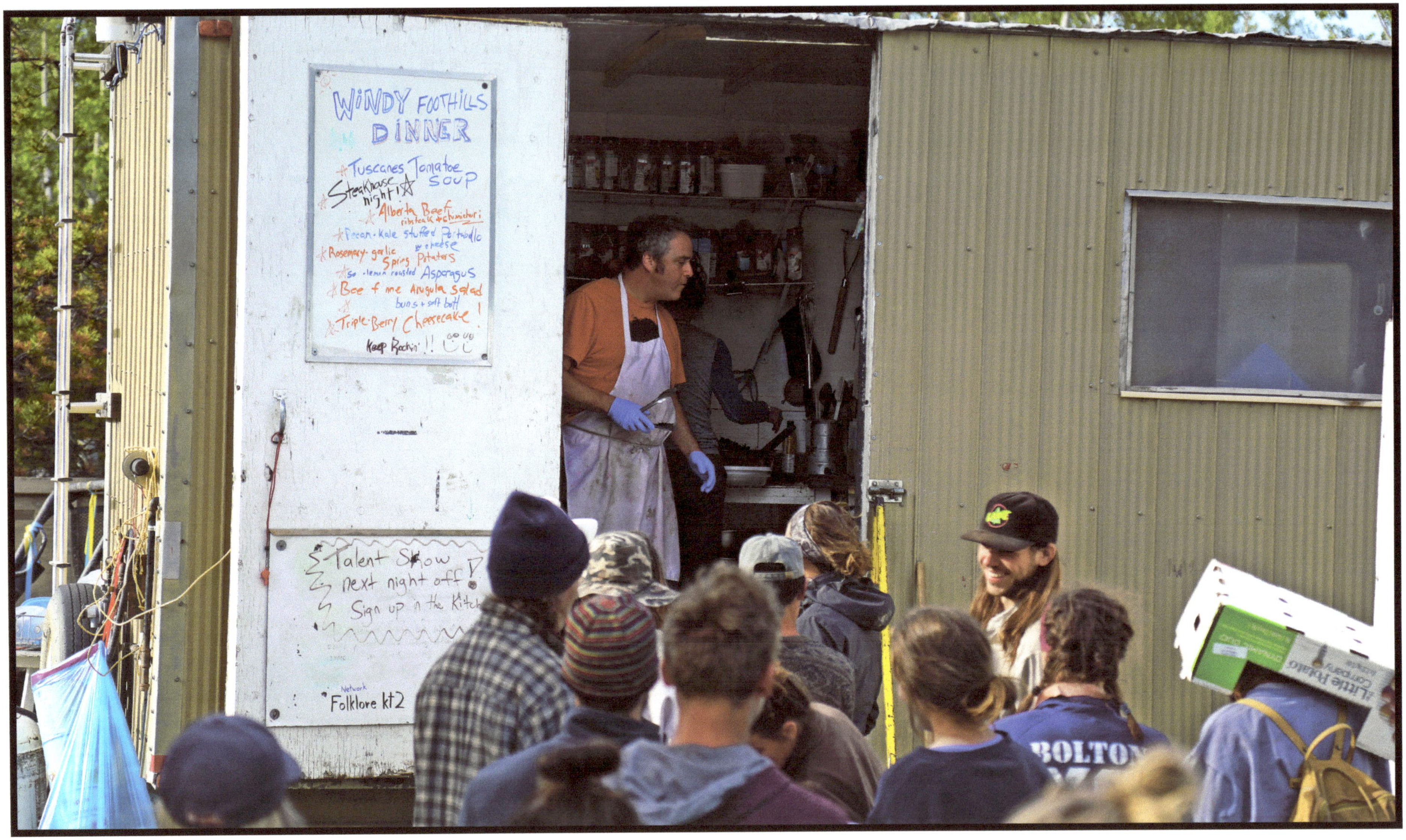

For most planters, dinner is the best part of the day.

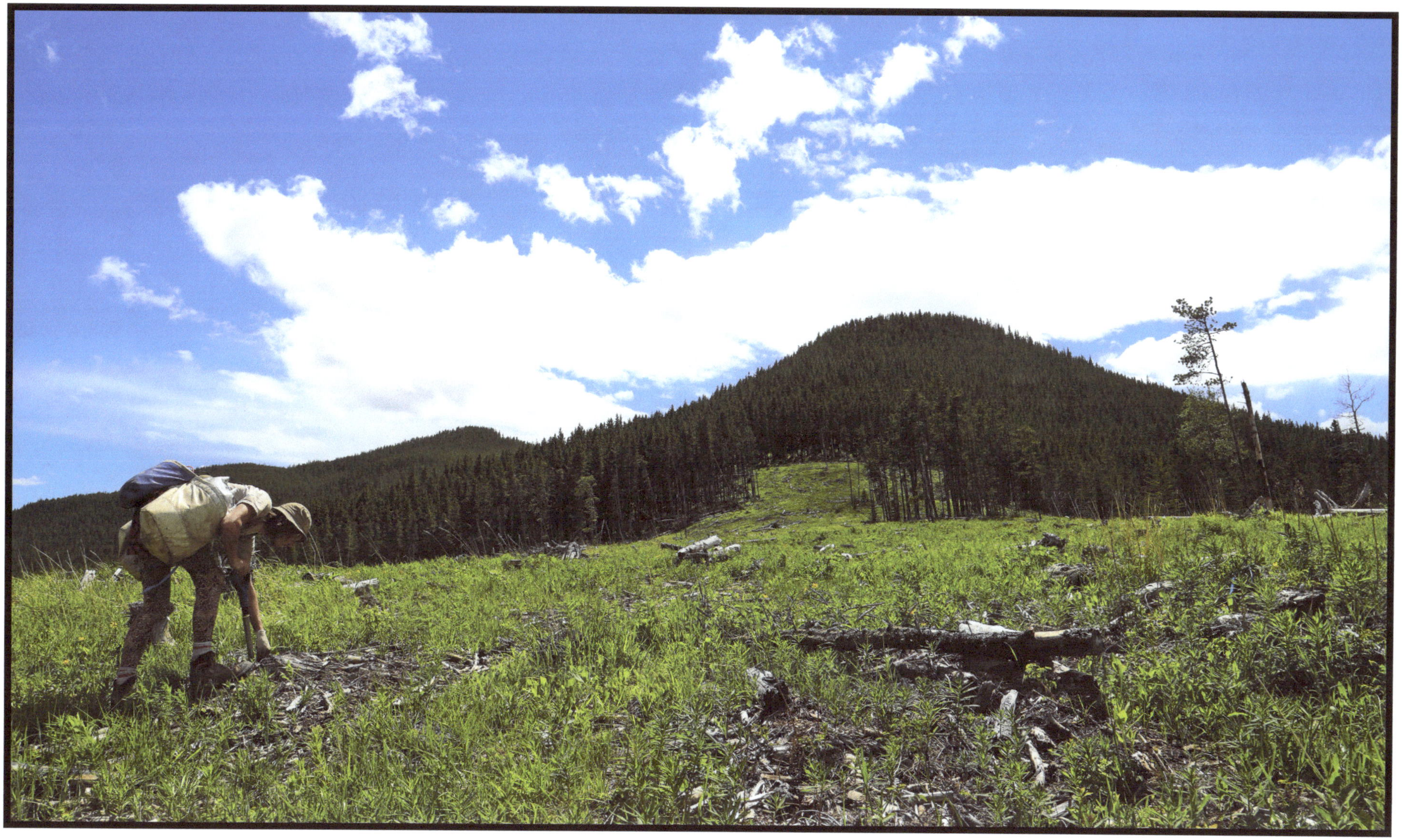

Although sunny days can be hot, they're usually a lot more enjoyable than rain.

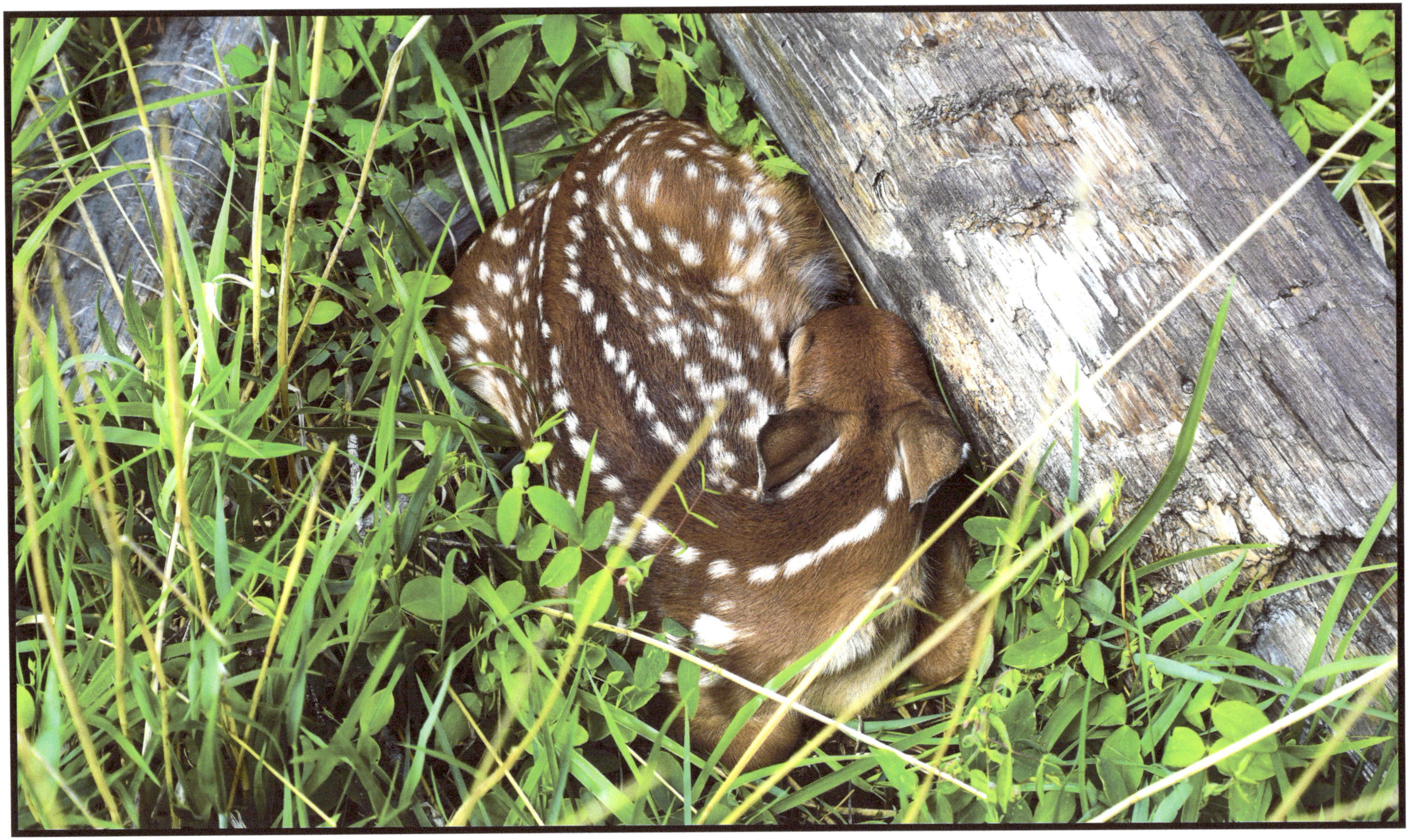

Be careful where you're walking. You never know what you might run into when you step over a log.

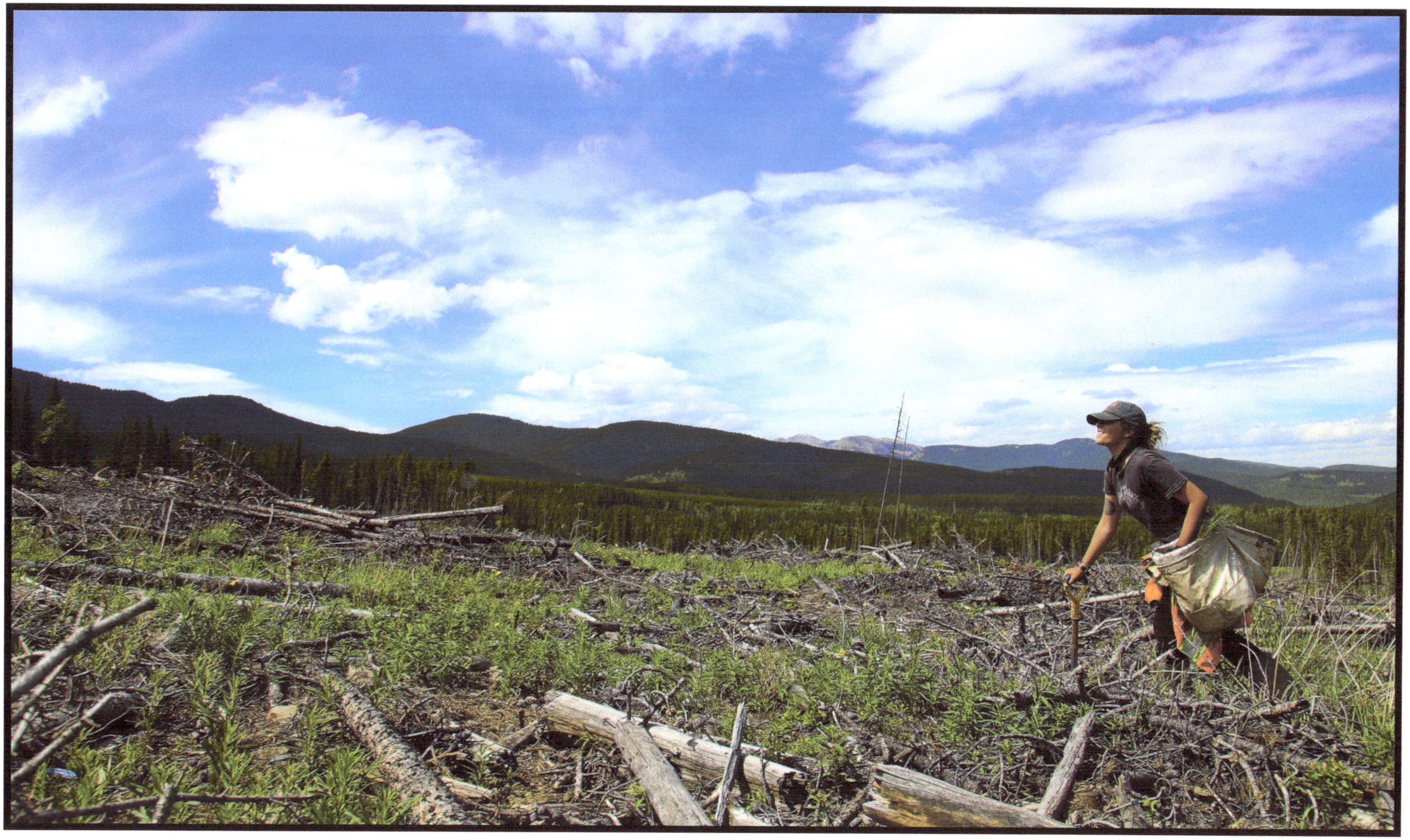

Karen looks like she's enjoying herself. Karen attended Lakehead University where she obtained her honours Bachelor of Science in Forestry. She is now working full-time in forestry at a mill in British Columbia.

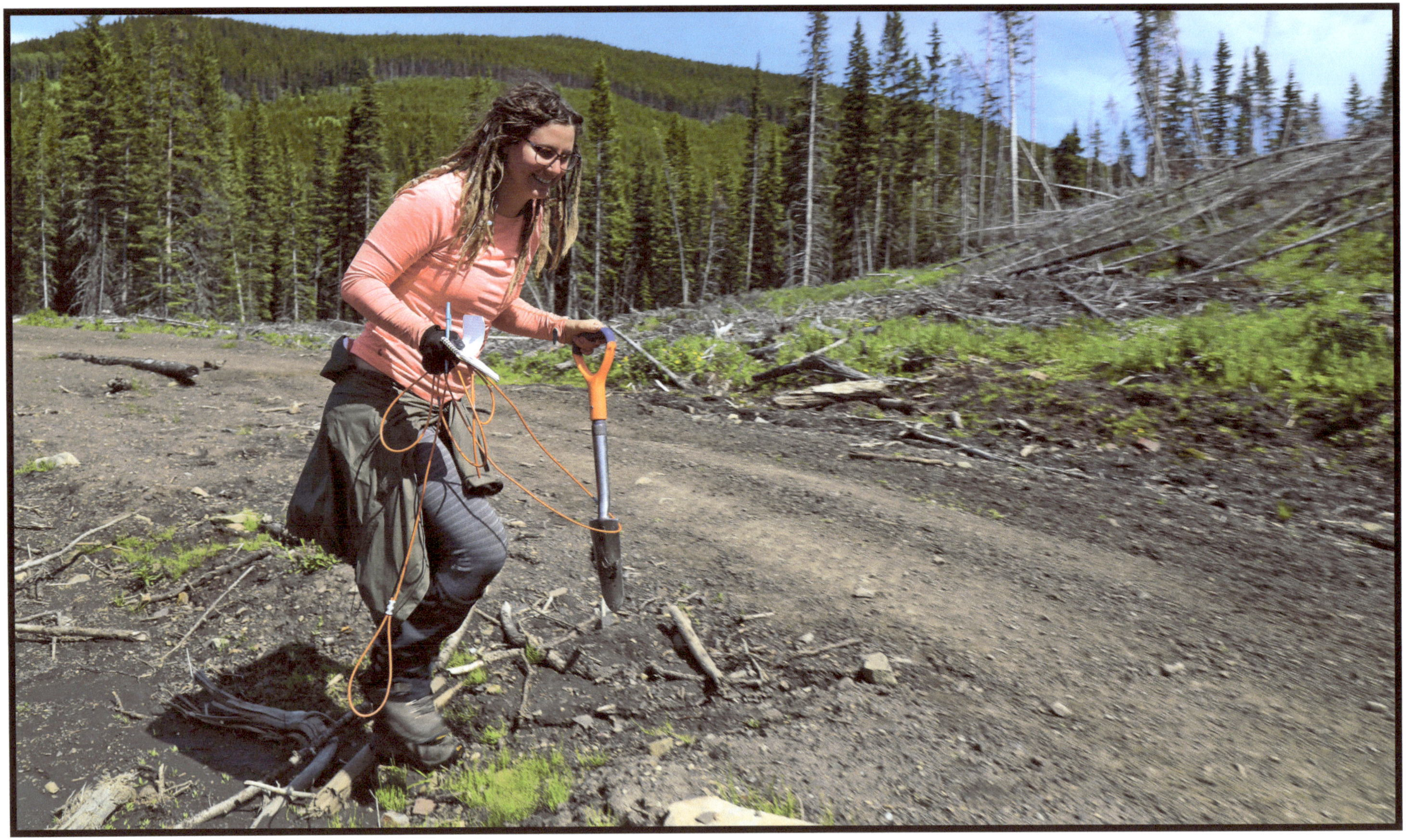

Based on her smile, Patricia must be finding some good quality and density in the plots that she's taking.

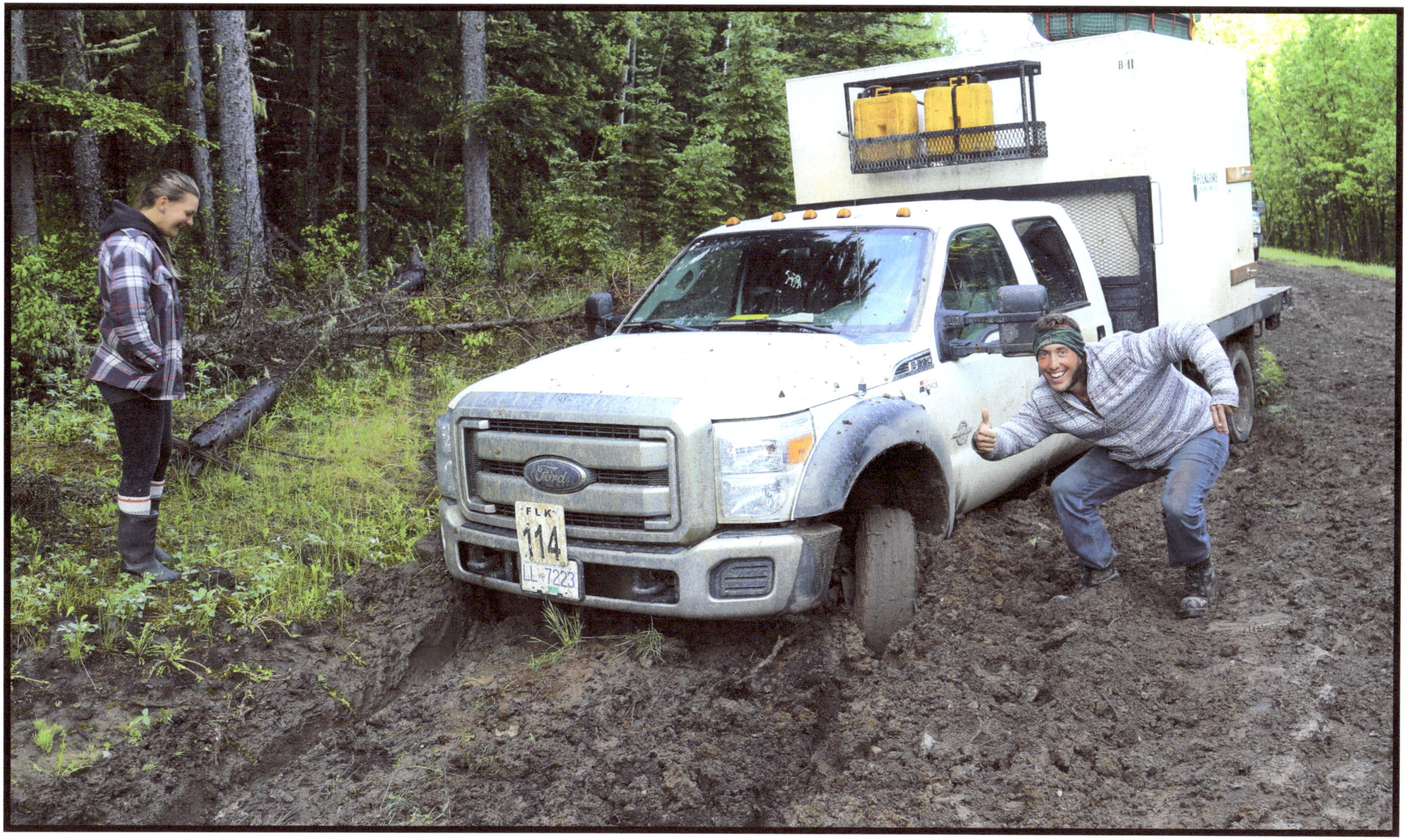

Unfortunately, things are not going well for this truck. If you're ever having a hard time finding a good parking spot in a city, remember this photo.

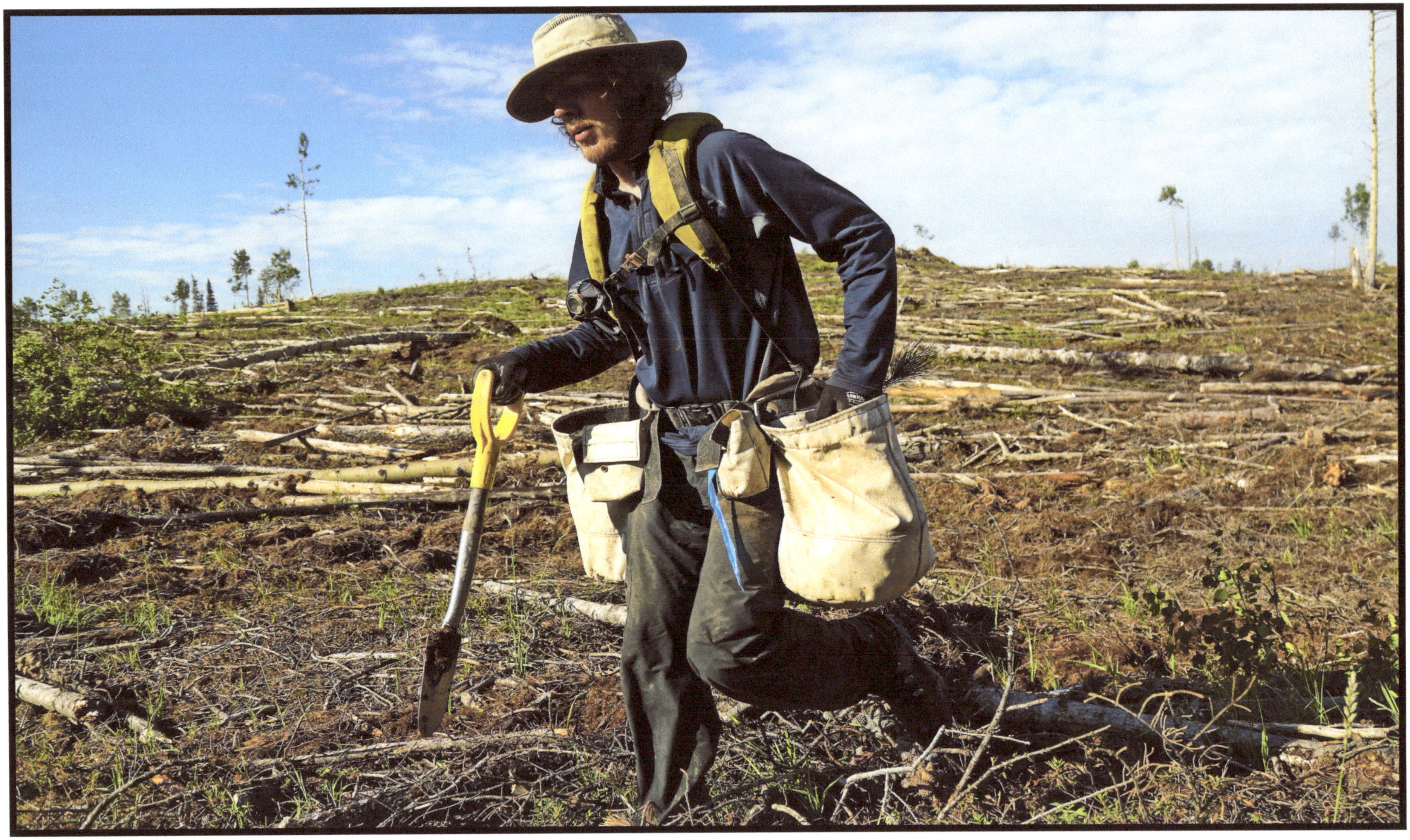

A sun hat is an essential piece of clothing for keeping sunlight out of your eyes. It works much better than sunglasses. Shane is one of our crew leaders, who really enjoys working outdoors. You won't catch him planting with sunglasses.

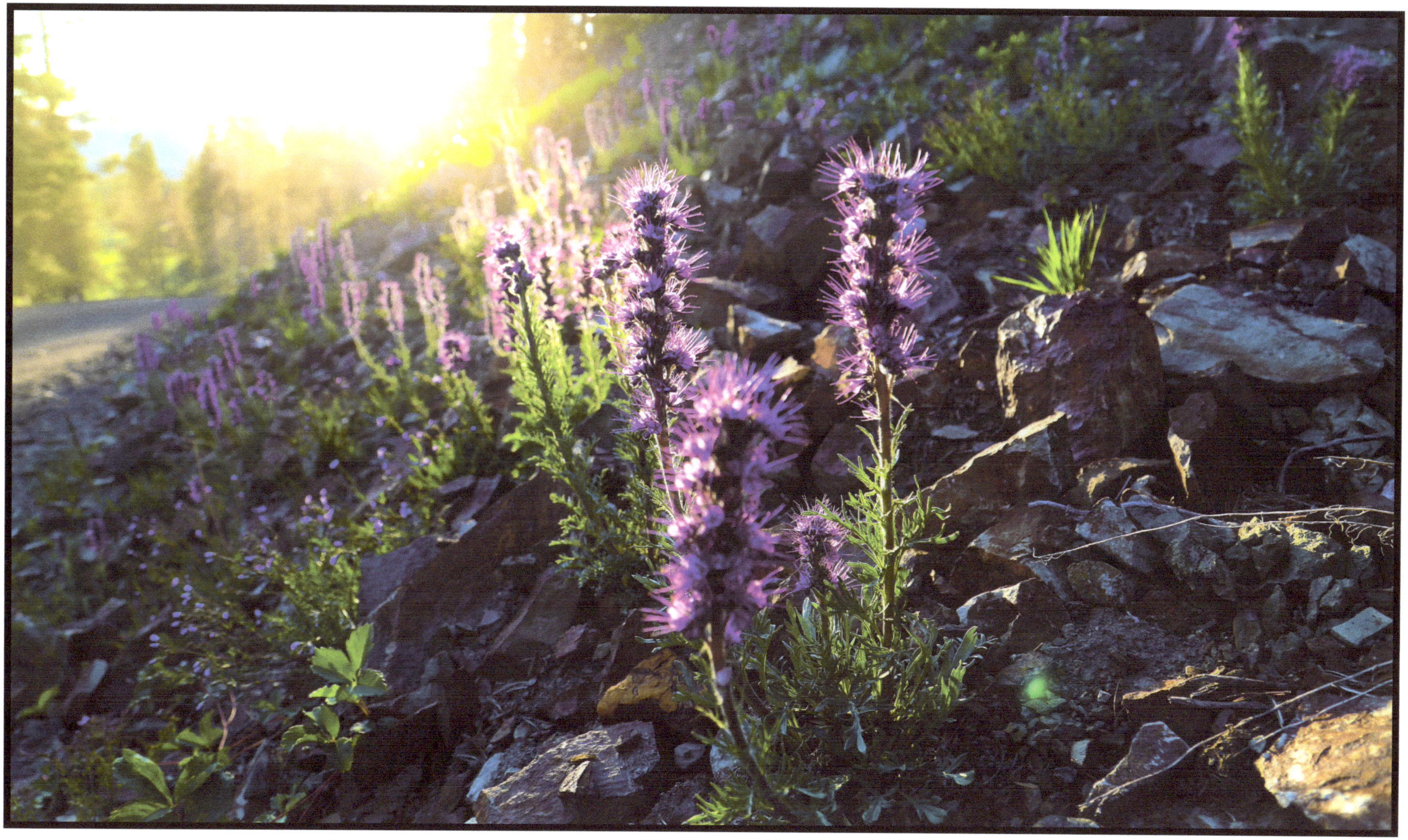

Nature can really be quite beautiful, if you stop to look around.

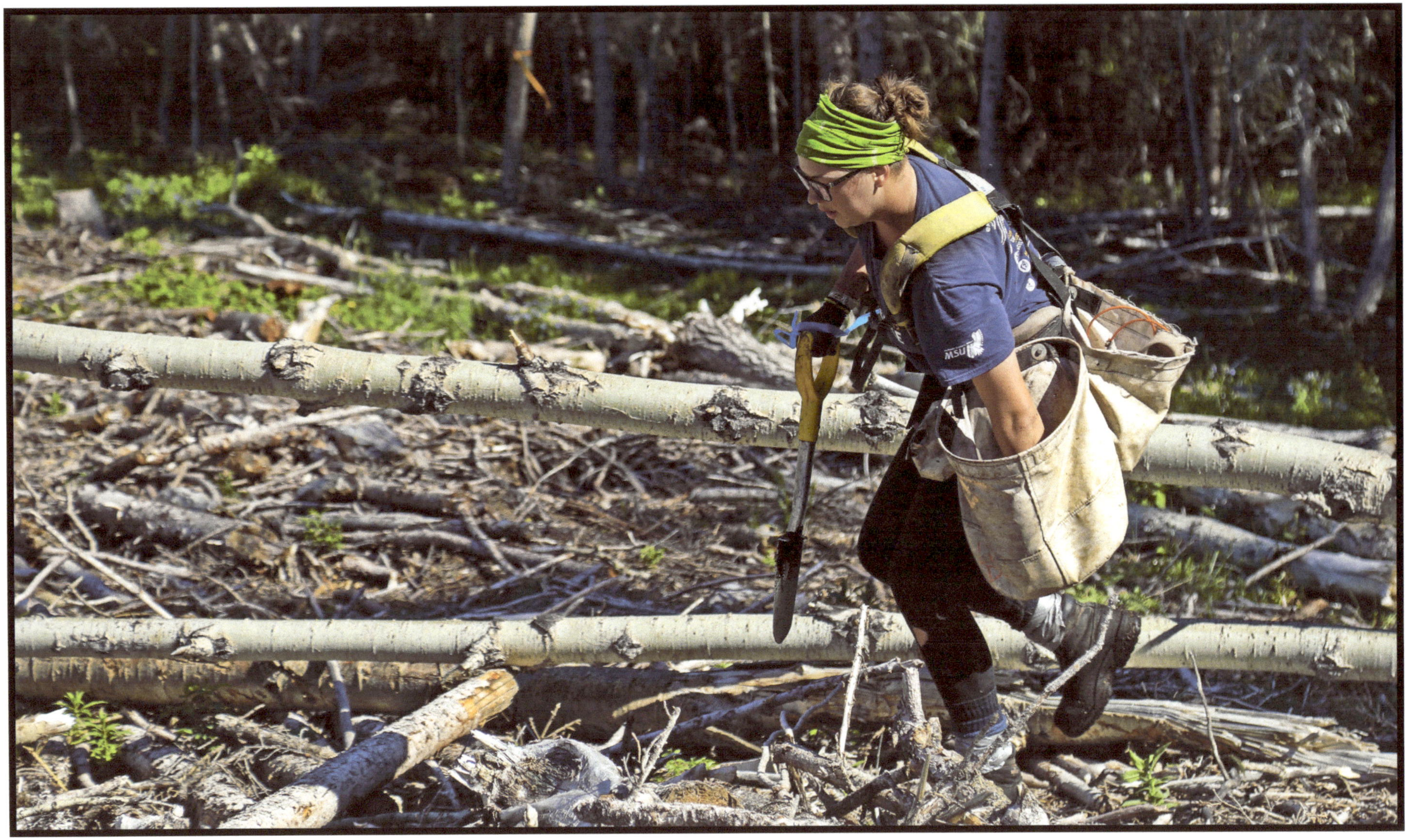

Liz is originally from Toronto. She is a graduate of McMaster University.

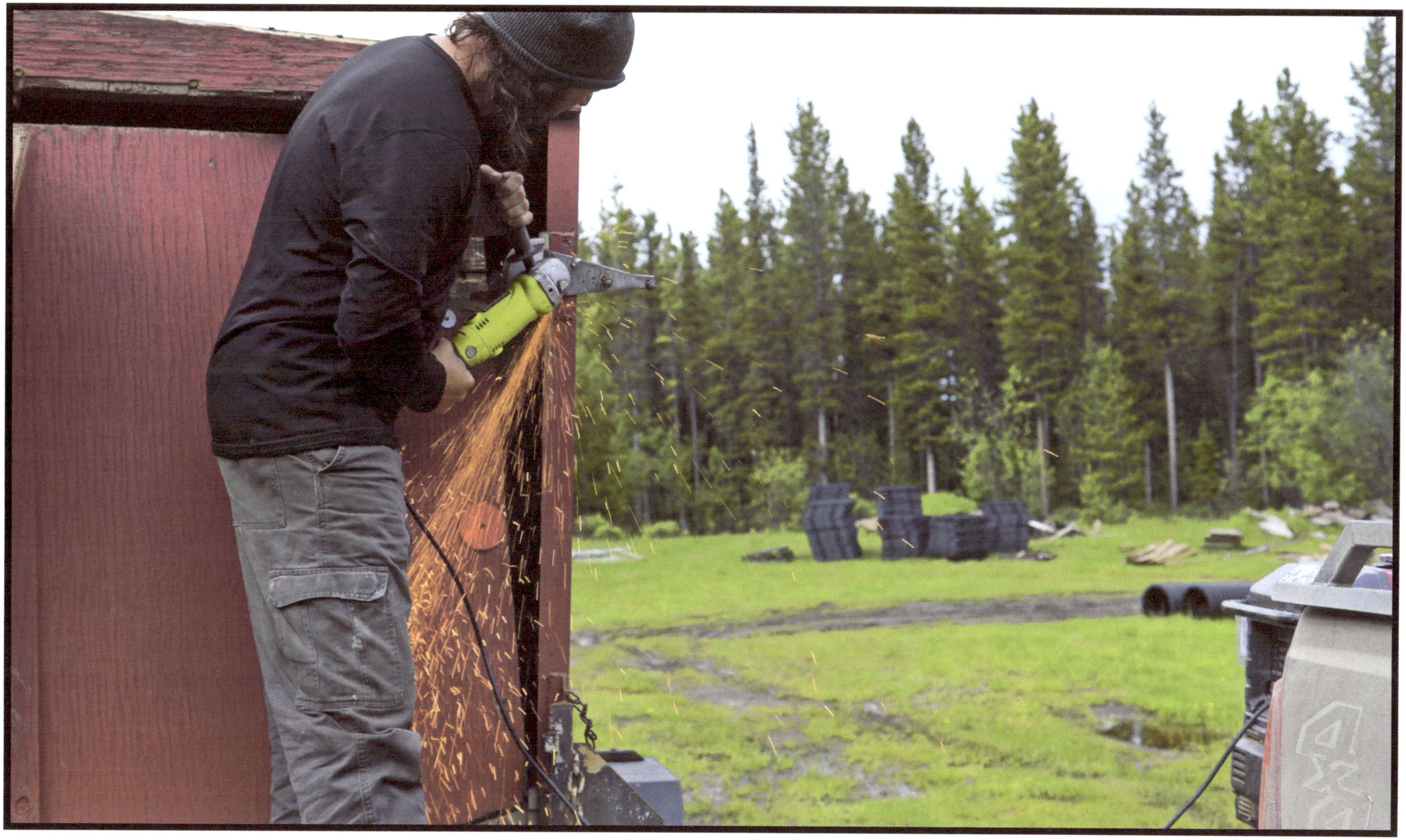

It is common to have to perform makeshift repairs while in the bush. Versatility is an asset.

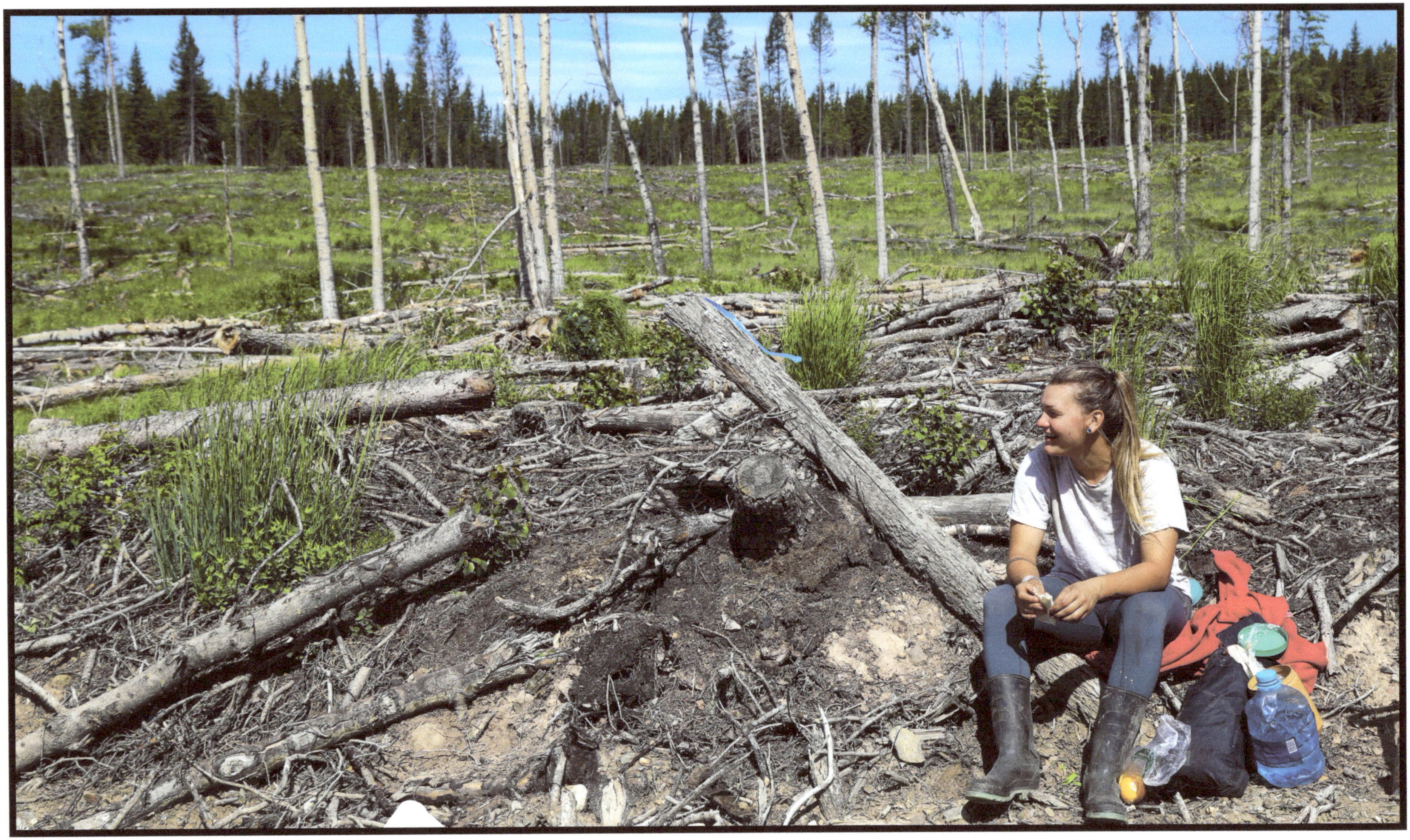

Shonda looks happy as she sits and waits for the truck at the end of the day. Shonda's long-term goal is to complete a masters degree in sports psychology, then to follow a career path helping with post-concussion recovery for professional athletes. As a hobby, she reads a lot of books about neuroscience and sports autobiographies. She also plays hockey, and spent four successful years playing in Alberta's AJFHL.

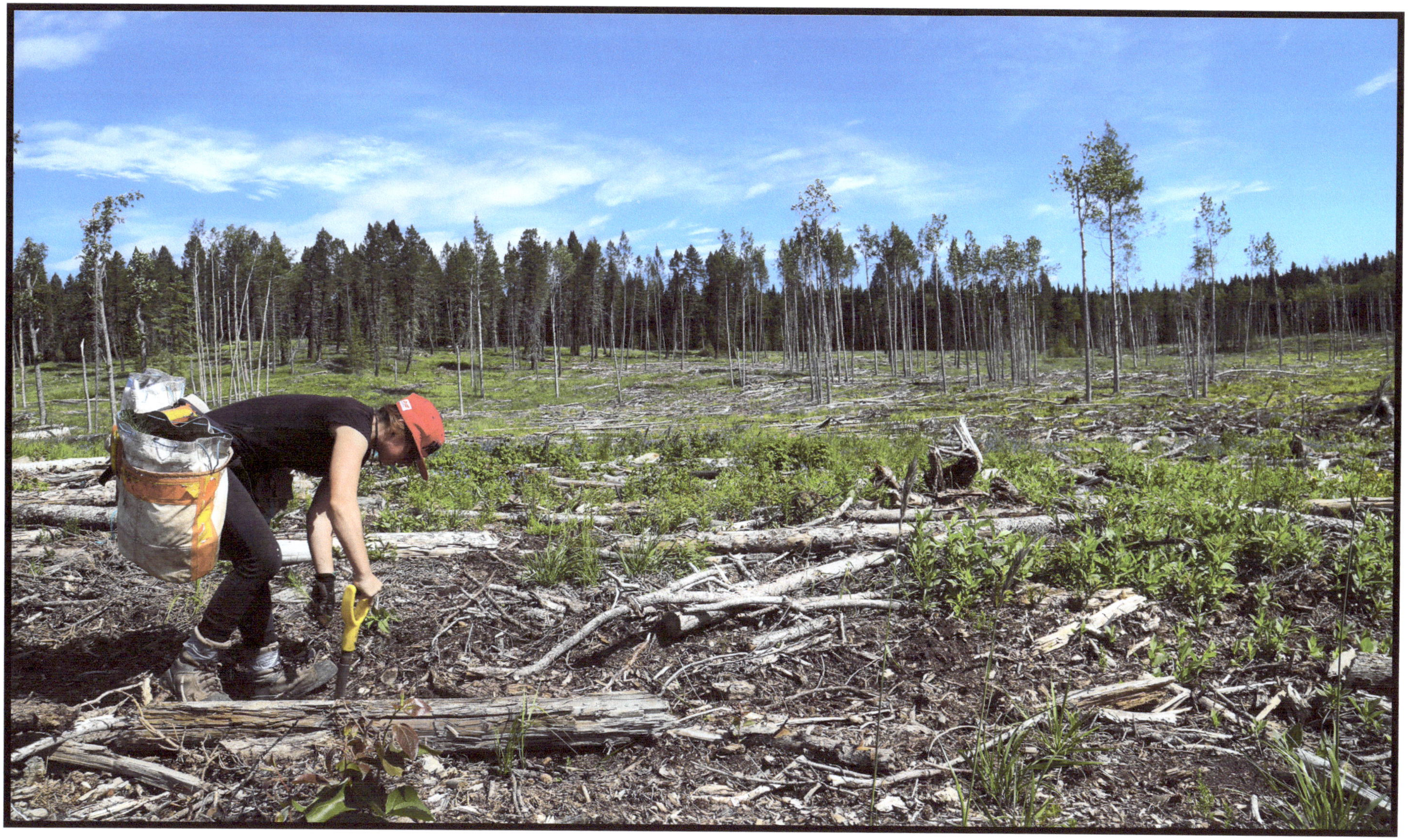

Cami also studied Neuroscience at UBC, and Psychology at Carleton. Originally hailing from Ottawa, Cami spends a lot of time travelling around Canada, especially in coastal areas. Her hobbies include aerial silks, static trapeze, pole dance, hoop dance, and burlesque.

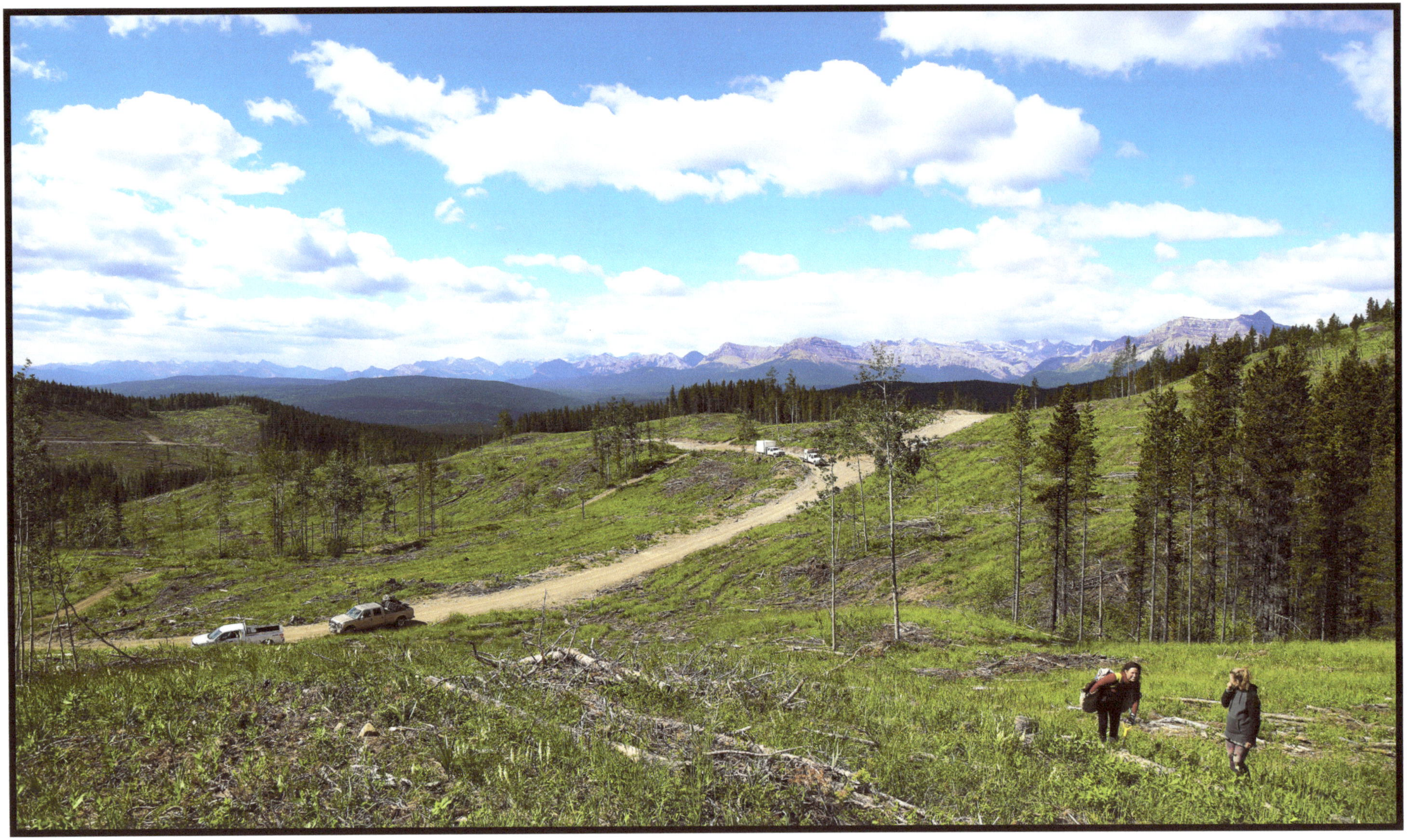

We sometimes get to work in areas with quite a view.

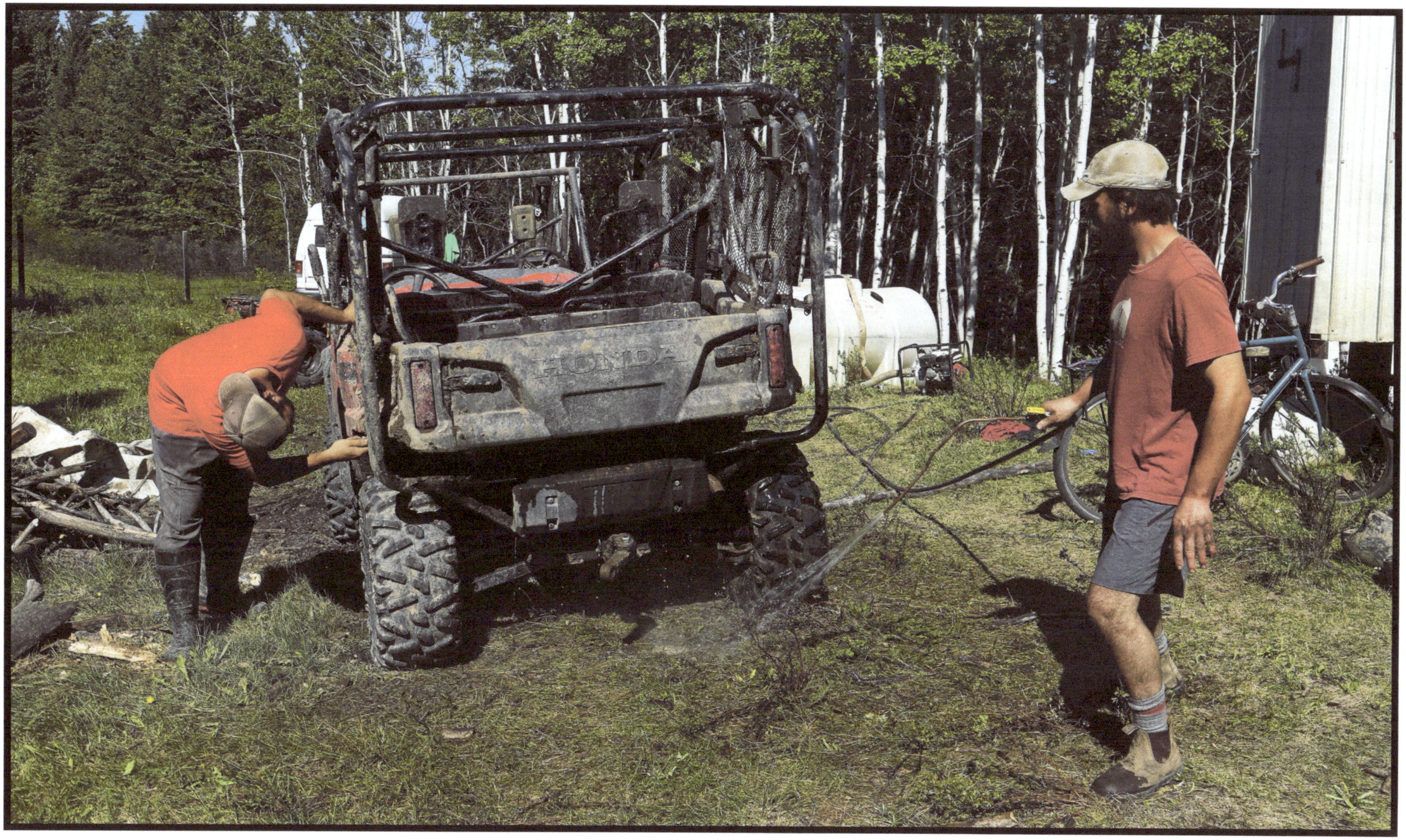

These foremen know the importance of keeping their equipment clean and doing regular maintenance.

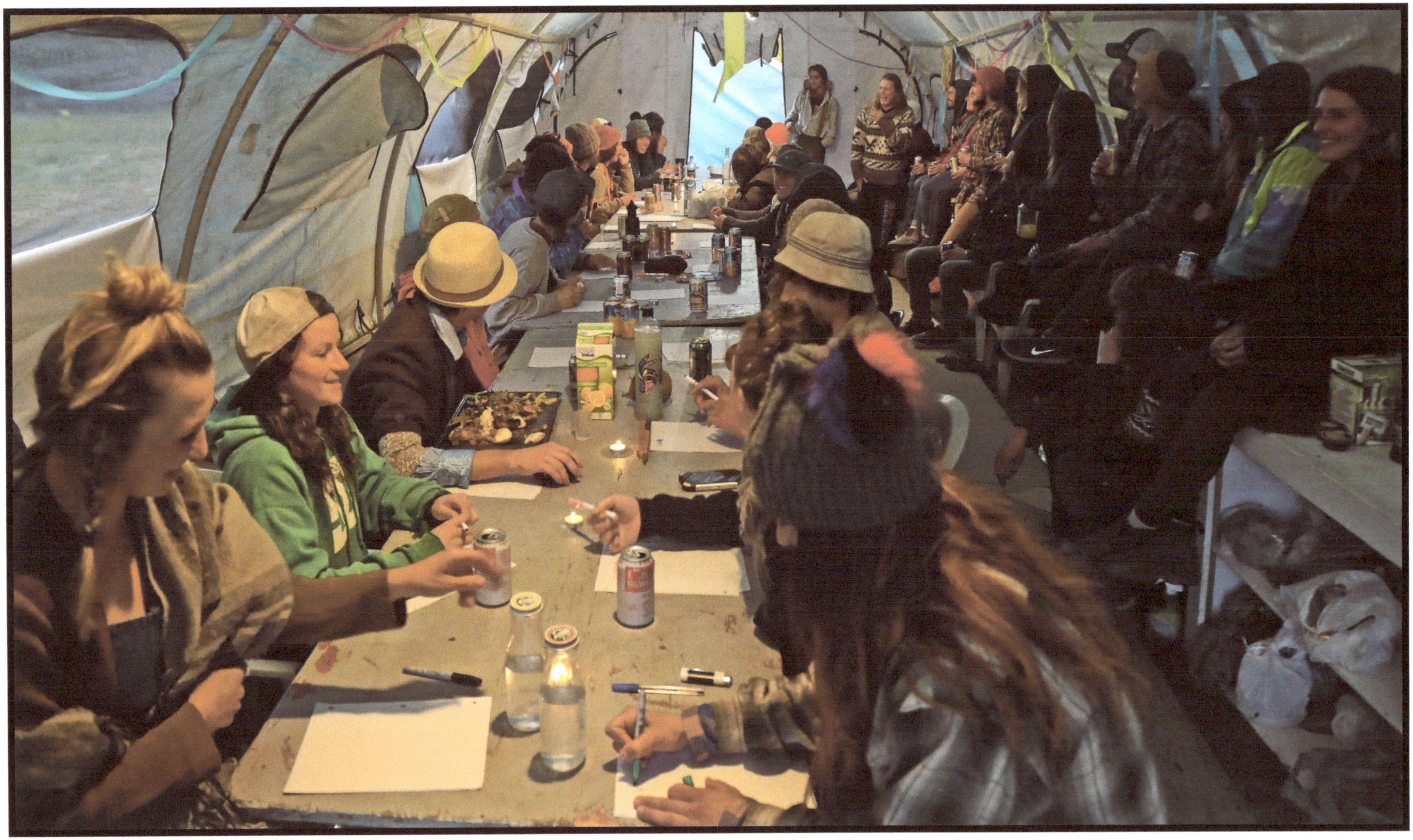

The annual "Newlyweds" game on a night off.

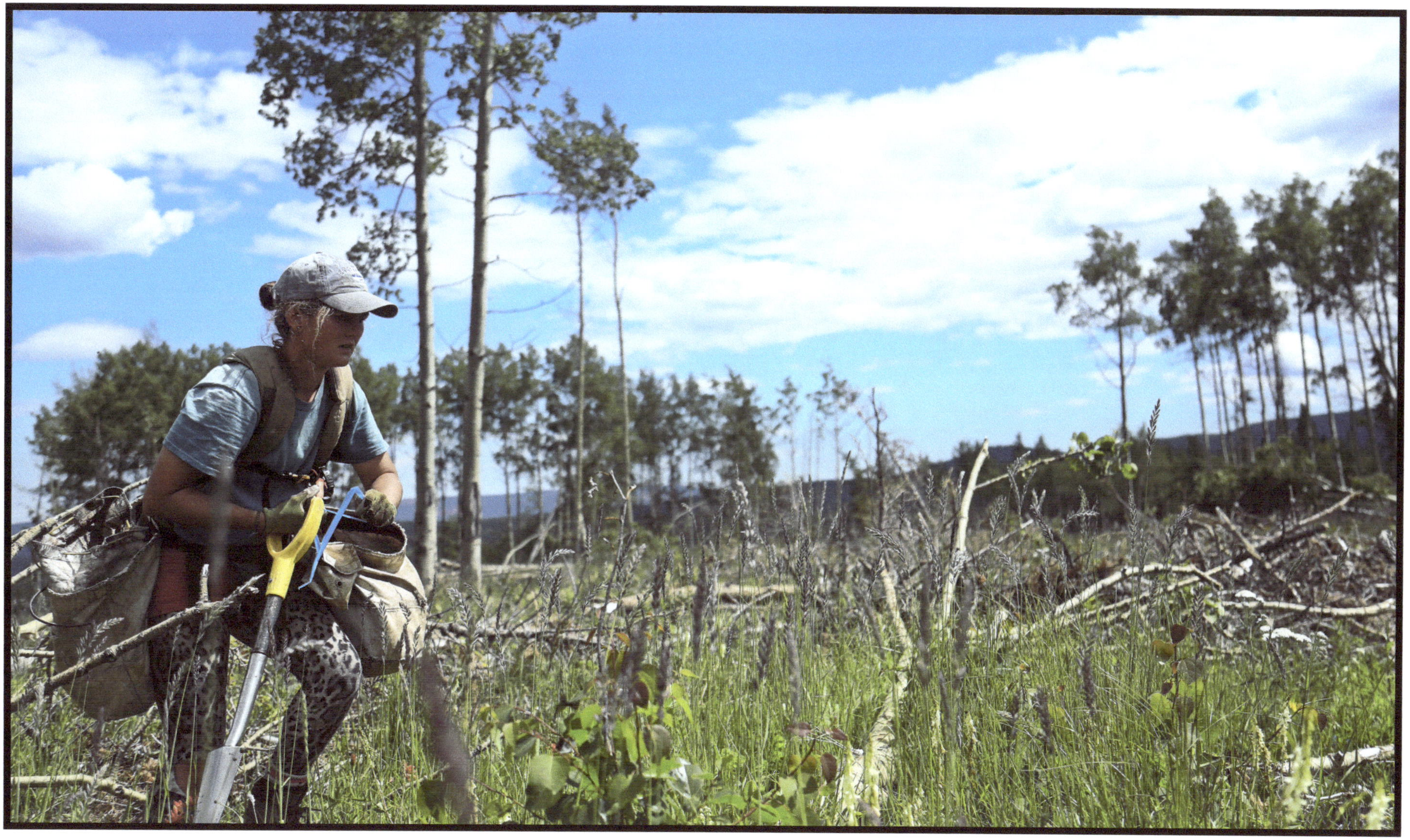

Georgina's hometown was Toronto. She attended Queens University, where she earned a Bachelor of Science in Geography and a subsequent Bachelor of Education. She is currently working as a supply teacher, but intends to move into a full-time teaching position shortly. When Georgina has free time, she enjoys camping, canoe tripping, and recreational sports.

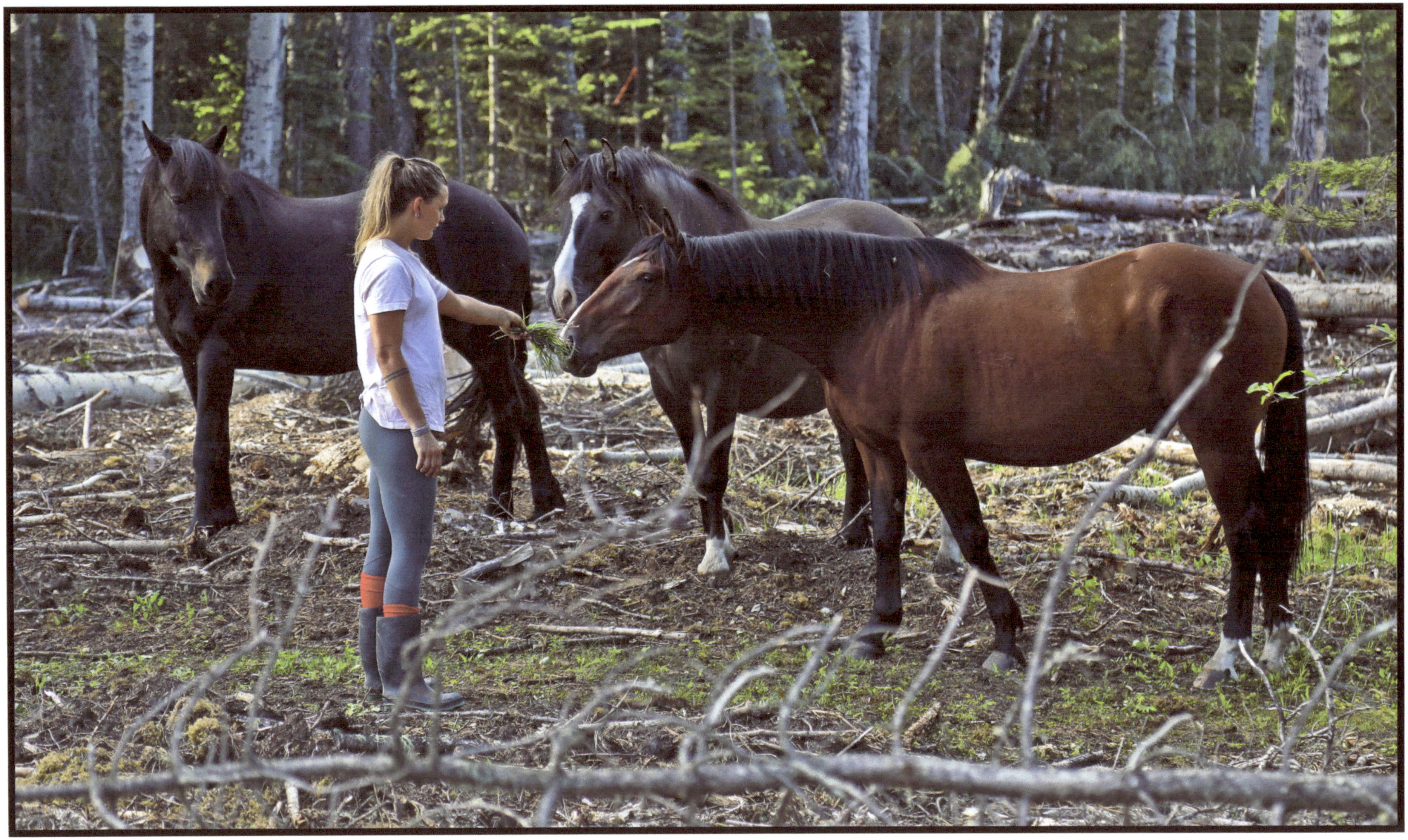

Having grown up in Barrhead, Shonda is a natural horse whisperer. I'm pretty sure that these feral horses wouldn't let any other planter get this close.

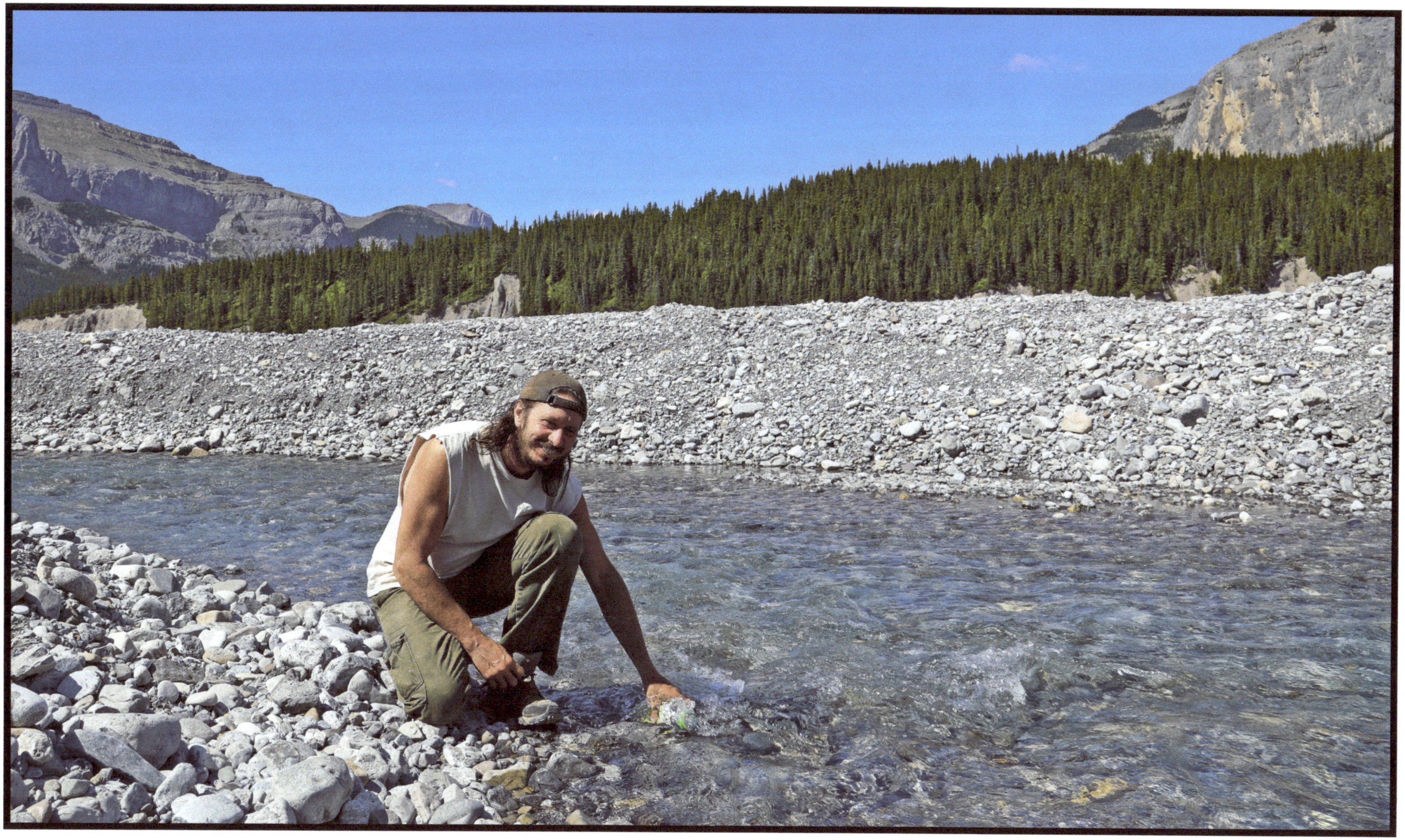

They say not to drink out of streams. They say that you'll get beaver fever. I find it hard to pass up a clean stream like this one.

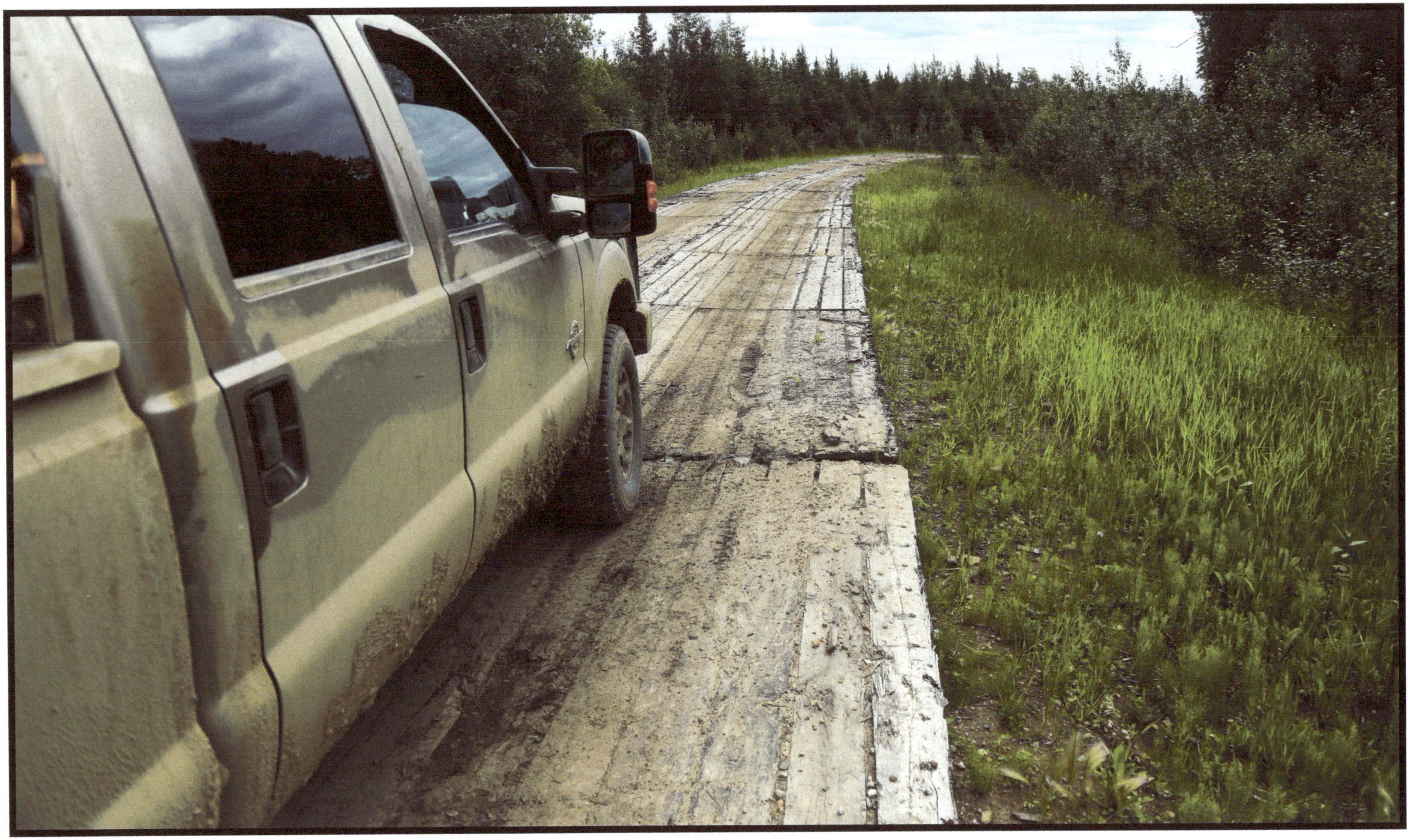

When we're working in areas where oil and gas infrastructure is in place, we sometimes get to take advantage of the money that Oil & Gas throws at a problem. It's not very common to see rig mats covering a muddy road like this, but it definitely makes it much easier for our trucks to move around.

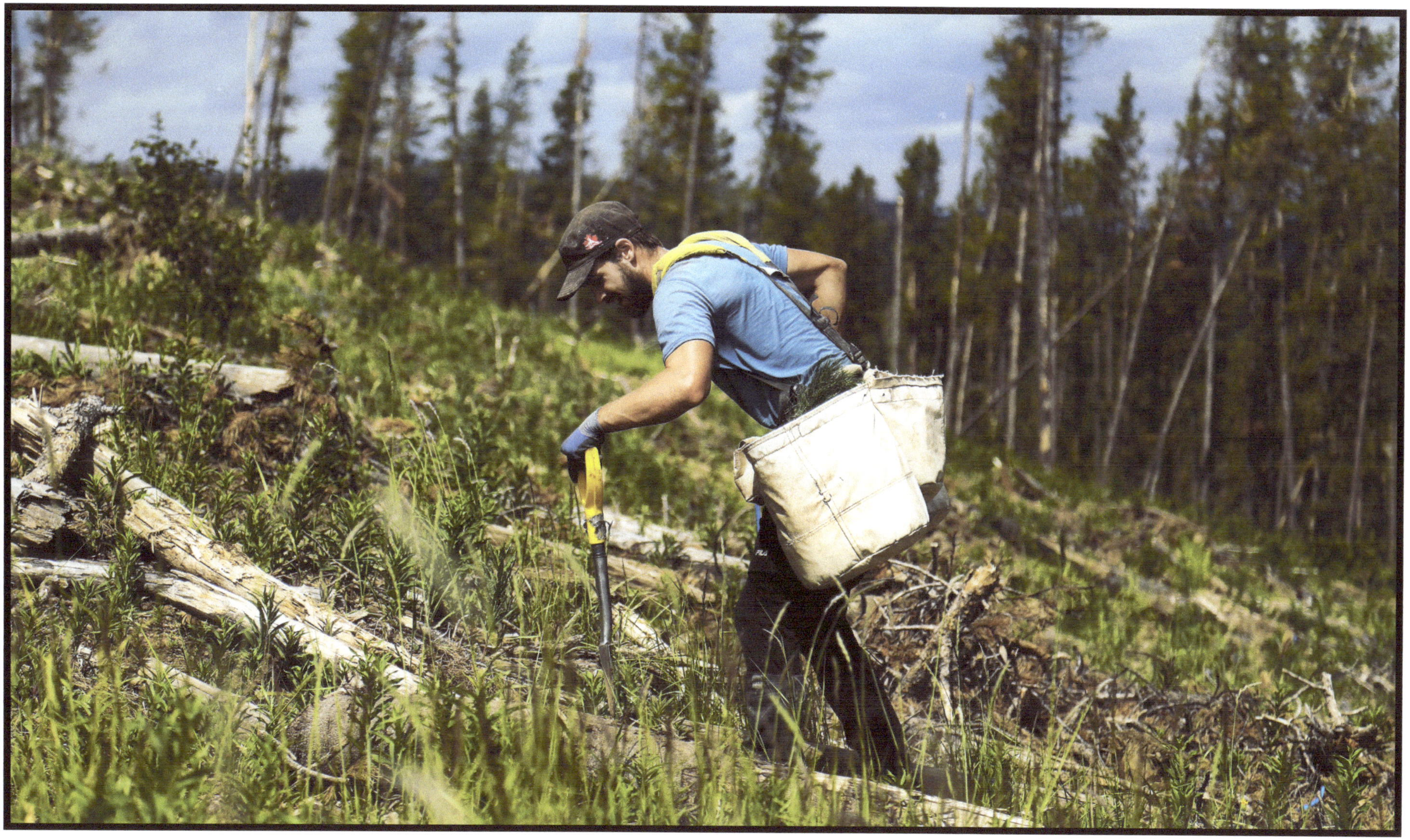

Francis loves cross country skiing and cycling, and has completed a program in Physical Education from the University of Quebec in Montreal (UQAM). One of Francis' many long-term goals is to become the head coach of a cross-country ski club.

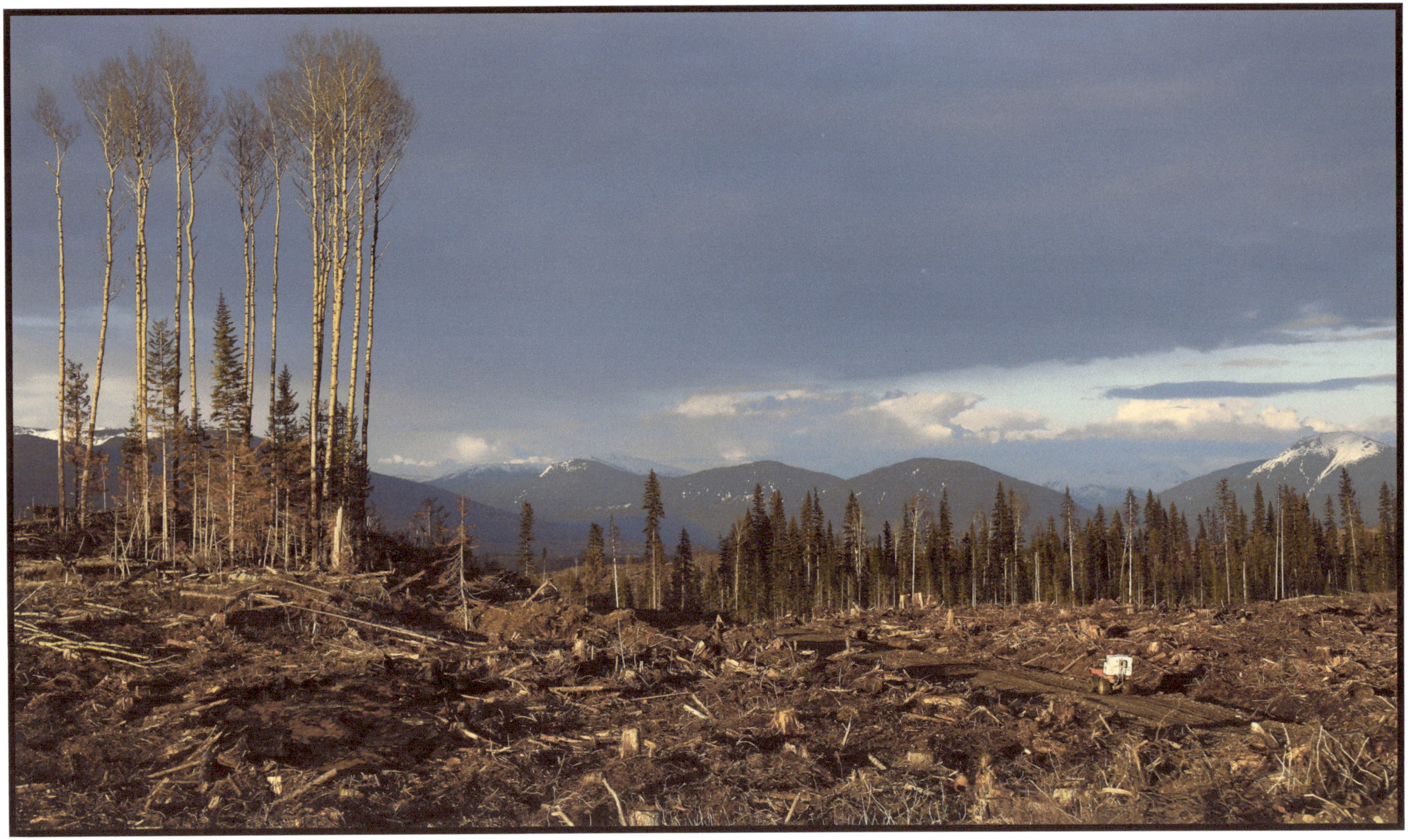

Sometimes, there is great beauty to be found in a cut-block.

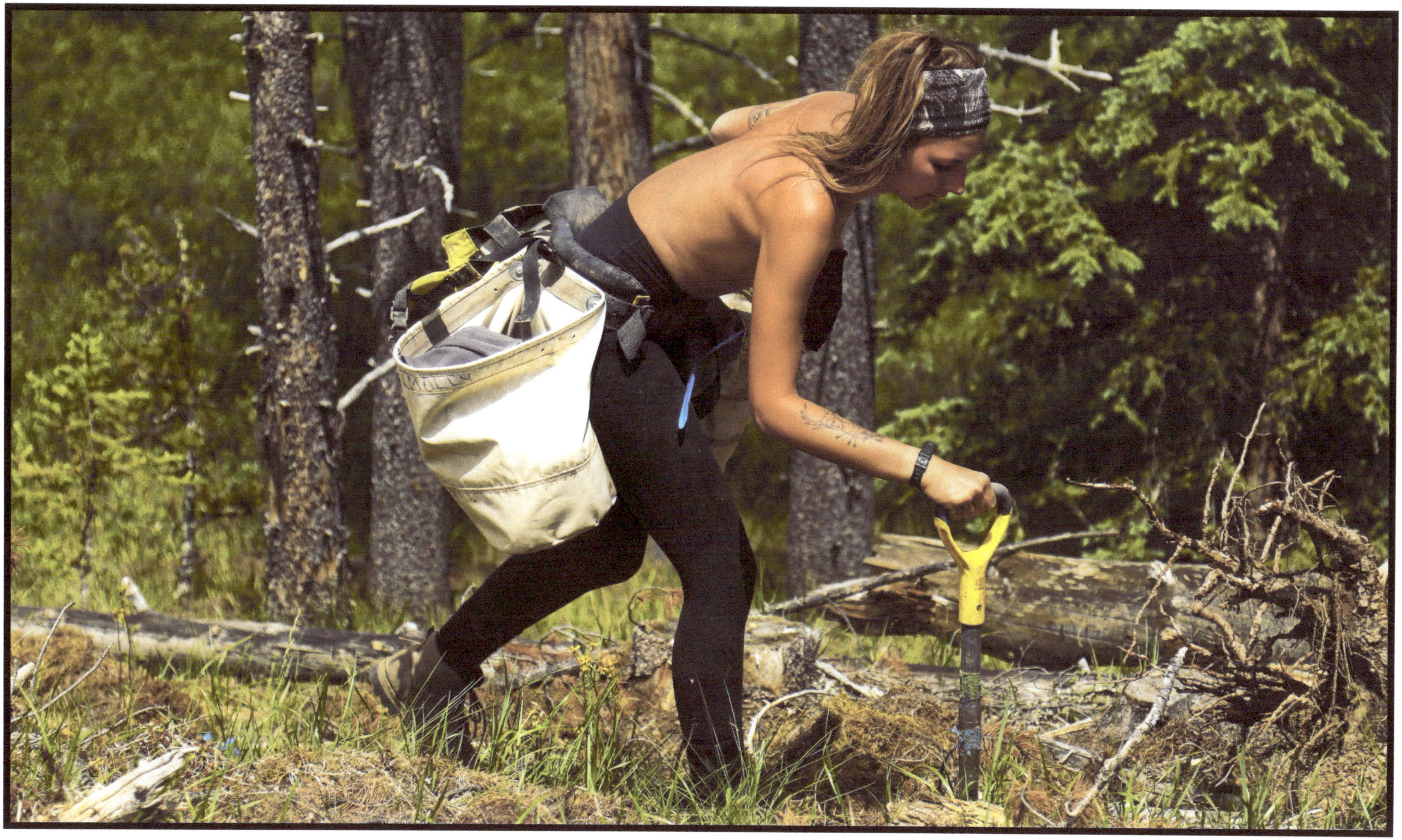

Remember to always wear sunscreen.

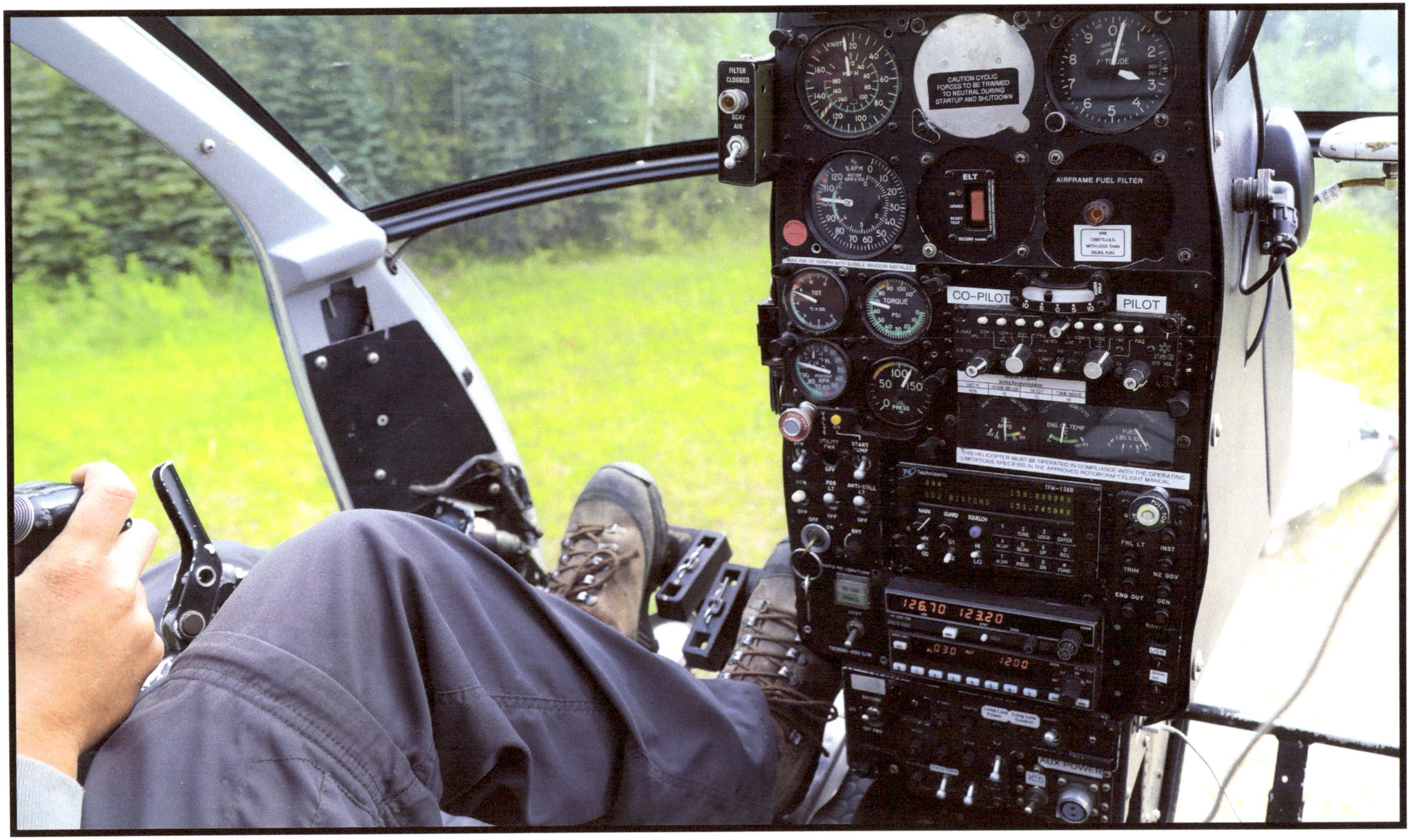

If you think your job is complicated, try flying a helicopter.

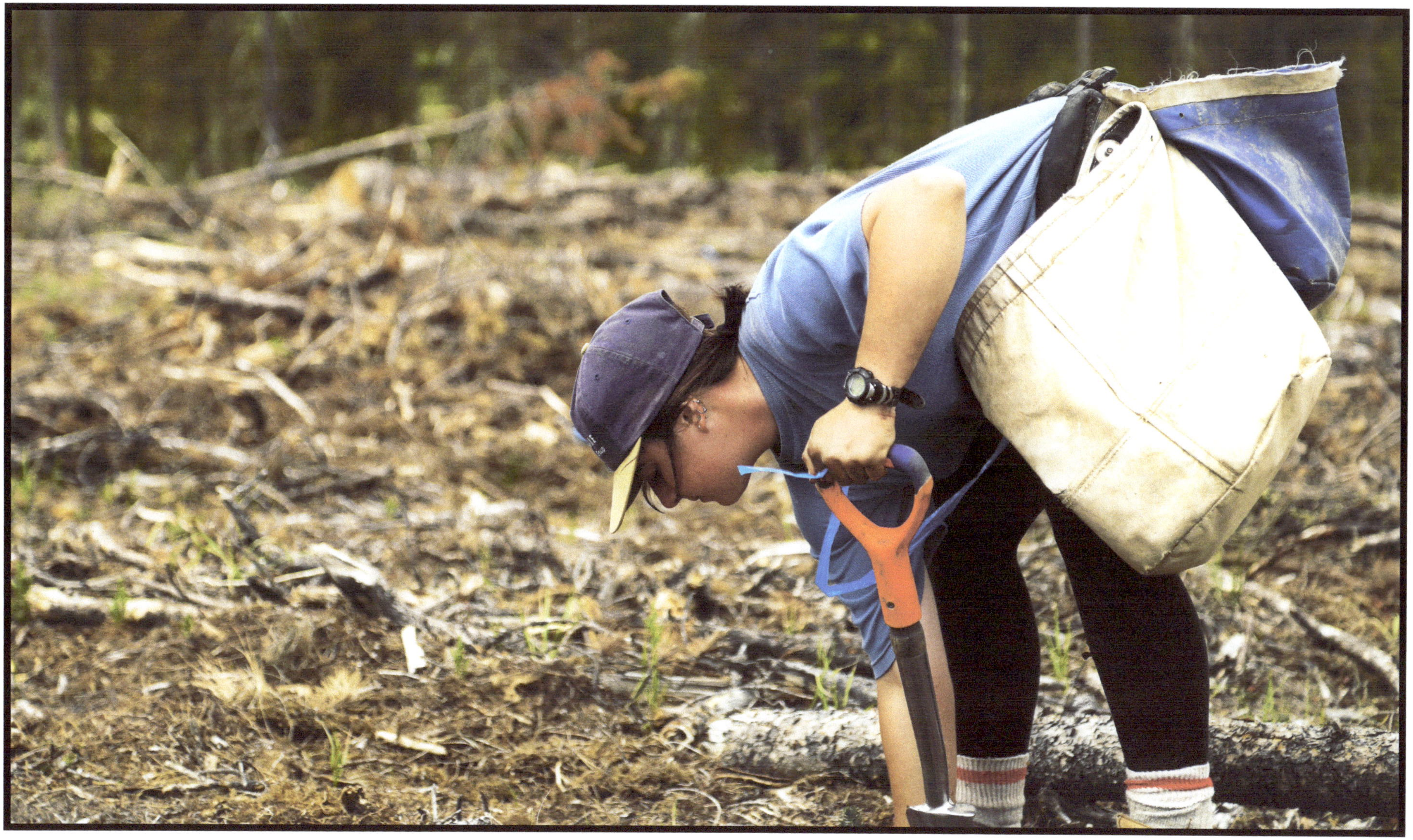

Sam, who grew up in Toronto, loves planting.

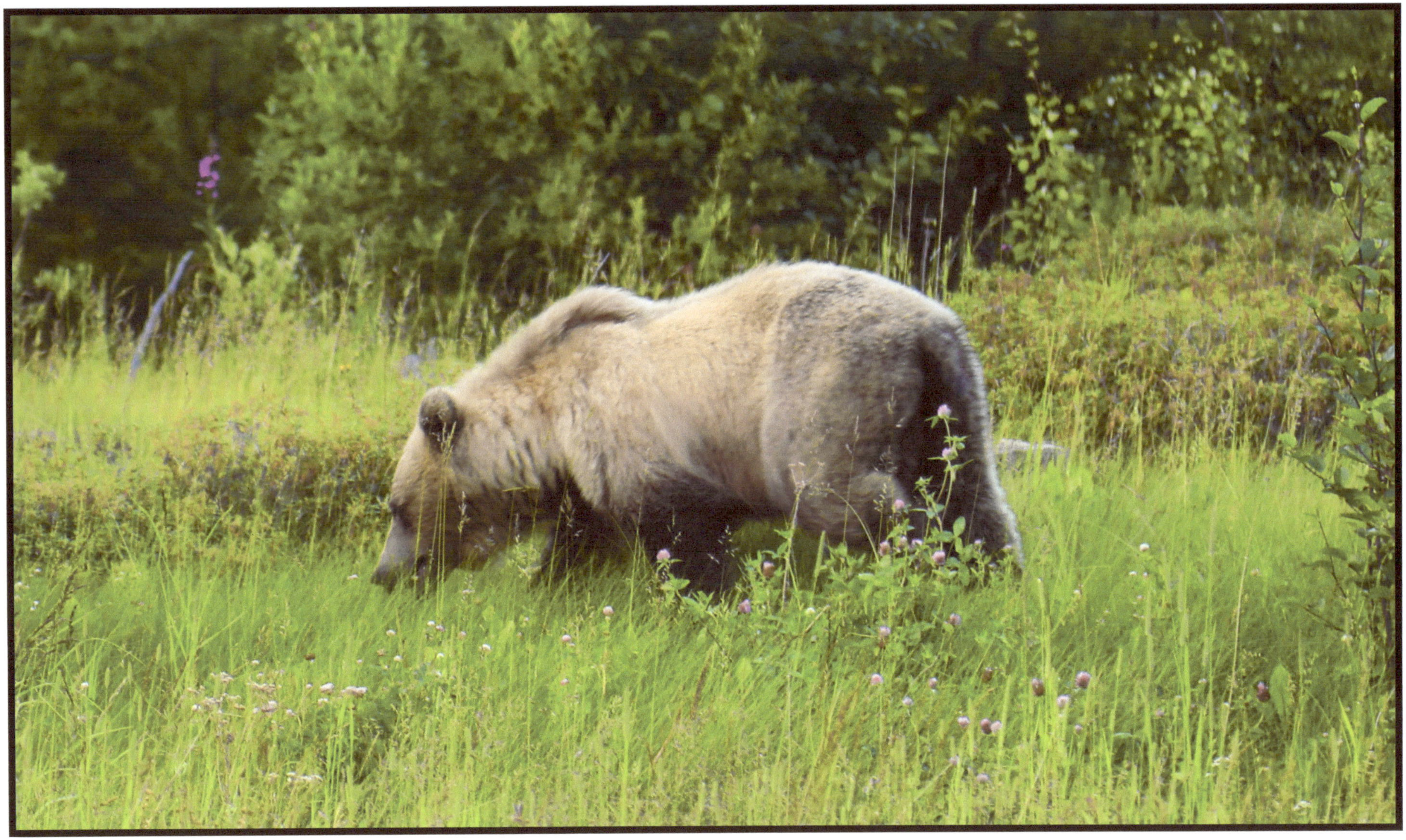

This grizzly bear is a rather magnificent specimen. I rarely see bears with fur that is this light in colour.

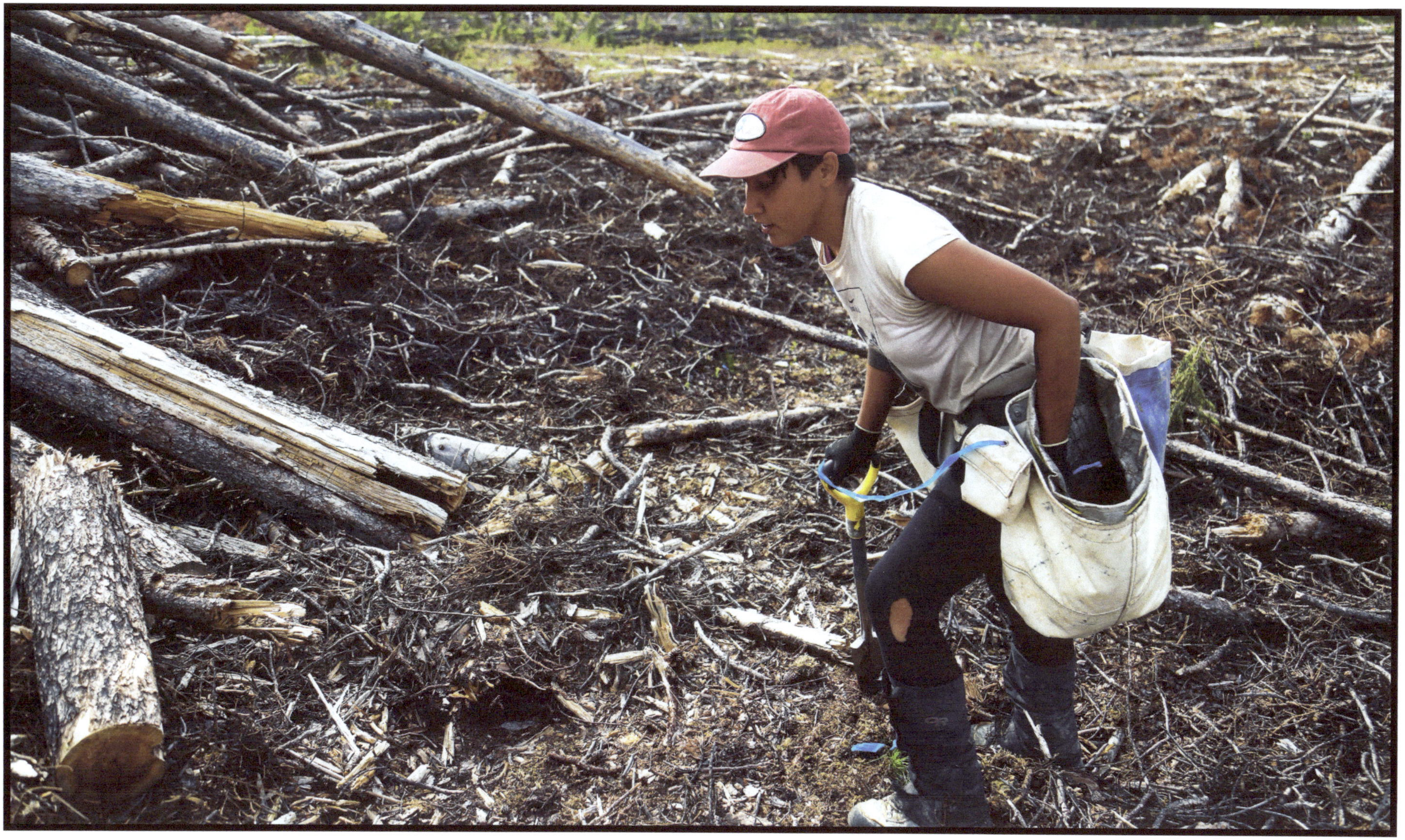

Planters often use small pieces of blue flagging tape to mark their trees. The blue colour makes it easier to see the trees once the grass and other competition starts to grow, and helps to ensure that we hit our density targets.

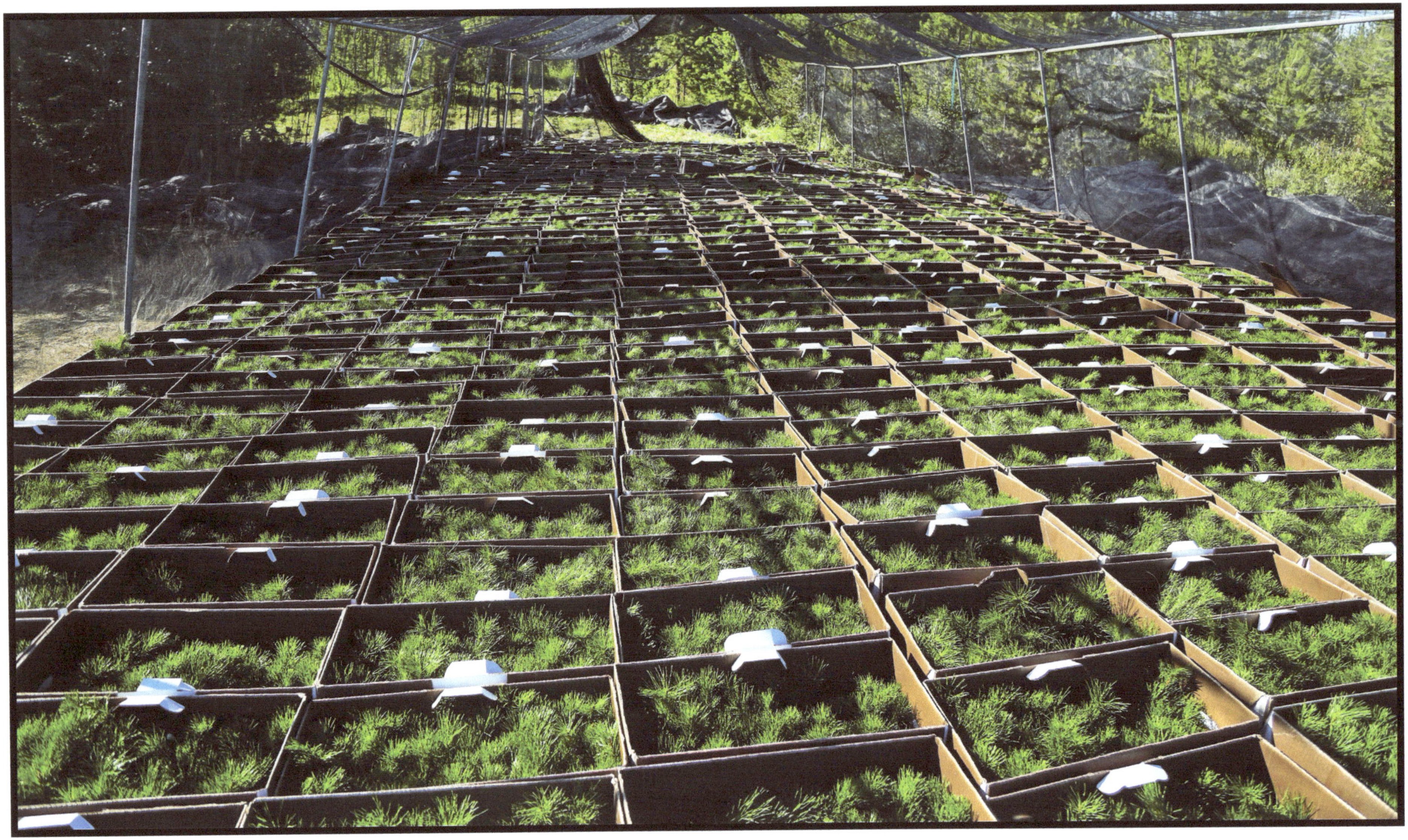

This shade tent contains boxes of trees with 225 seedlings apiece. One thousand boxes arrived in this shipment, which works out to almost a quarter million trees. That's about three days of work for my camp early in the season, when everyone is still getting into shape, but only about two days of work later in the season when people are at full production.

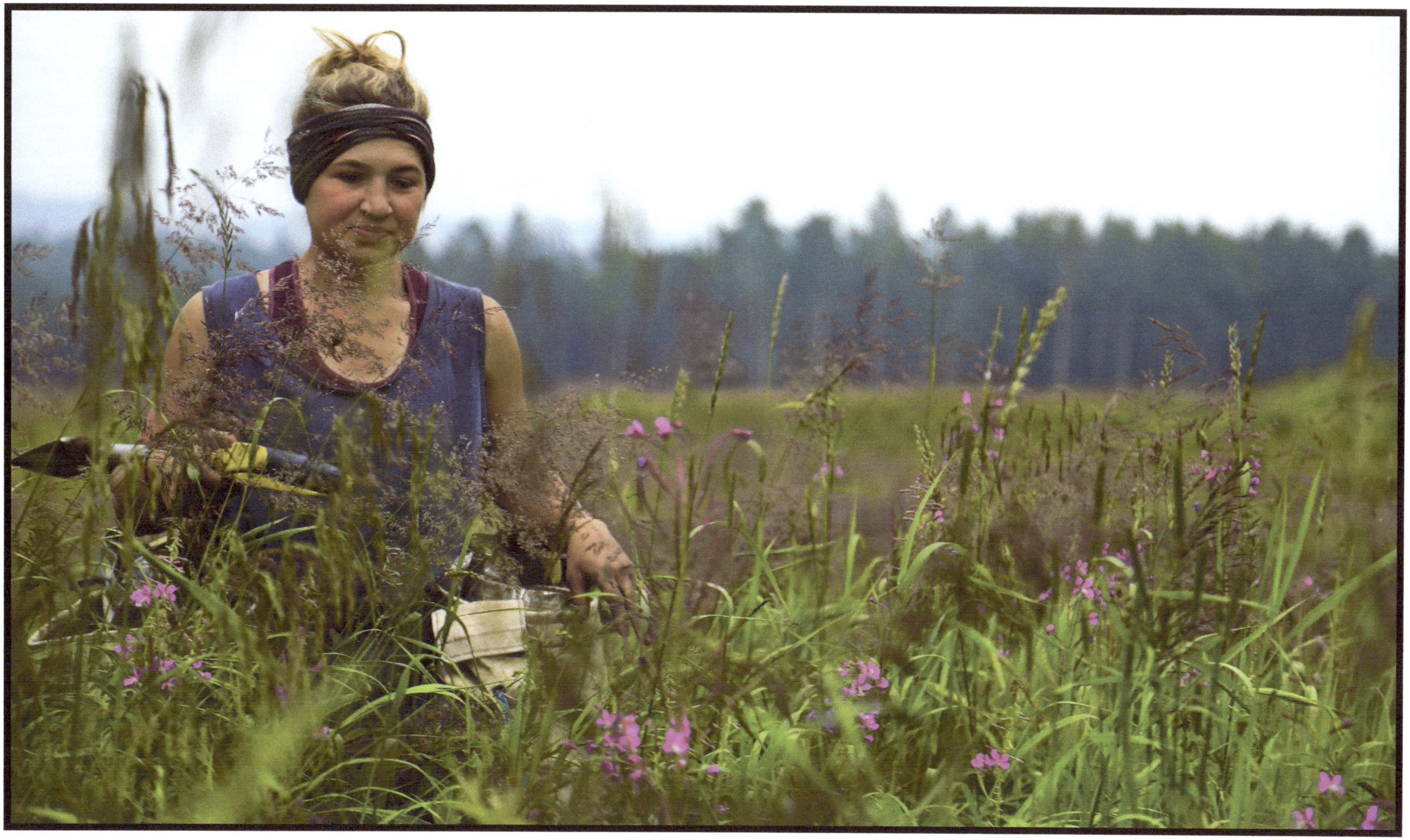

Ania, who spends most of her year teaching in Ontario, was almost lost in the tall grass on this block. This was a "special mission" block, when two of us went out to plant a number of logistically challenging blocks that had been deferred too long, allowing the grass to get out of control. Thankfully, we were only planting the burn areas on each of the blocks, but it was pretty hard to find the burns in the tall grass.

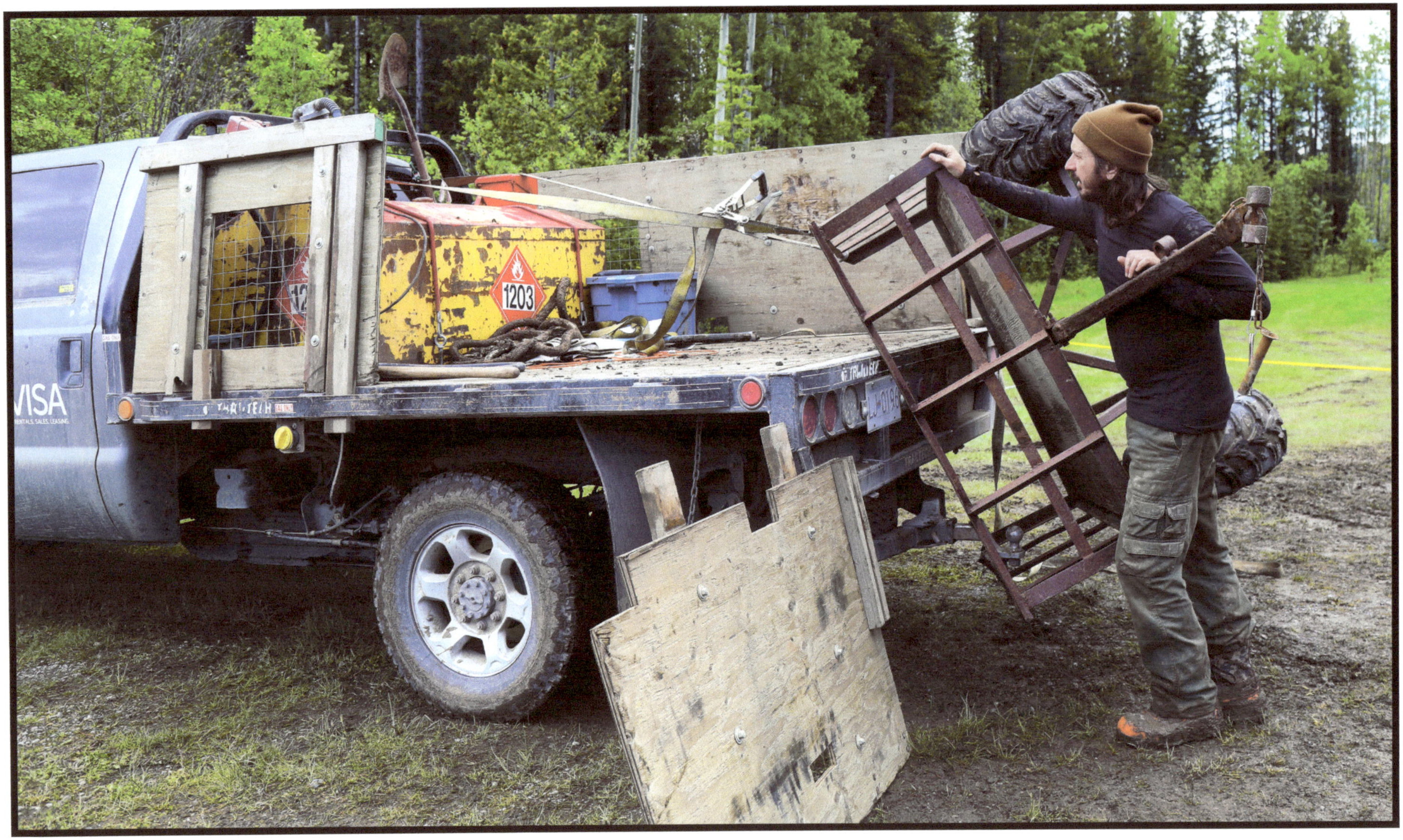

This quad trailer was much too heavy for me to load into the truck by myself. However, I wanted to use it as an example to demonstrate that nothing is impossible if you set your mind to it. I had initially considered backing the truck up to some sort of ramp, but I couldn't find anything suitable. The solution was to use some ratchet straps and a bit of heavy lifting to load the trailer onto the truck in a series of short steps.

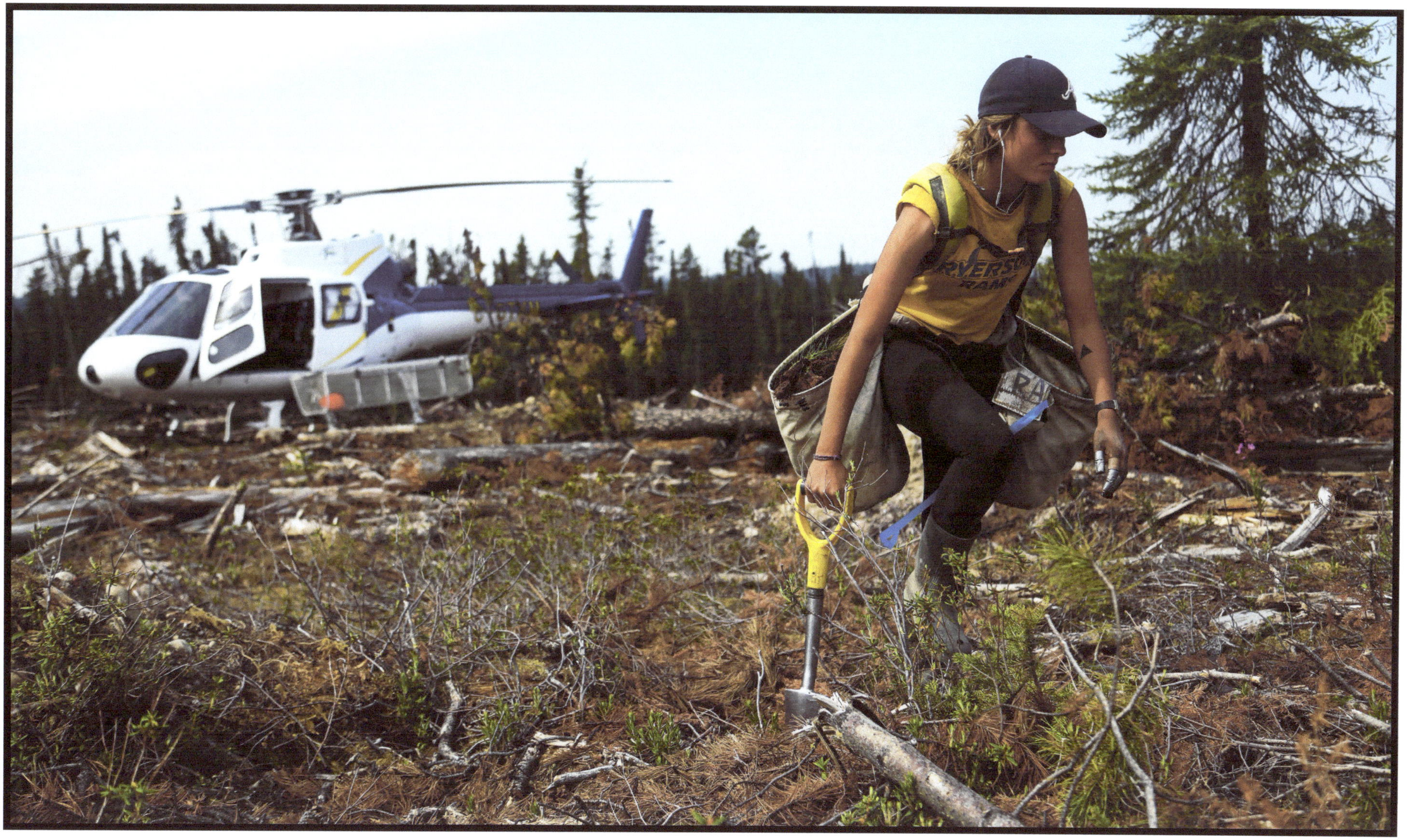

Becca seems entirely unfazed by the helicopter parked in her piece.

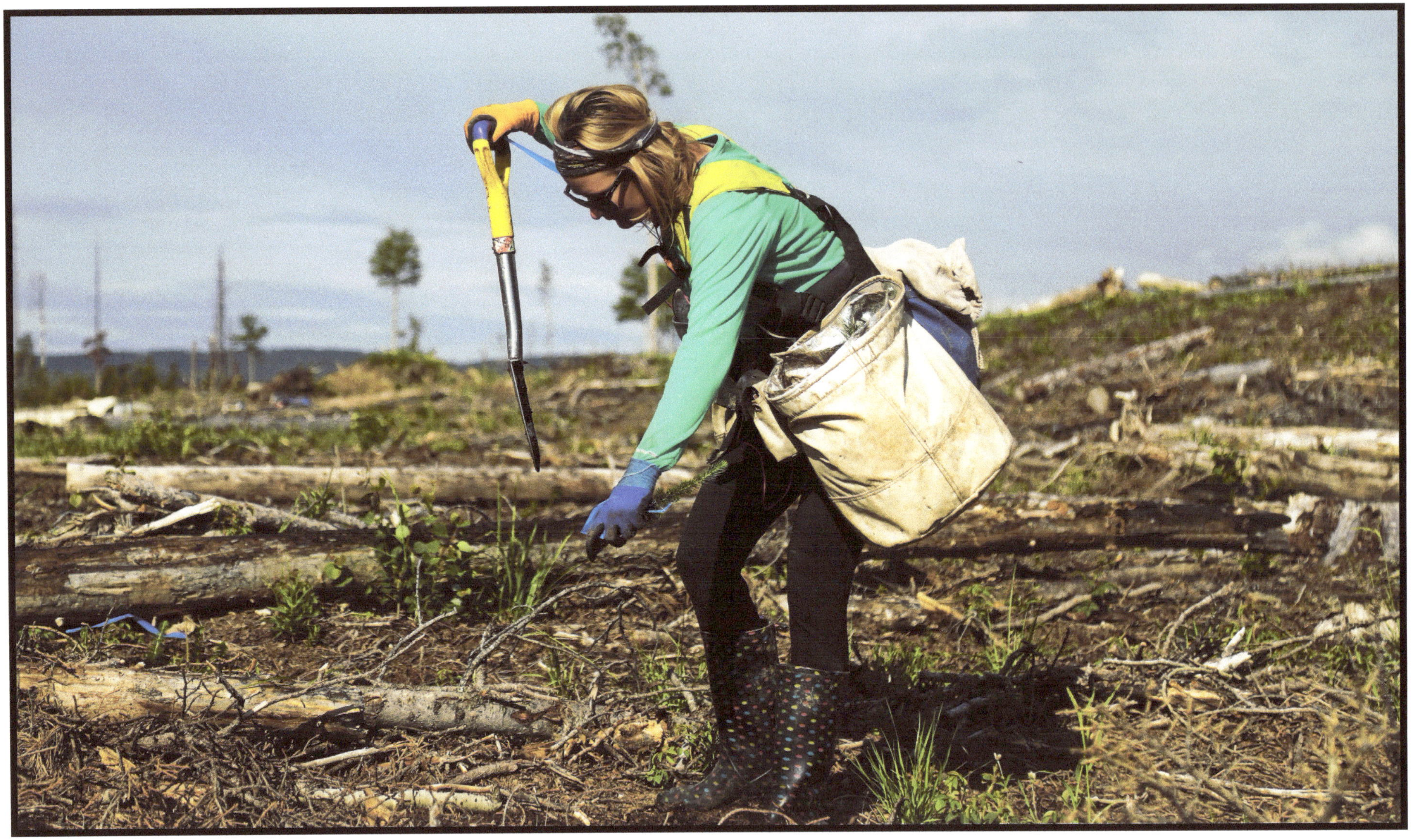

Katrina is from Oshawa. She studied History and Philosophy at Brock University. Her hobbies include singing opera, painting, and thrifting.

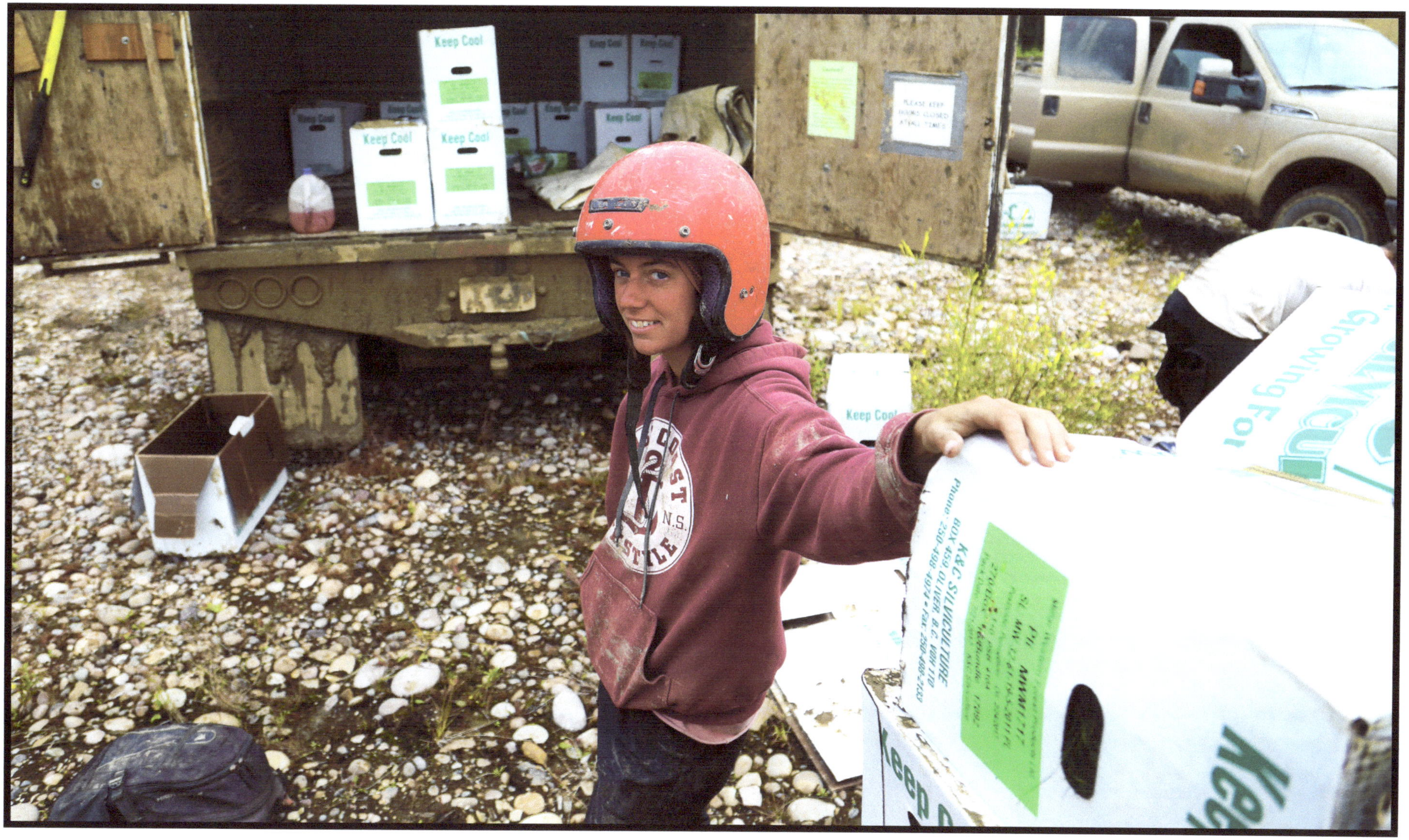

This is Jess, the crew leader from my earlier "the road ate your truck" story. Jess is from Nova Scotia. She's an energetic long-distance runner, and is currently studying to go into a career as a full-time paramedic.

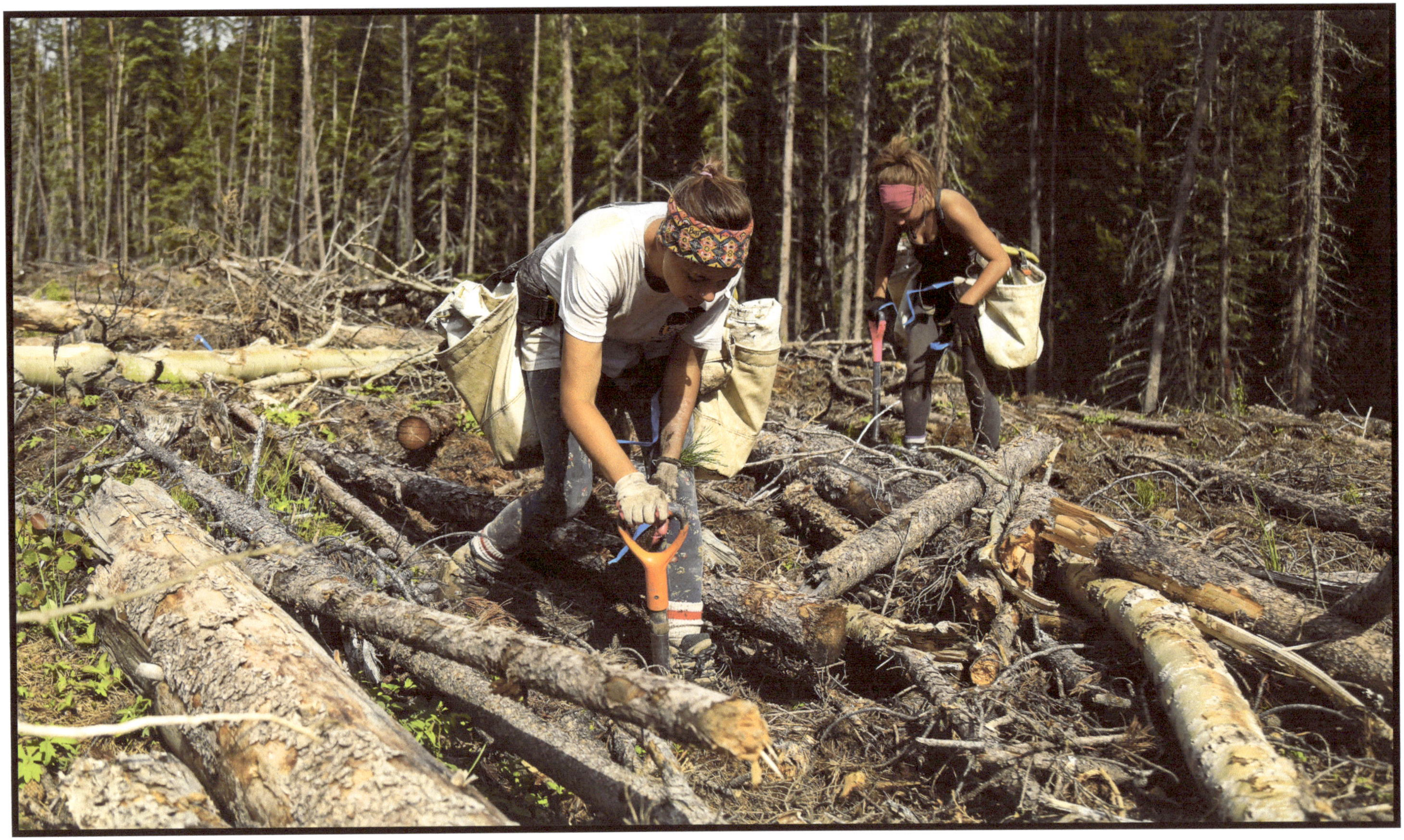

Some planters prefer to work alone, and others prefer to work in pairs. These two often worked together. Karla, on the right, is an Environmental Science student at the University of Winnipeg, doing a double major in forest ecology and biology. Karla likes to work out at the gym, nerd out on anything related to science, and drive her truck through a lot of mud. That seems ideal for tree planting.

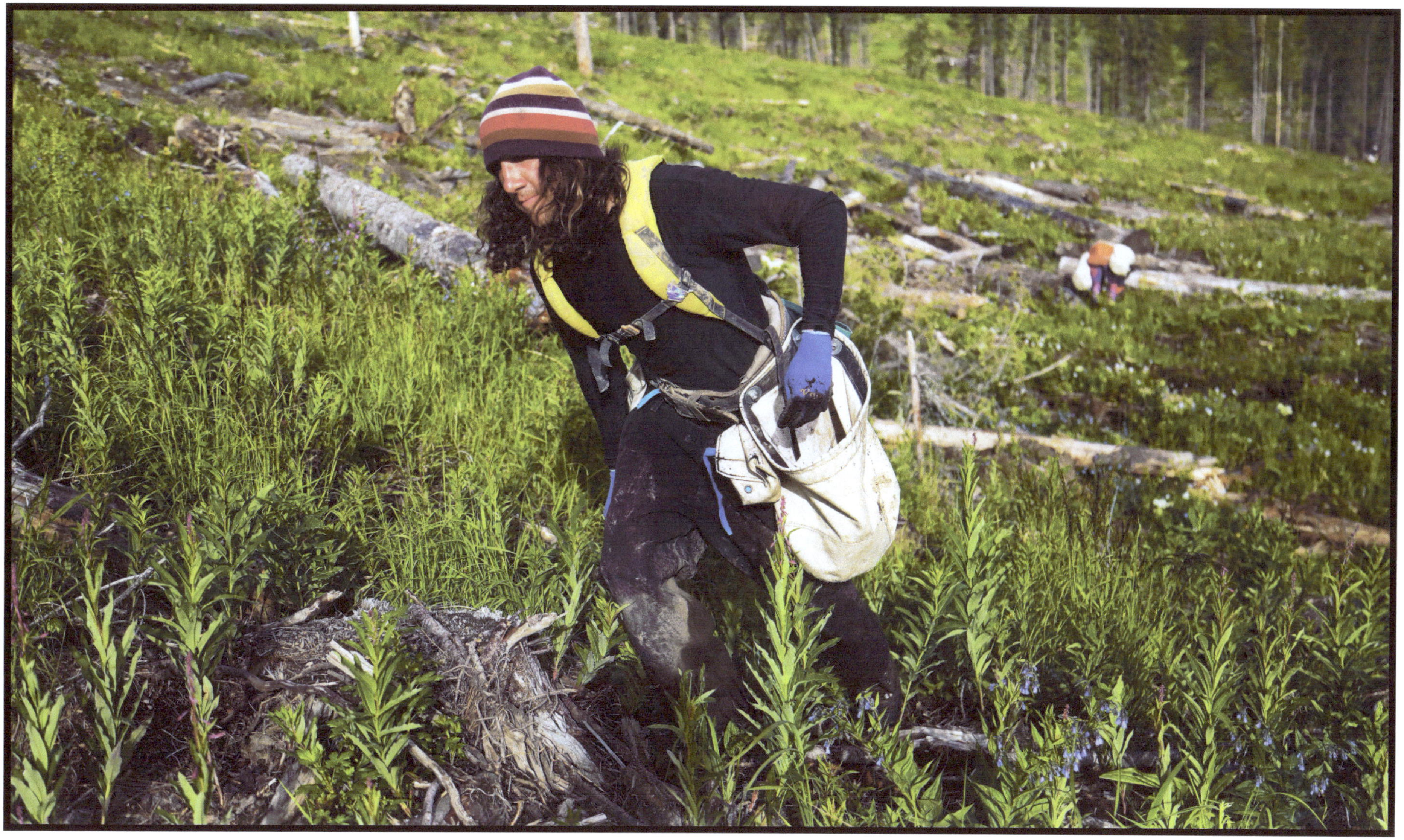

Abe, who grew up in New Brunswick, studied biology at Mount Allison University (with a focus on ecology). He enjoys planting, and thinks that he will probably pursue a long-term career goal in the forestry sector.

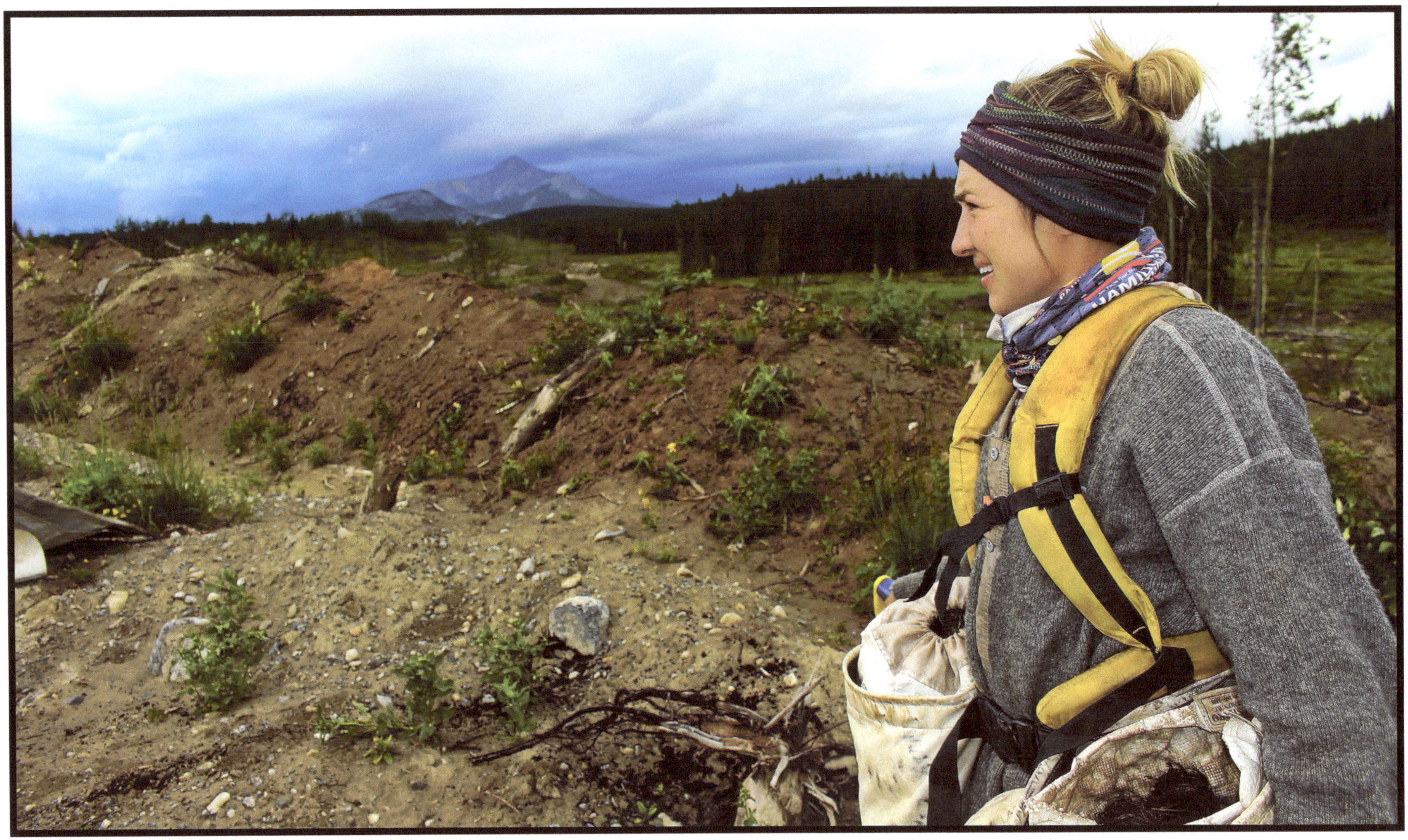

The lighting in this photo was quite unique. This photo was taken after dinner. The skies were still fairly light behind us, but there was a pretty nasty storm approaching from the mountain that you can see in the distance. We ended up suffering through about twenty-five minutes of intense rain while the storm passed overhead.

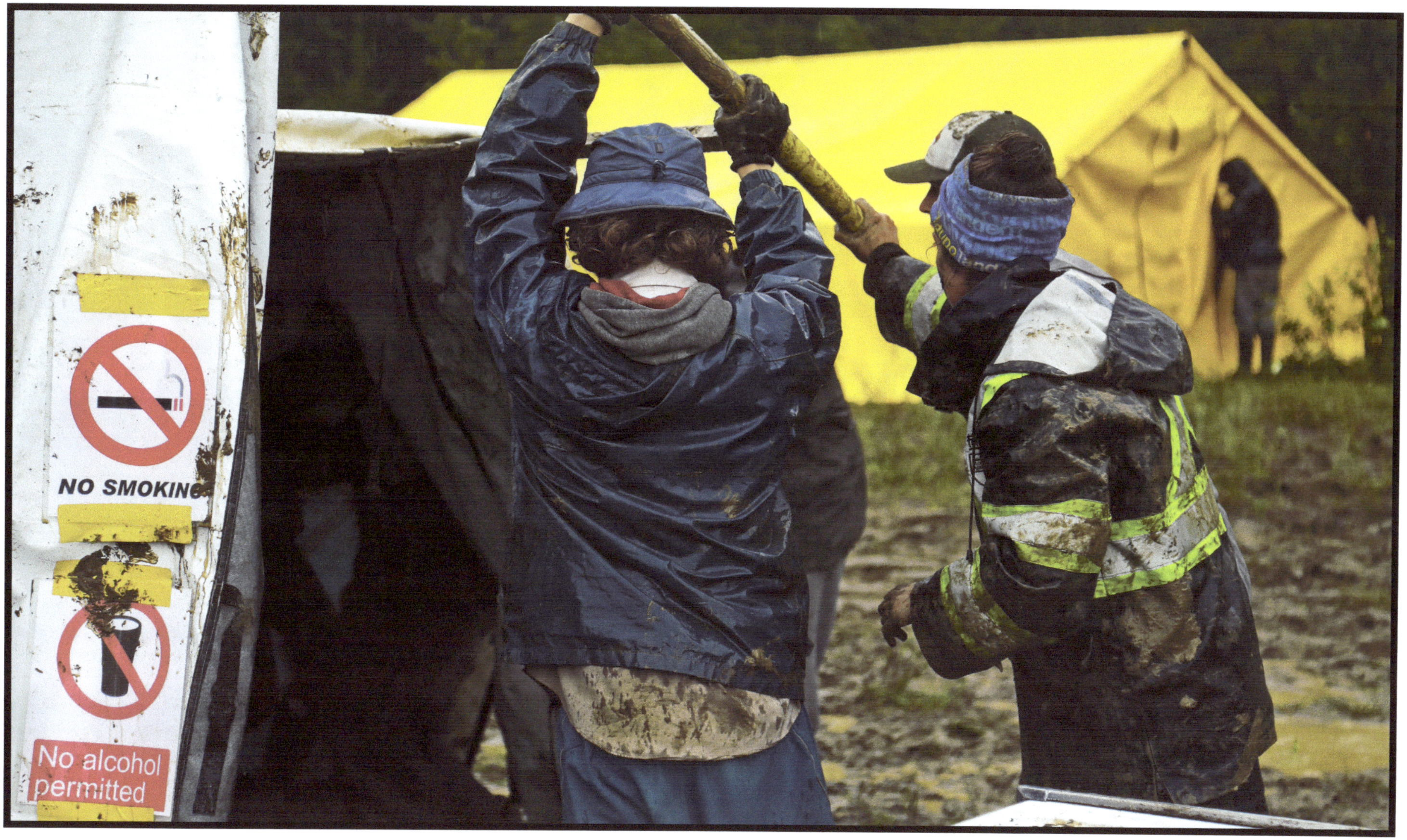

Rain during a workday on a block is usually bearable. Some planters grow quite accustomed to it, and don't mind it at all. However, setting up or tearing down a camp in the rain usually leads to a bit of a challenge.

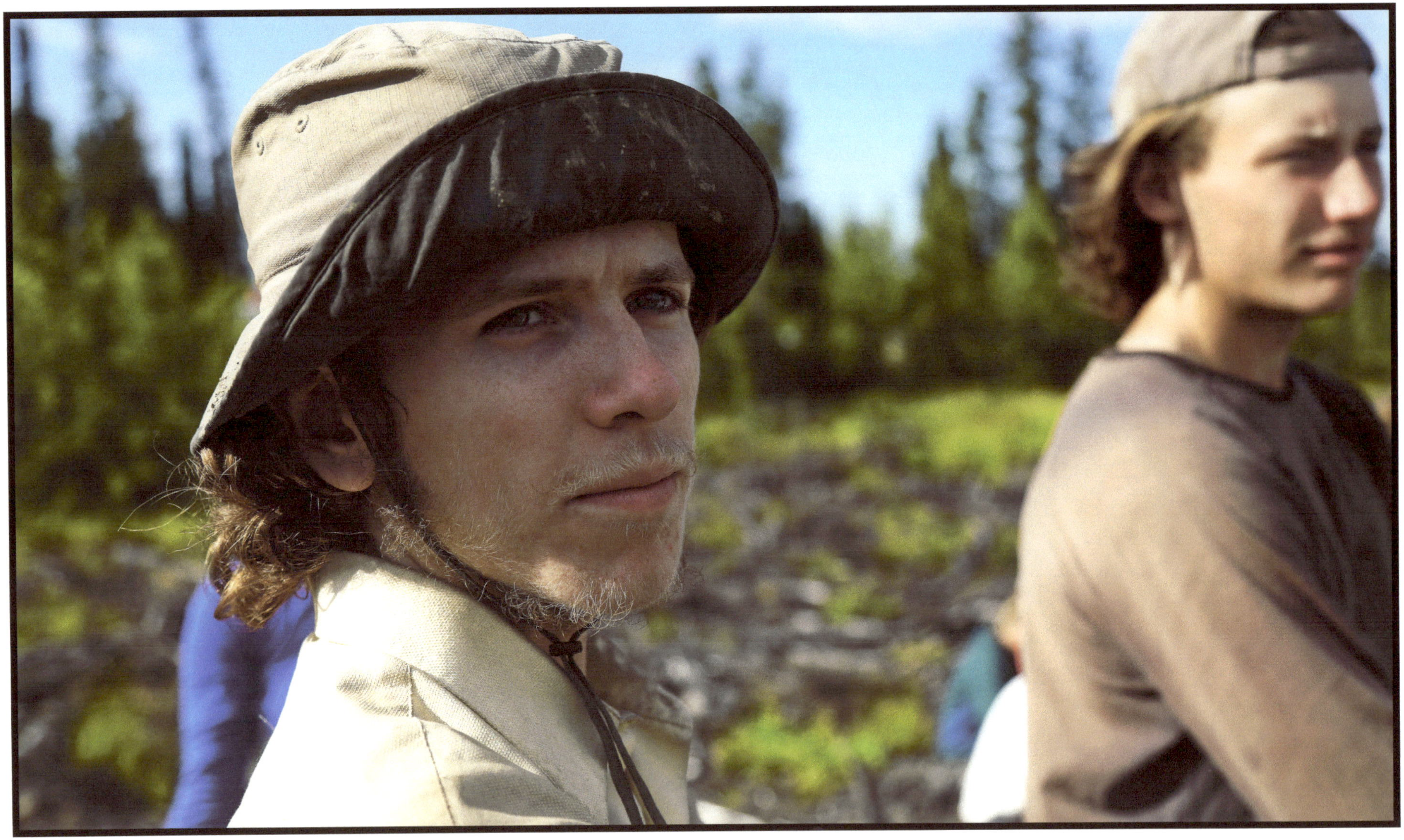

Liam is a journalism student at Carleton University, who plans on travelling and freelancing once he finishes his degree. His interests include writing fiction, rock climbing, ball hockey, and planting. He writes poetry and short stories, and is currently working on his first novel.

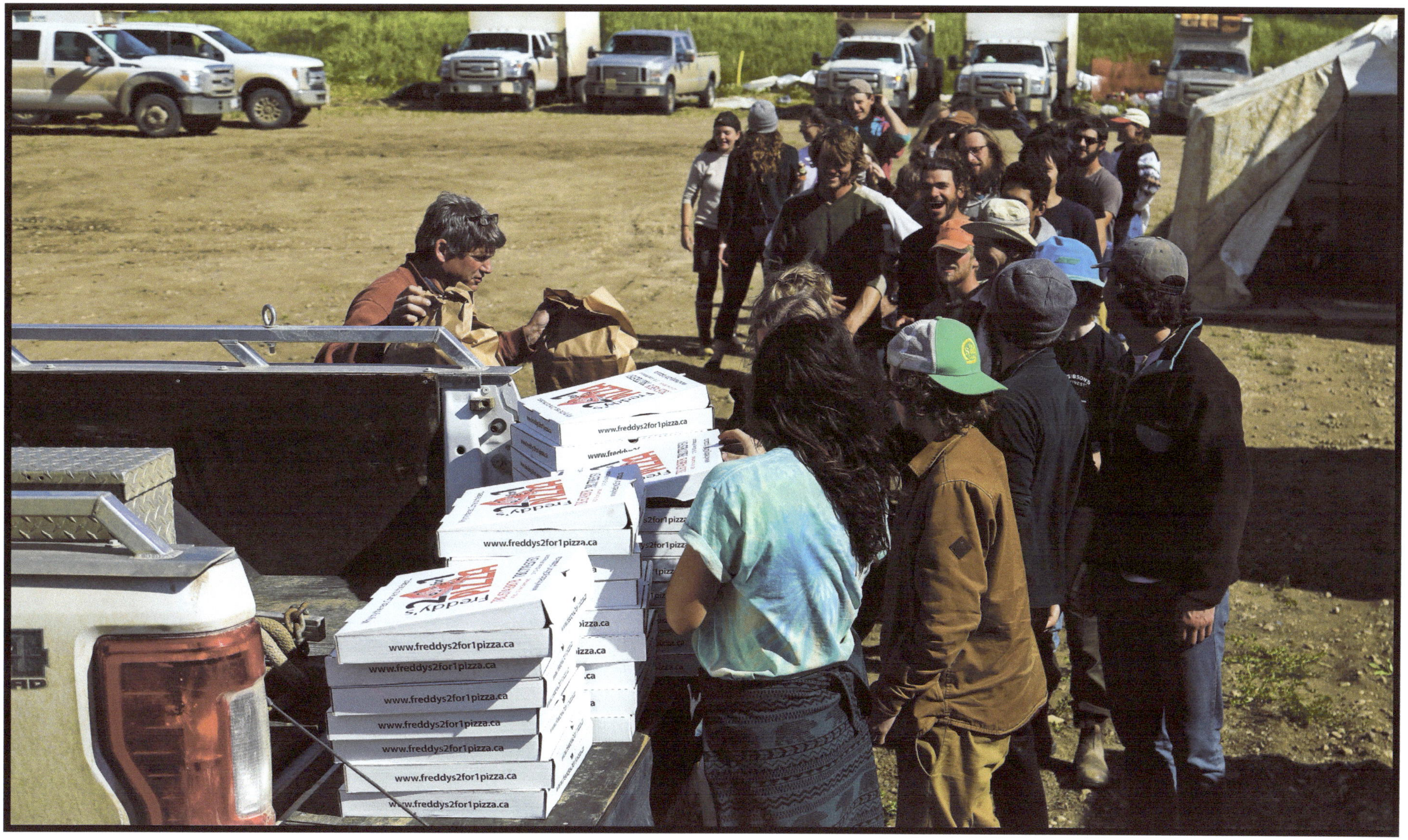

Jim gave our camp a pizza party. We ate a lot of pizza.

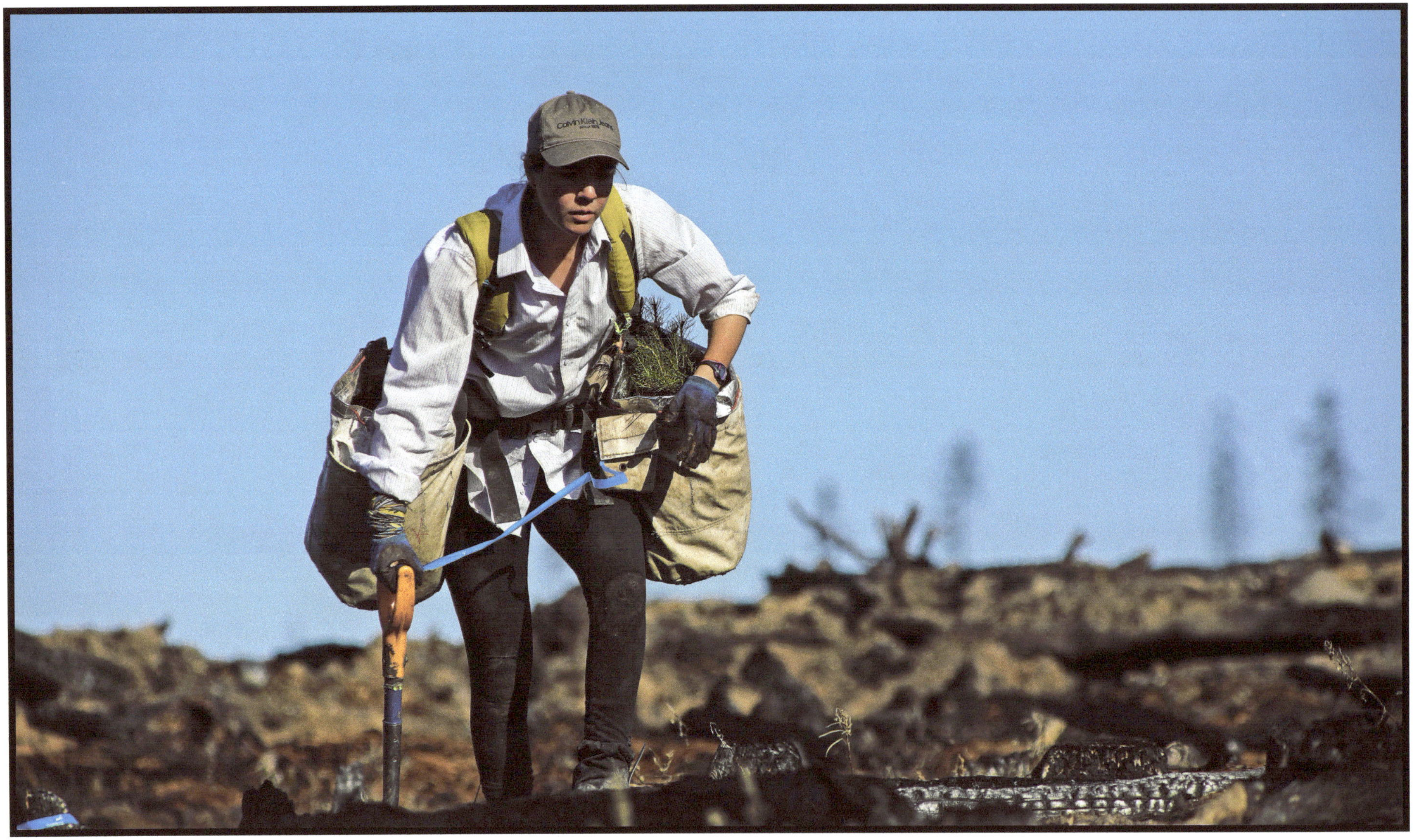

Jessica is from Montreal, and went to Concordia University to do a Masters of Environment program. Her long-term career goal is to seek to work with the Nature Conservancy of Canada. In the meantime, when she has free time, she loves to go cross-country skiing.

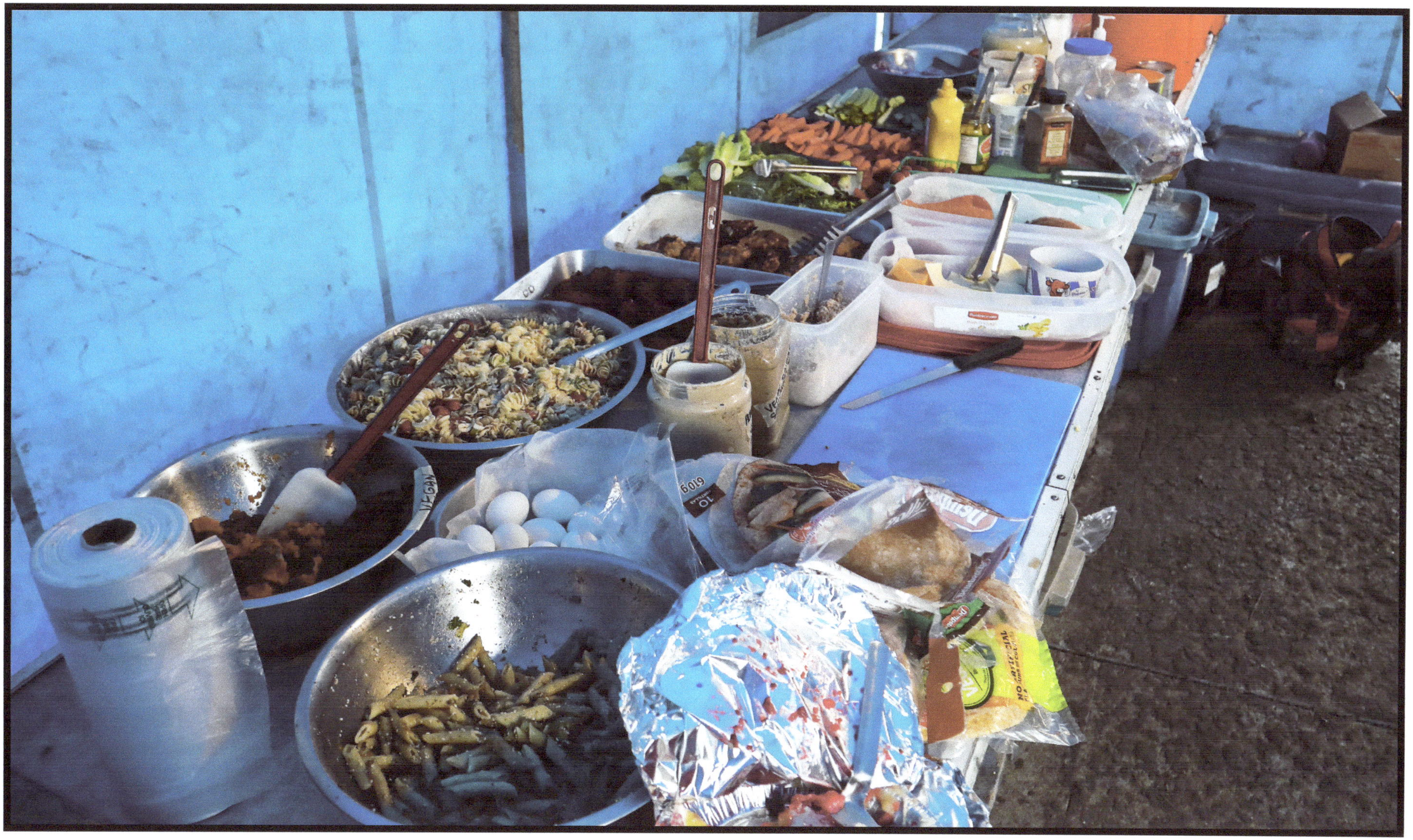

Tree planters eat a lot of food. This is the lunch table, at 5:30am, before the planters start to build their lunches. In an hour, it will be almost completely decimated.

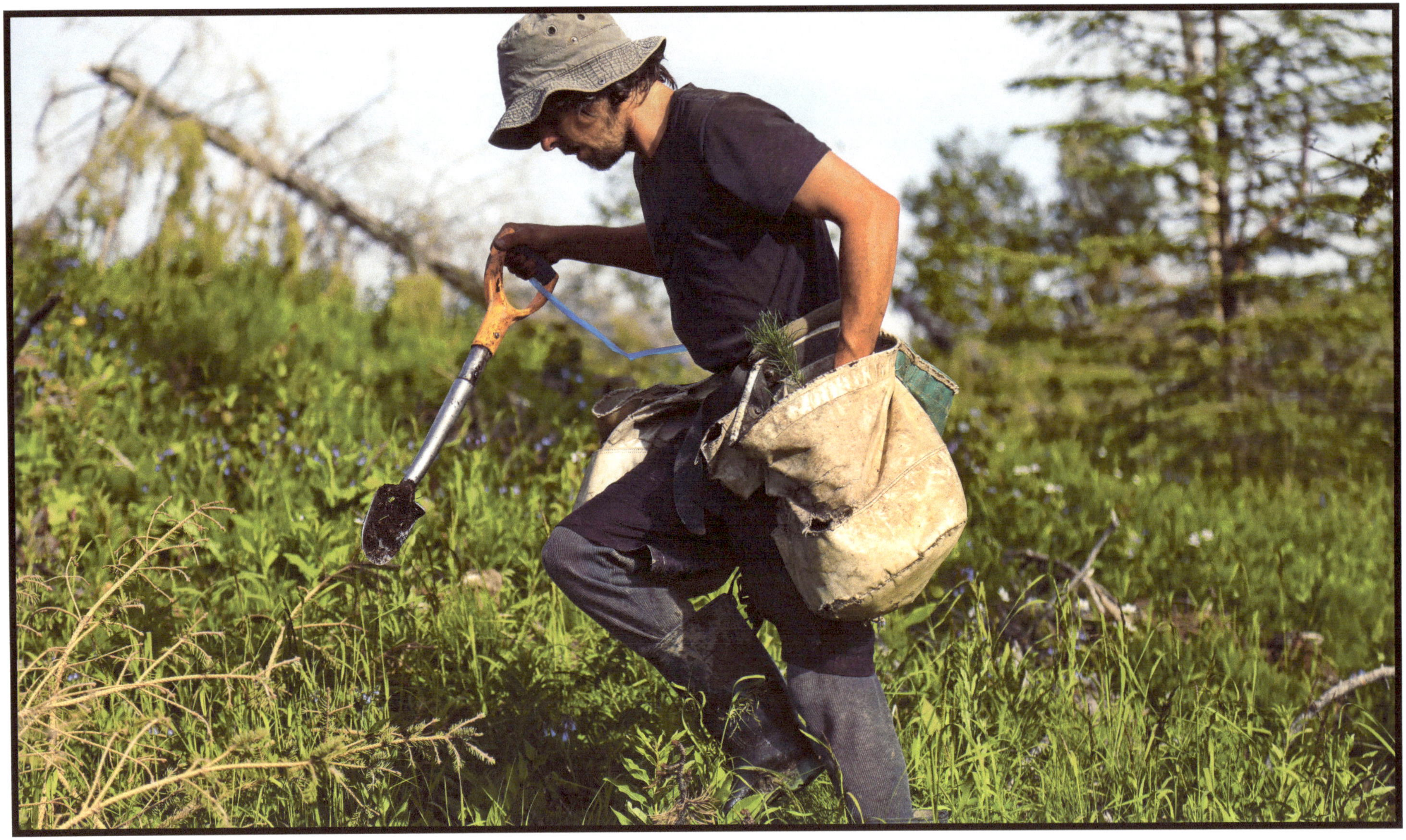

Peter majored in Philosophy and Political Science at Concordia in Montreal. Peter enjoys the meditative aspect of planting, the freedom that the job provides, and the spiritual growth that results from overcoming hardships in their various forms. Focus is key.

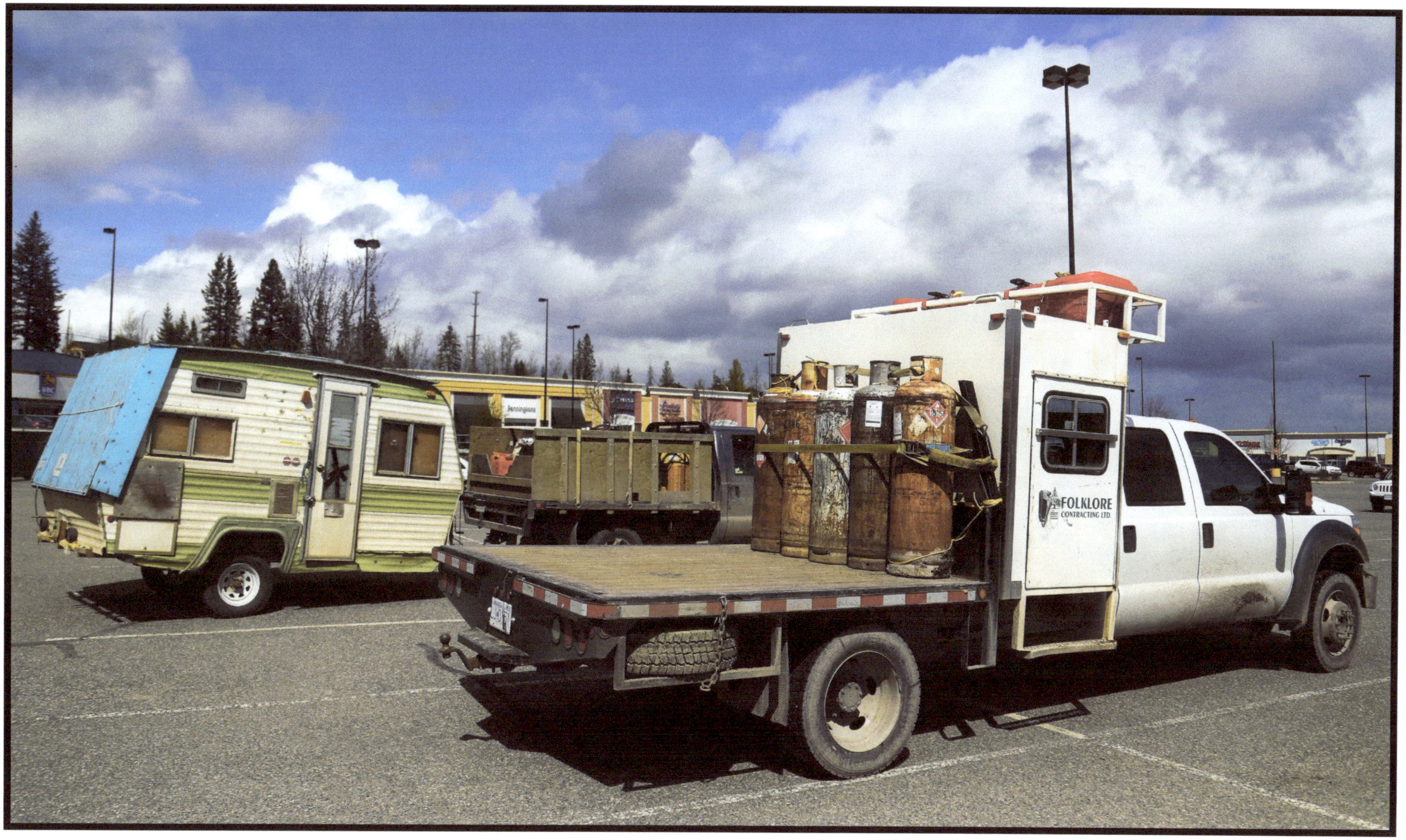

If you're driving around British Columbia, and you see strange looking trucks with ugly camper trailers, chances are good that they belong to a tree planting crew. The crummy in the foreground is carrying large propane tanks to power the ovens and hot water heaters.

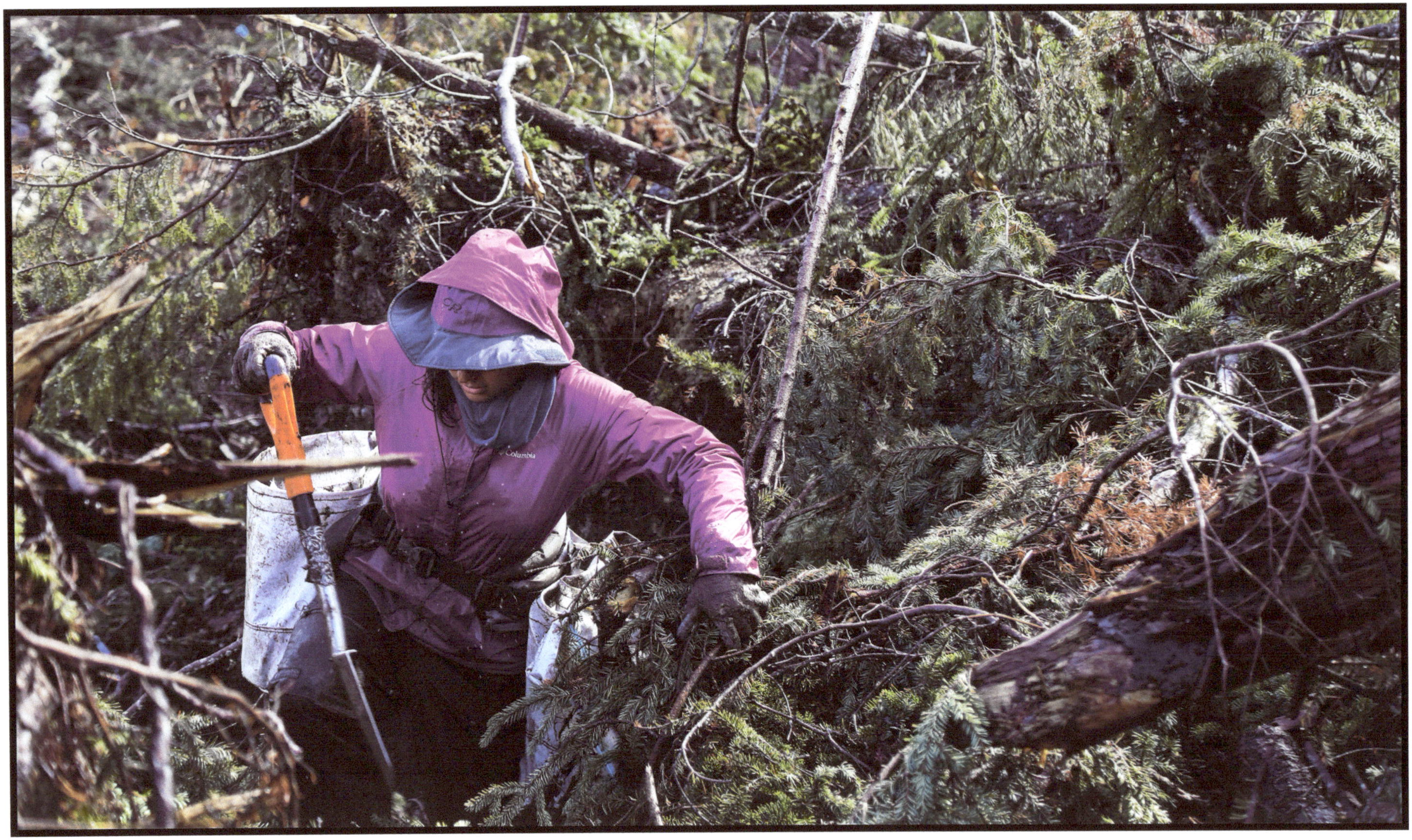

Planters often talk about personal challenges. The slash in this piece looks quite challenging.

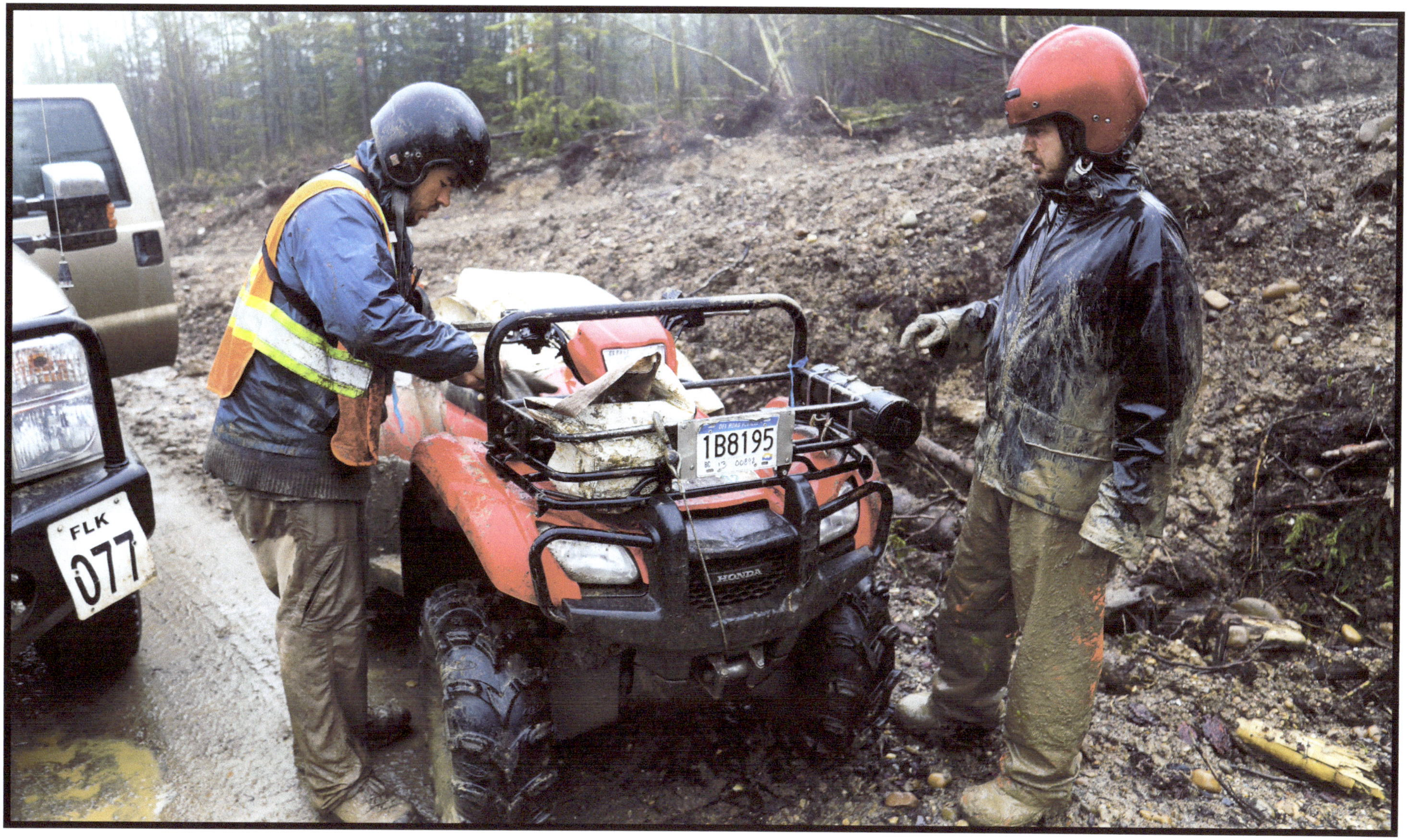

When it rains, planters have the advantage of constant movement to help keep themselves warm. Crew leaders and tree runners don't have that same luxury.

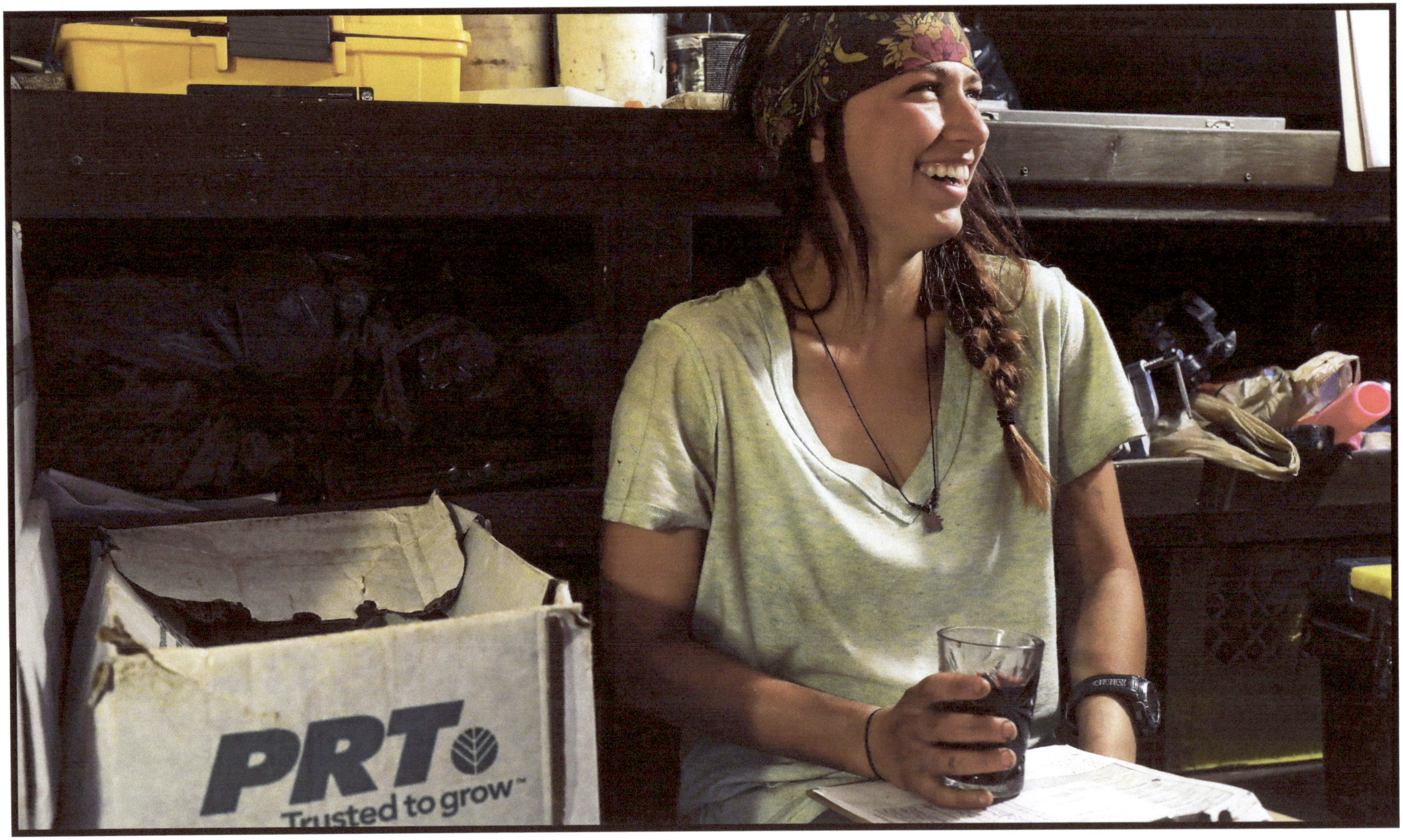

It's refreshing to see someone who laughs this much while working on seedling audits.

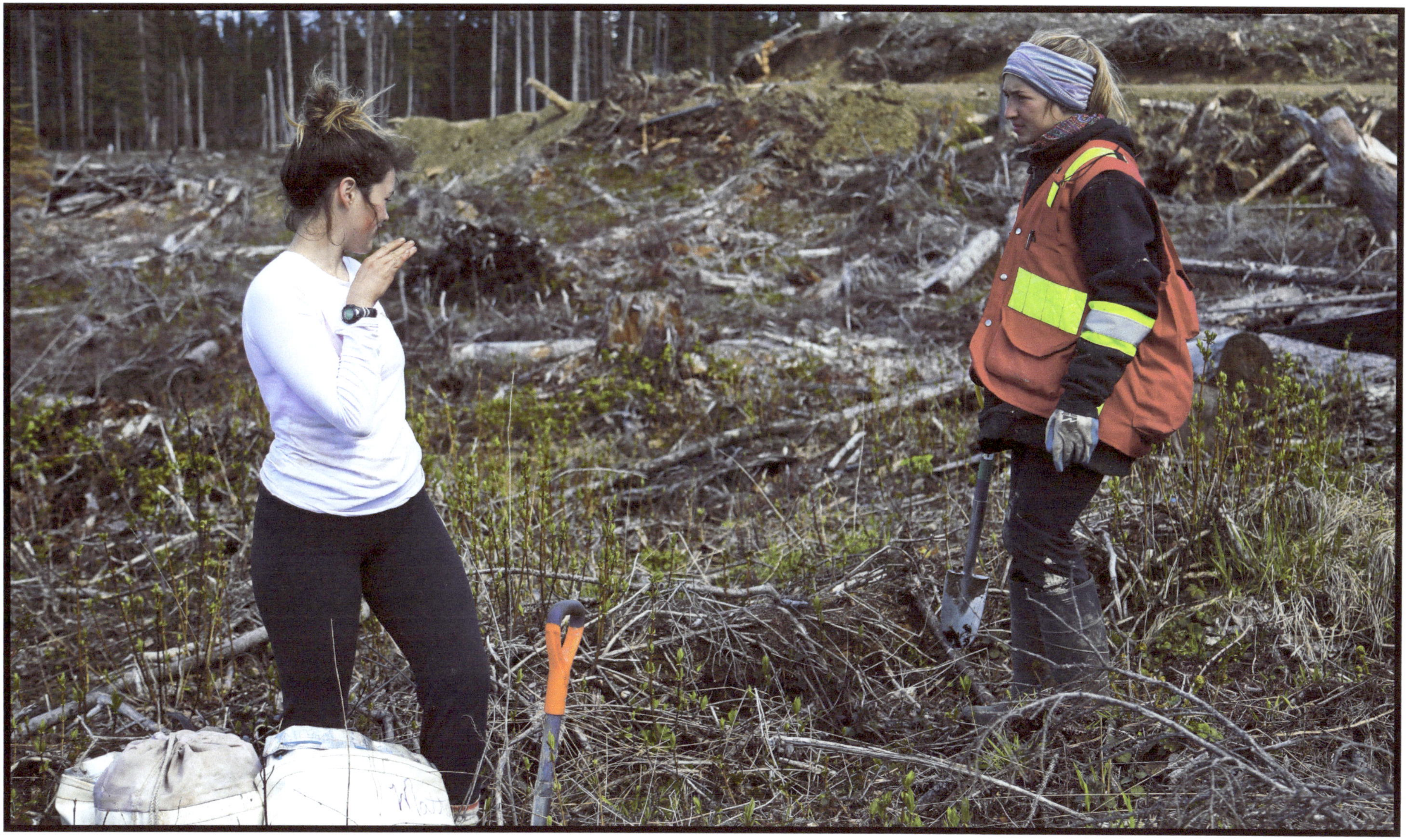

Our management team tries to spend as much time as possible working with planters to help them improve their planting skills.

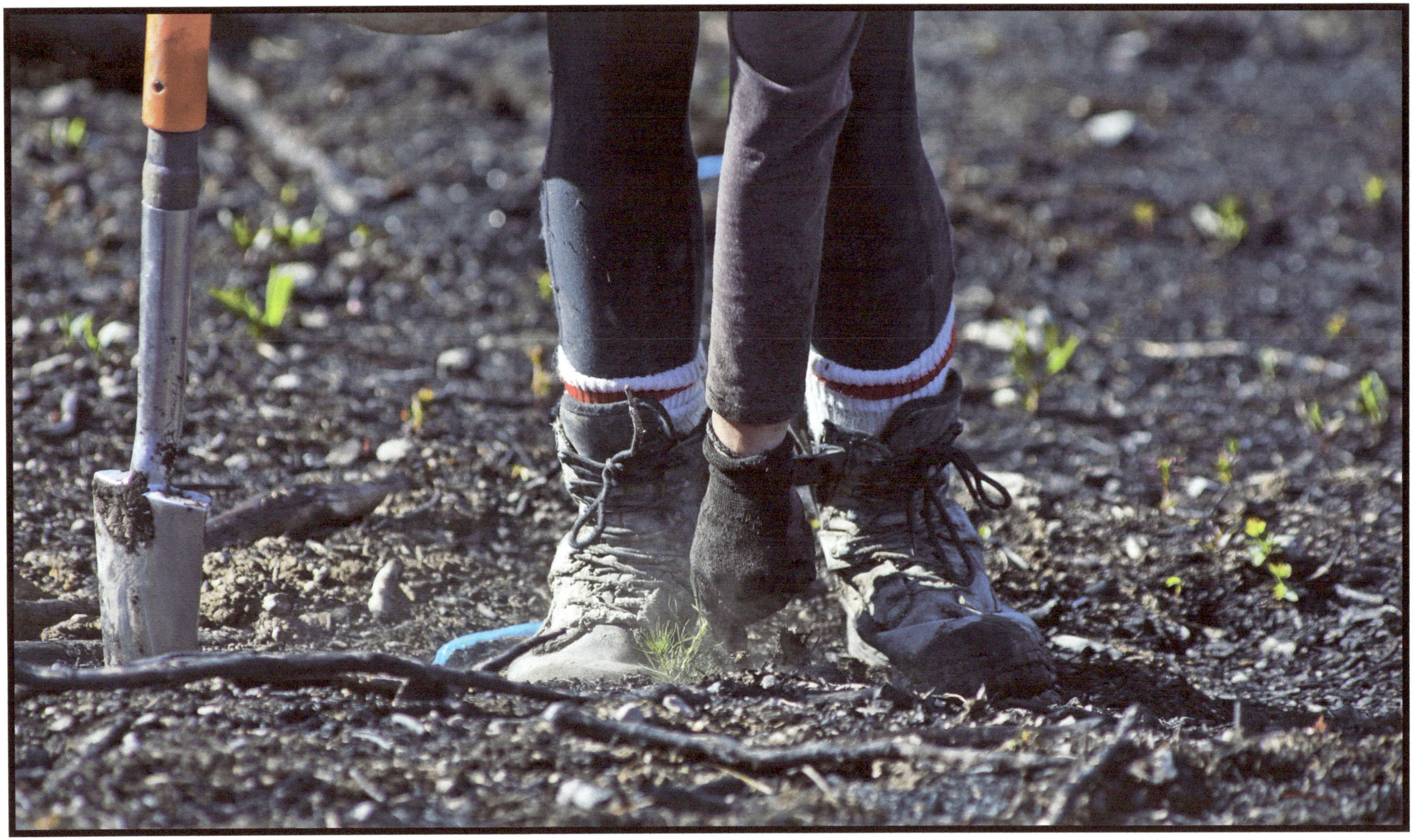

Welcome to your new home, young seedling.

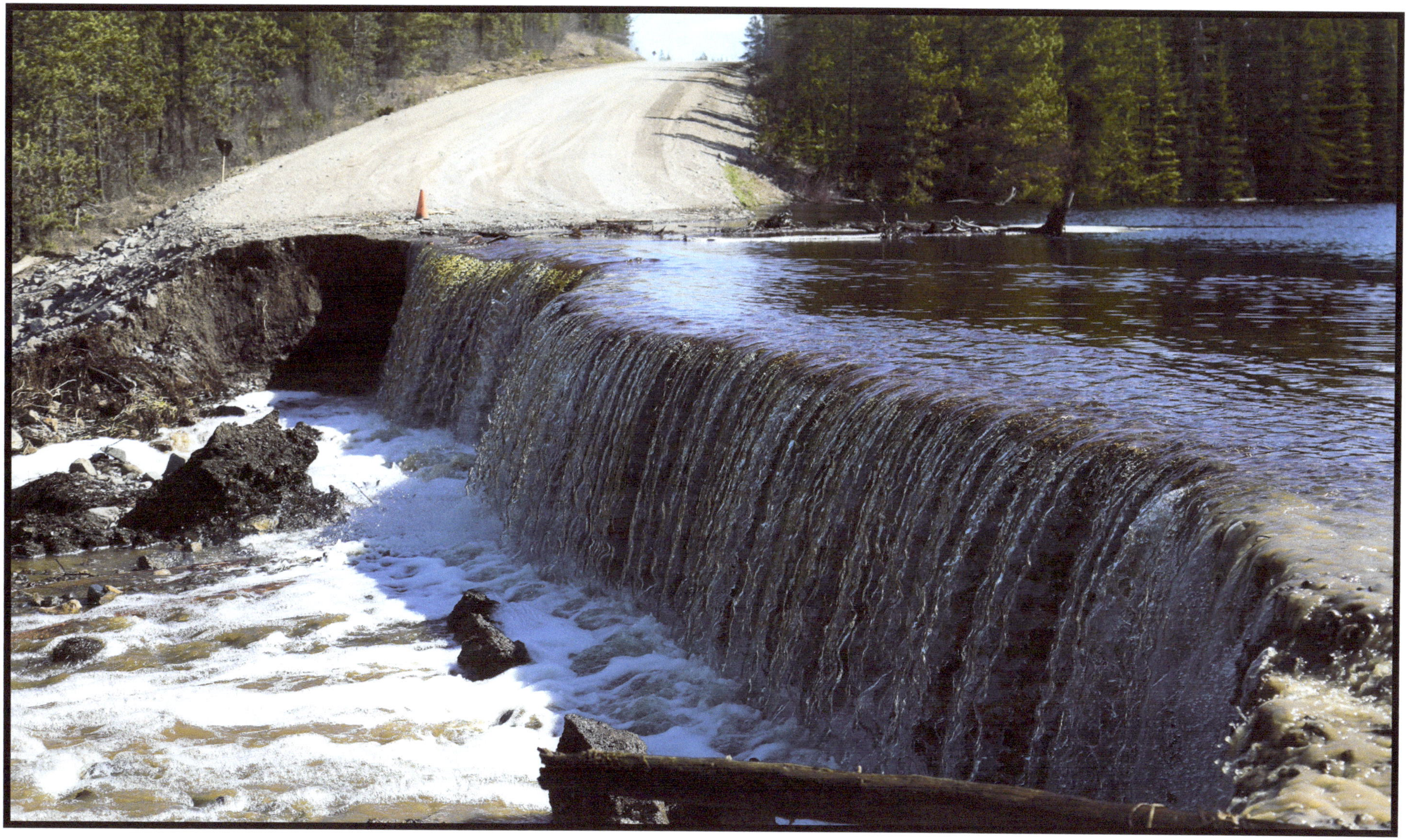

I've already mentioned that planters should learn to expect the unexpected. Having seen as many washed-out roads as we have, washouts are no longer unexpected. Canada's national animal, the beaver, causes a lot of headaches when the spring melt is happening.

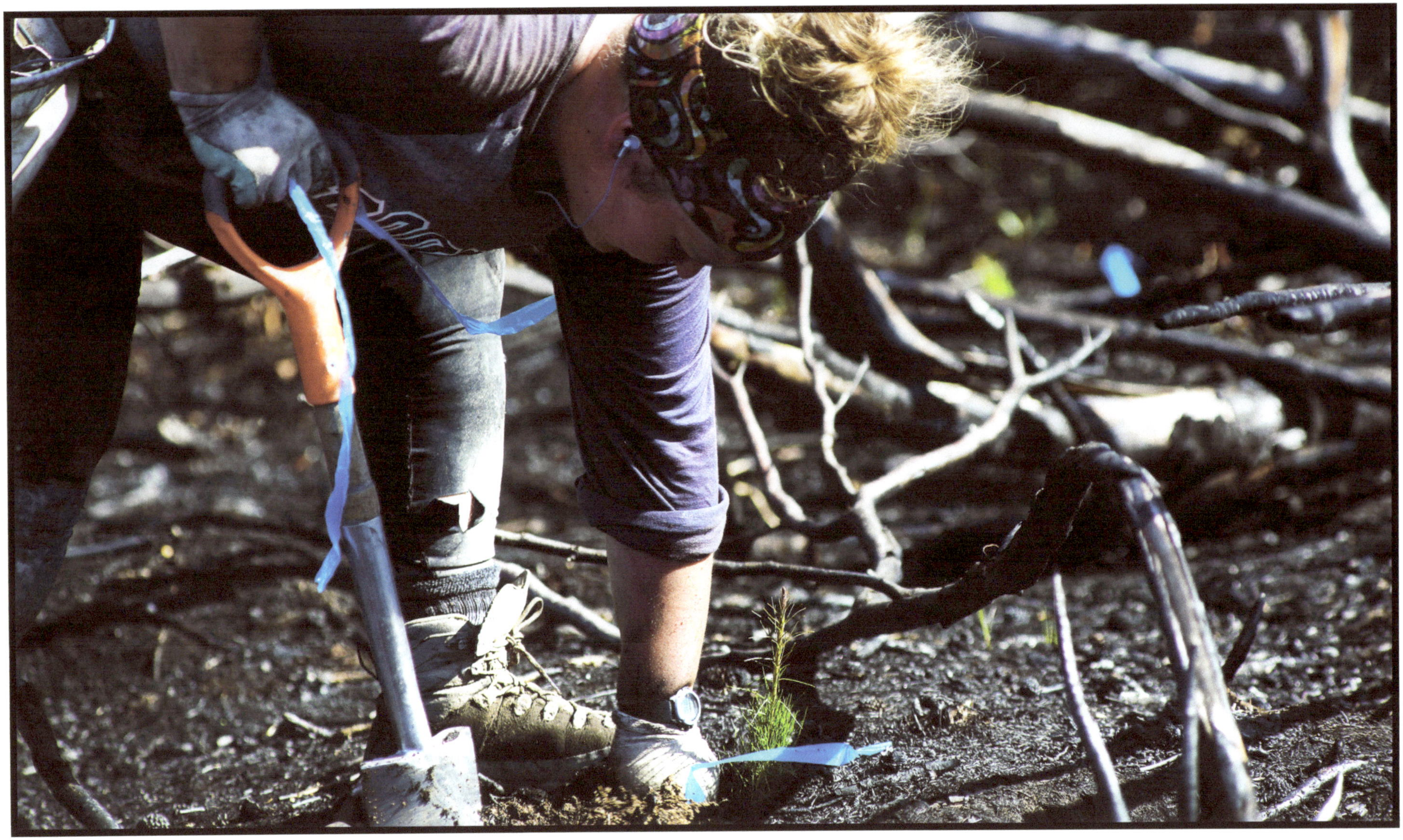

Jessica started her career in Ontario, then spent several years planting in British Columbia. She eventually intends to become a Registered Nurse, but in the meantime, enjoys working on painting and wood-working during the off season.

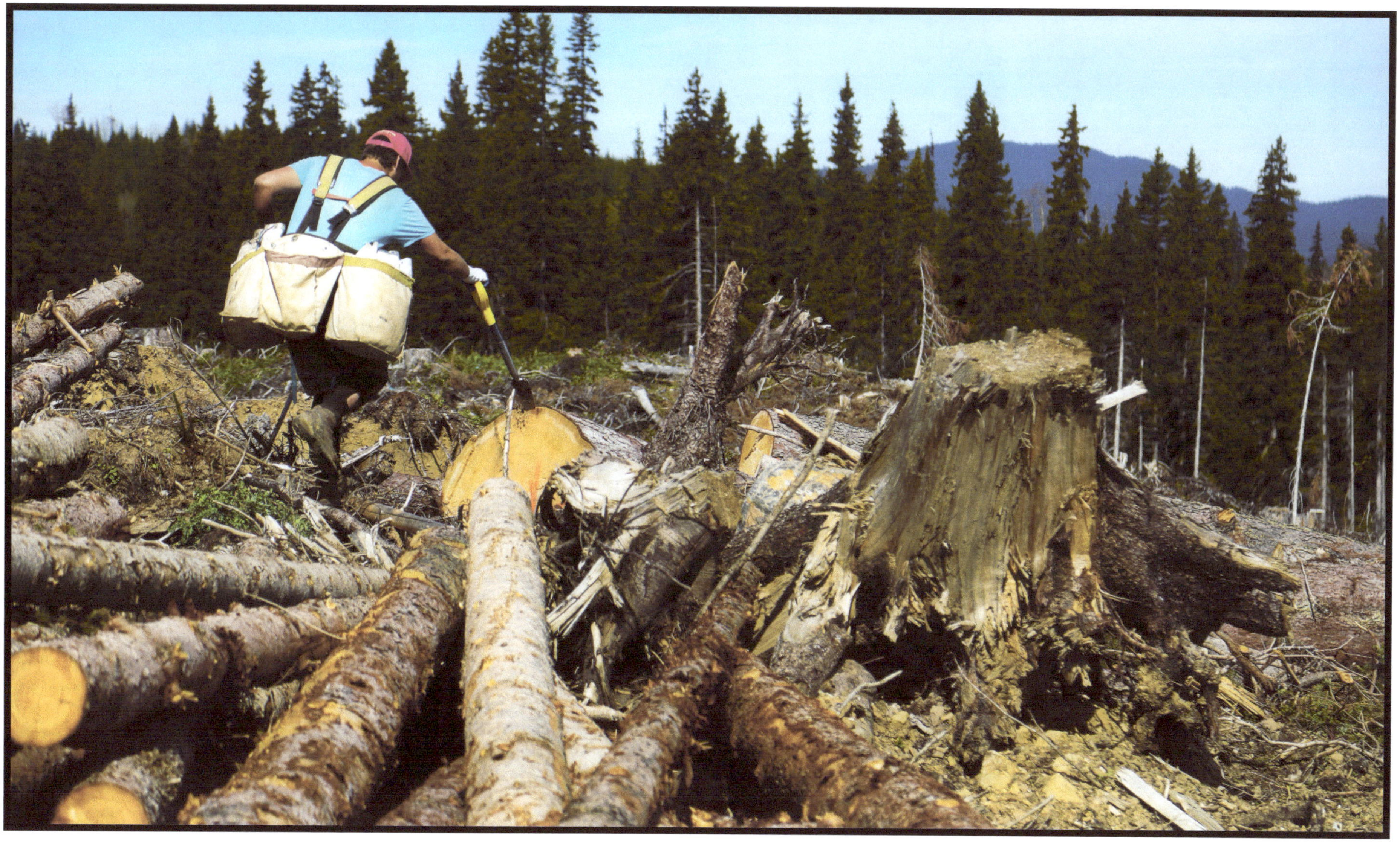

Logan doesn't seem to be slowed down by the awkward pile of logs in the front of his piece.

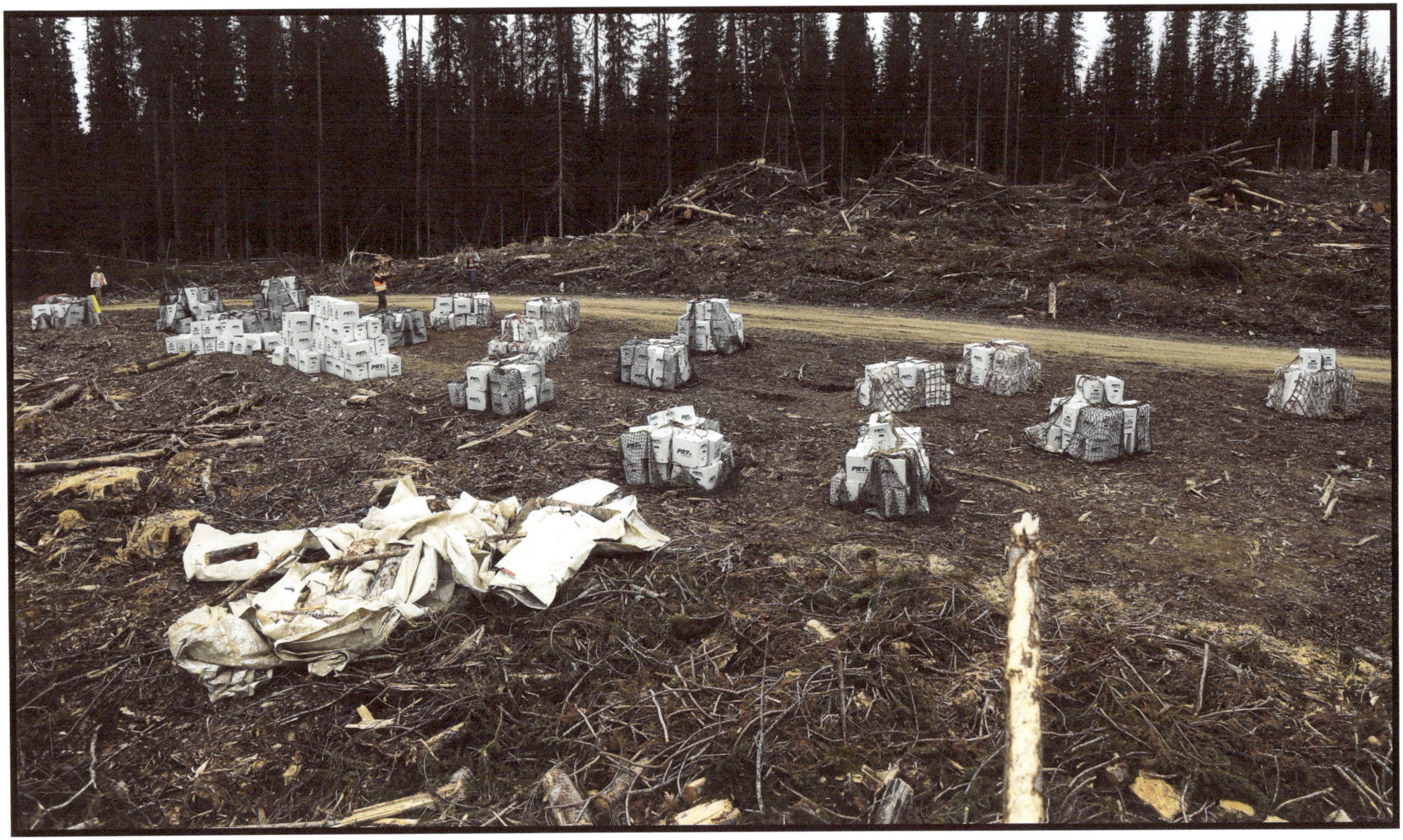

Here are some slings full of tree boxes, ready to be flown into planters' caches. It shouldn't take much more than an hour for the helicopter to fly all of these trees into their designated drop zones within the block.

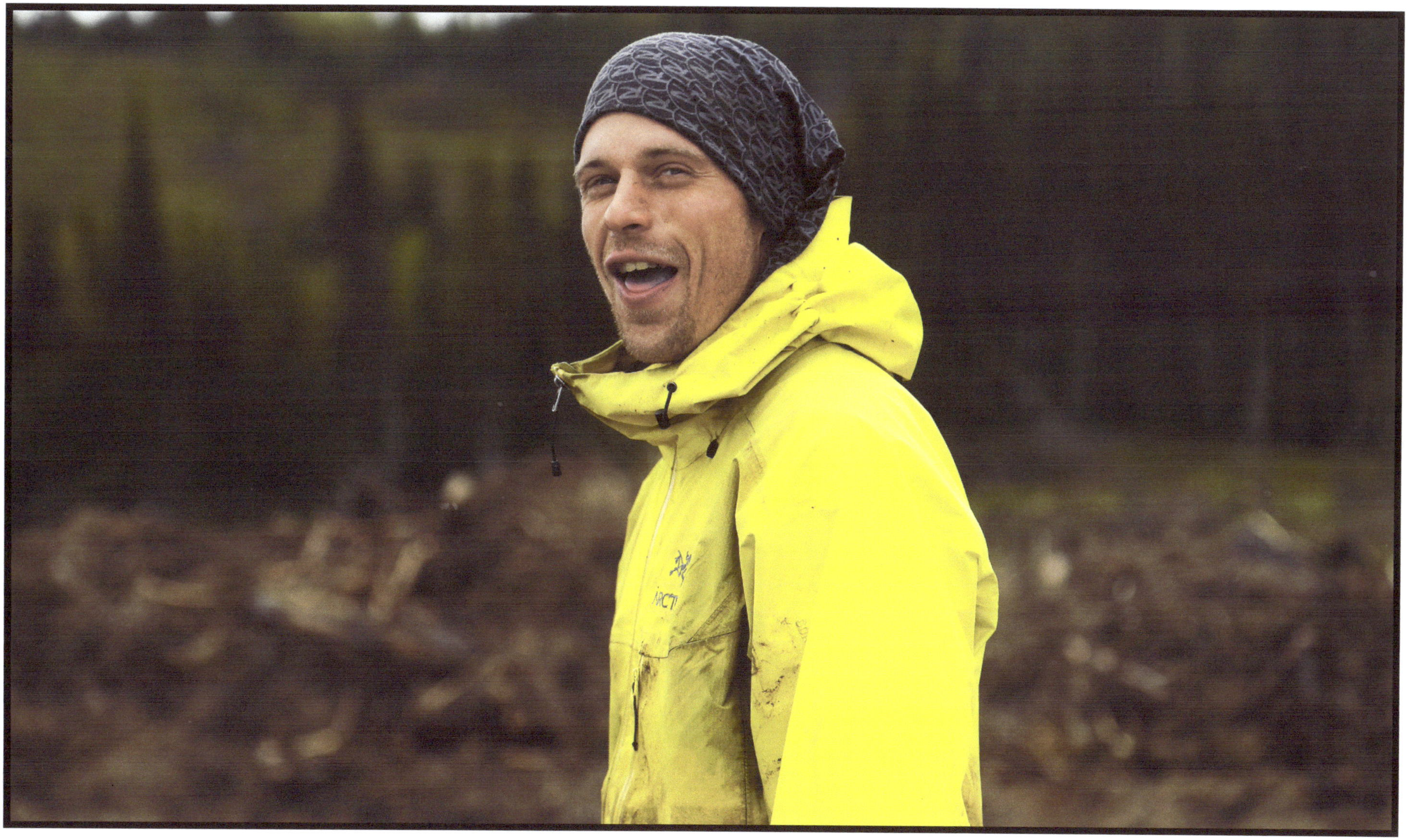

Riley is one of our most consistent long-term planters.

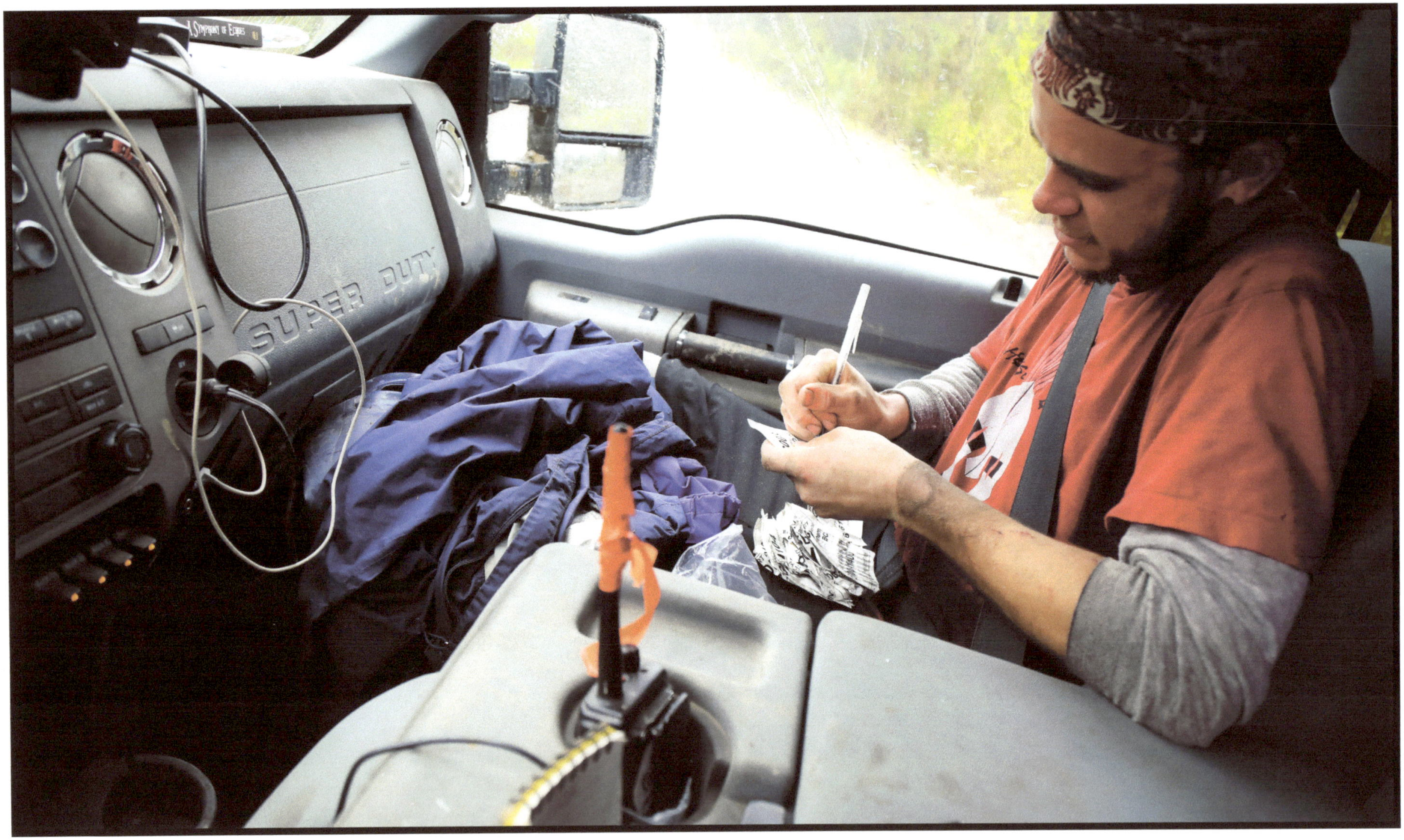

Aidan is counting through box-end stickers, double-checking his production tallies for the day.

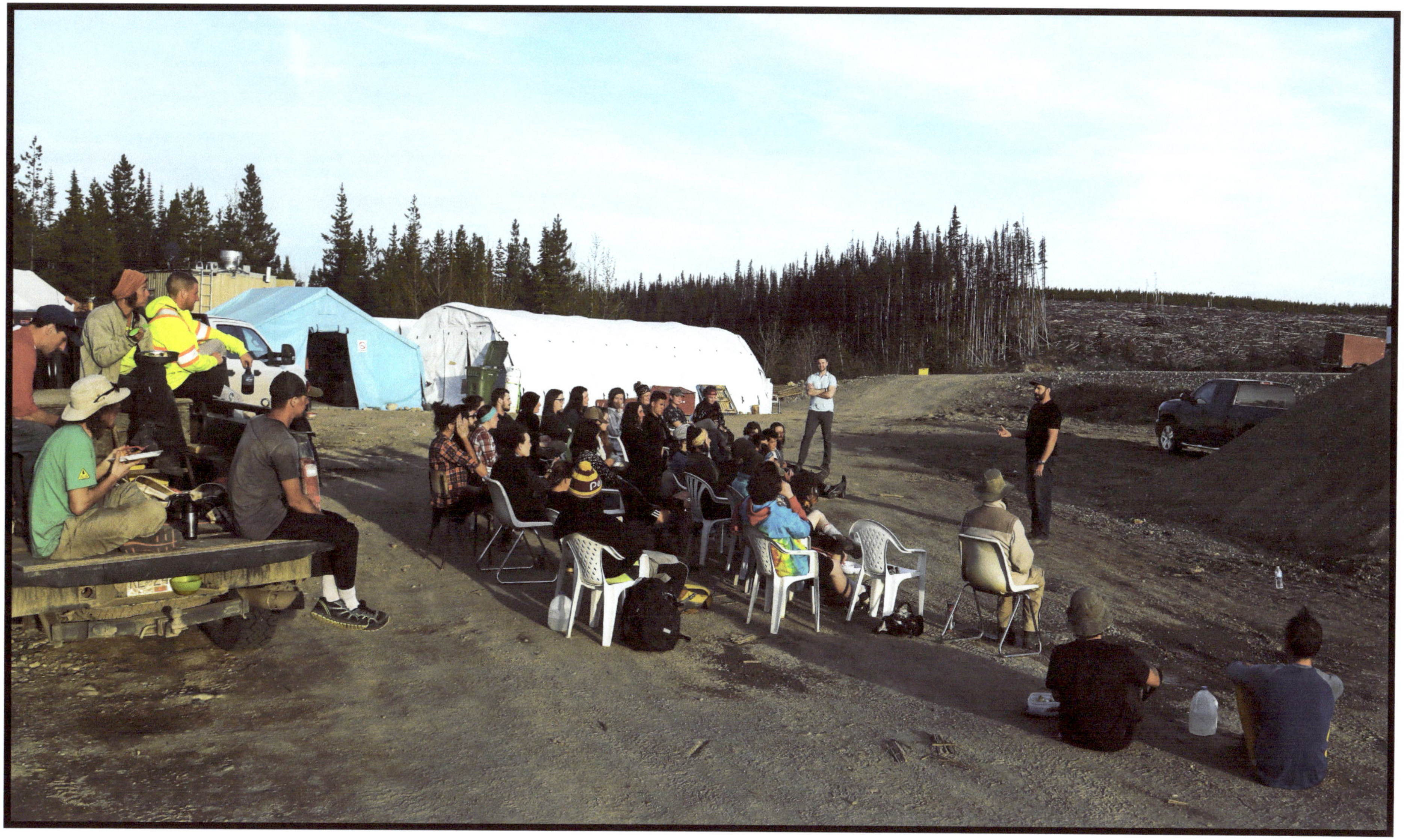

Total Physio is based out of Houston, BC. They have had an enormous positive impact thanks to a comprehensive program that helps mitigate or eliminate tendonitis and a number of other common planting-related injuries. A large number of planting companies now work with Total Physio. Their full program includes pre-season training for planters, in-house training sessions for management and first-aid personnel, plus site visits & presentations.

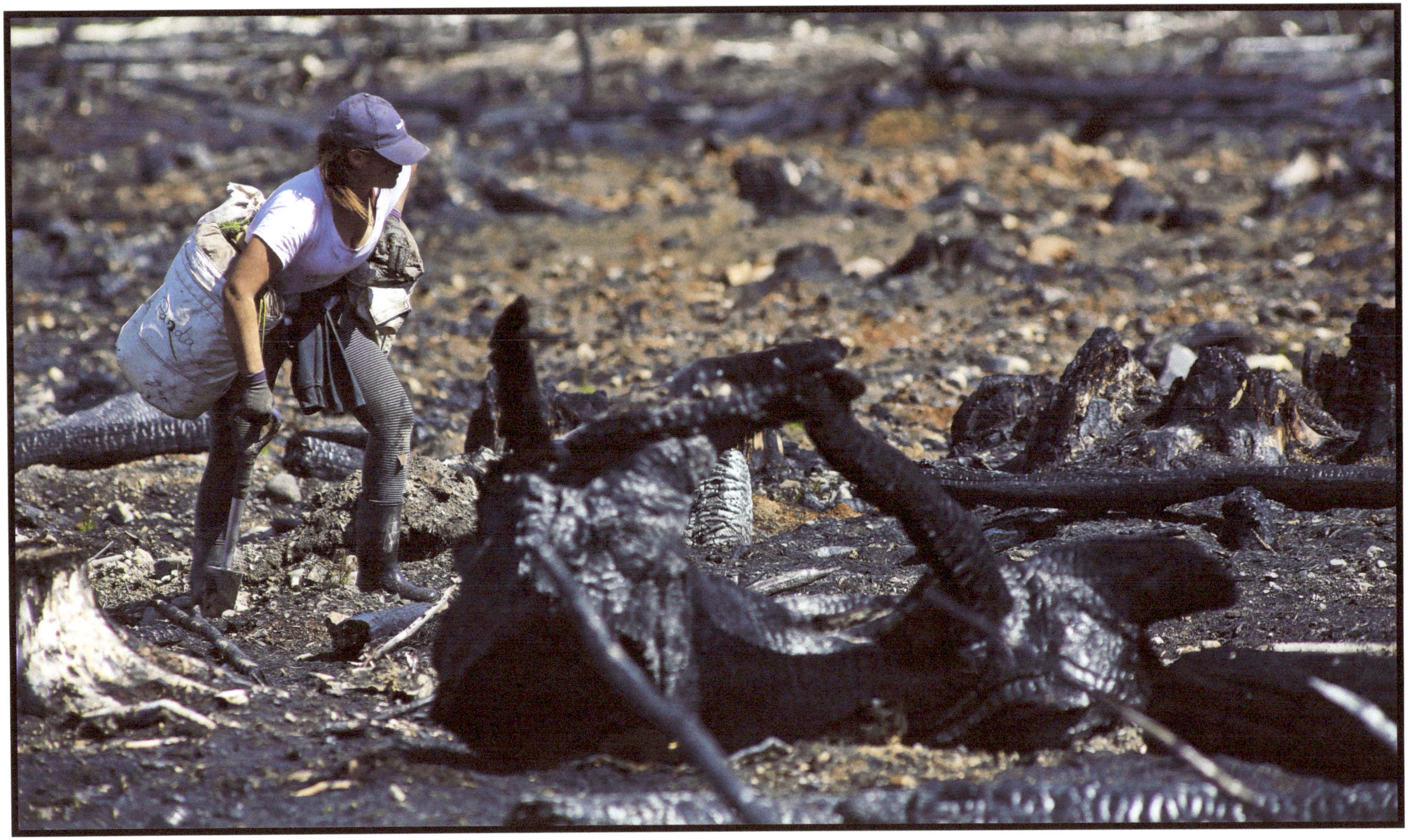

Amanda is from Victoria. The one thing that she misses the most while planting is the constant presence of the ocean. Amanda studied Anthropology at the University of Victoria, and her eventual career goal is either to become a First Responder (paramedic or fire-fighting), or to work within the prison system as a social worker.

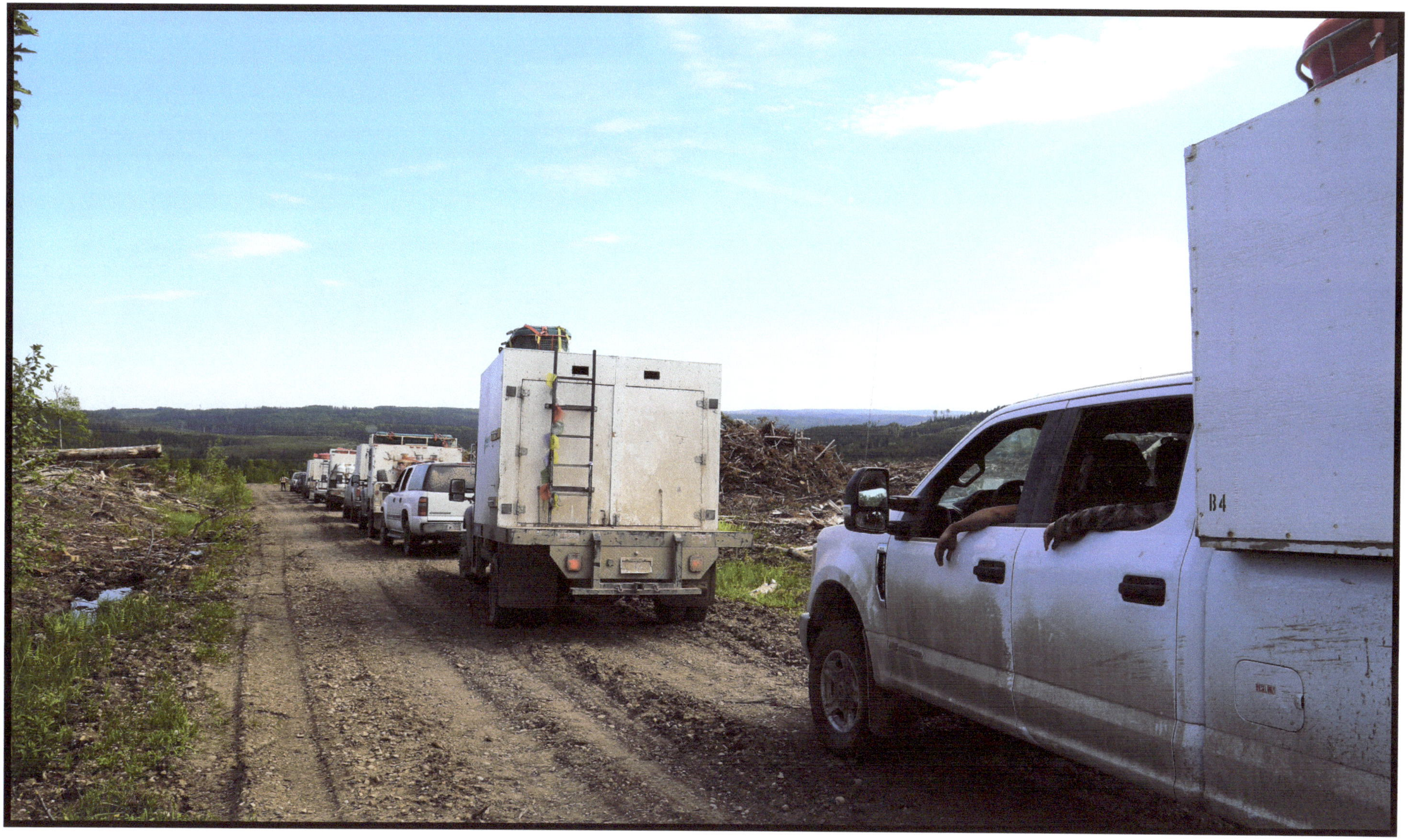

These trucks are all lined up along a block road during a pre-work conference for a new contract.

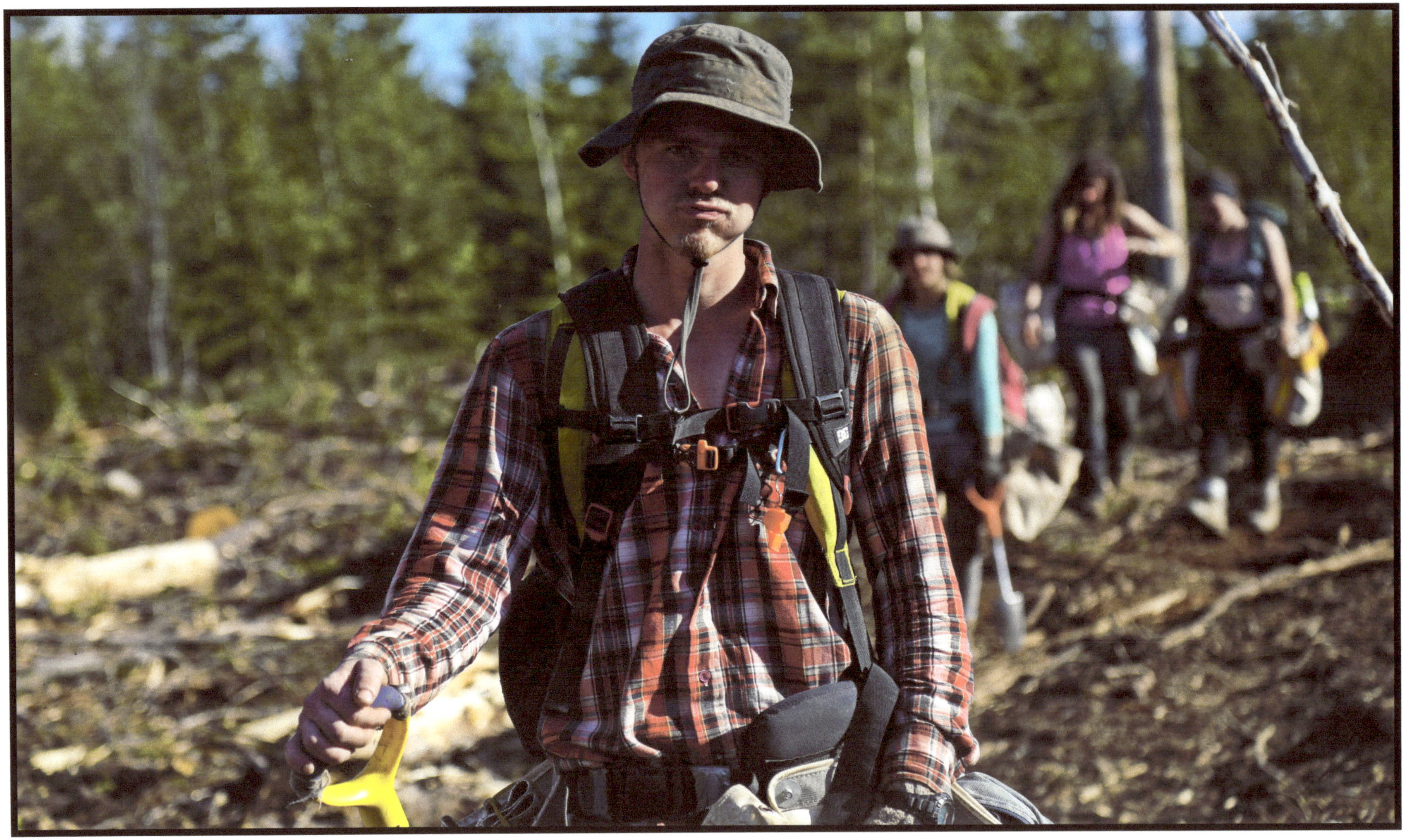

It's common for planters to have a lot less energy at the end of the day than they had at breakfast. But on a positive note, dinner is just around the corner.

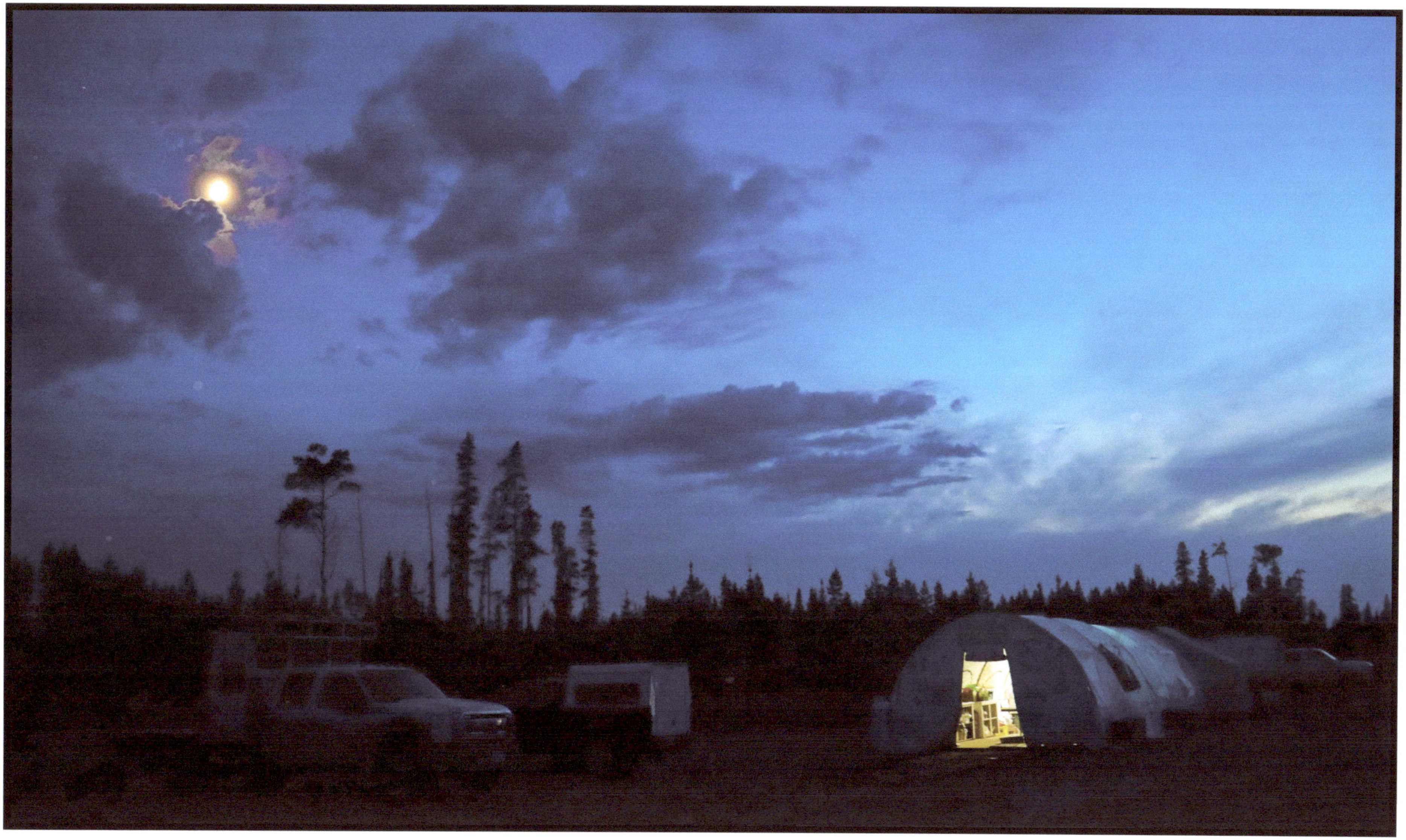

At this time of night, most of the planters are probably sound asleep in their tents. Some members of the management team are probably the only ones who are still up and at work in the mess tent.

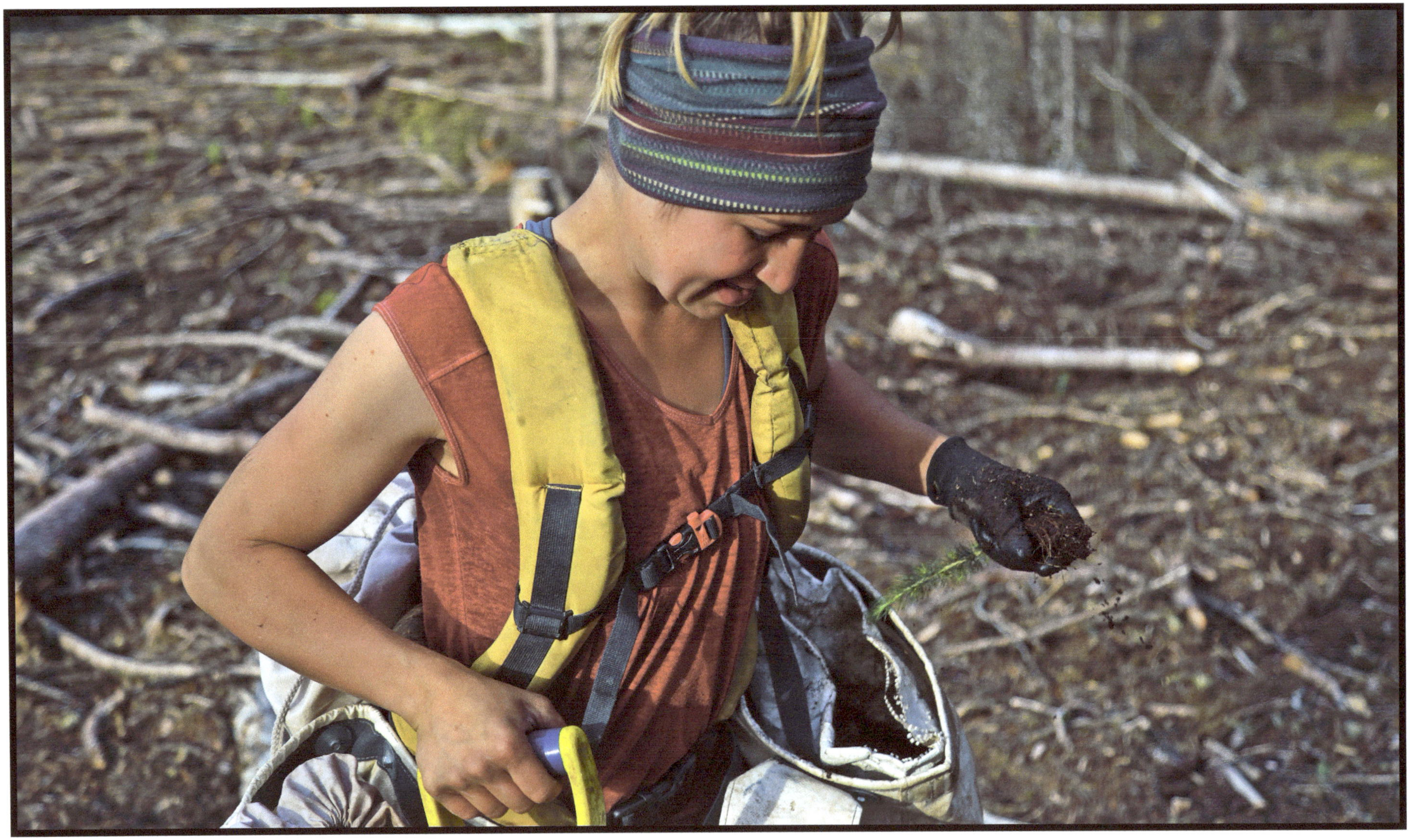

What really struck me about this particular photograph was the dirt flying off the bottom of the seedling. This part of the seedling is called the plug. The plug is typically quite solid due to the seedling's root network. However, the roots of the hot-lifted trees that we plant in mid-June often haven't had enough time to develop fully, so the plugs can be slightly softer.

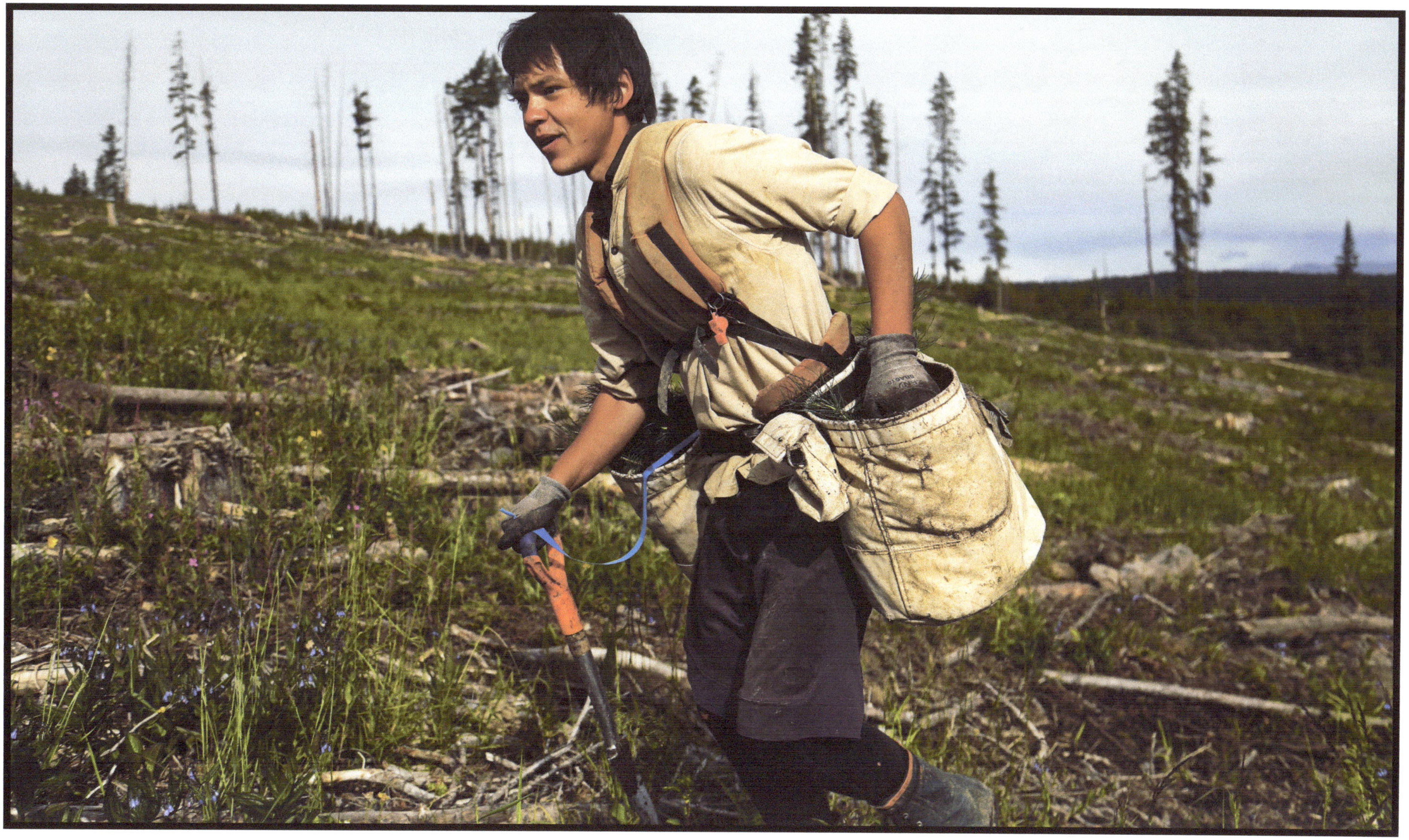

Steve is from Prince Rupert, although he studied at UNBC in Prince George. Steve notes that he loves people watching. Most planting camps are full of very unique people, so that's a great fit.

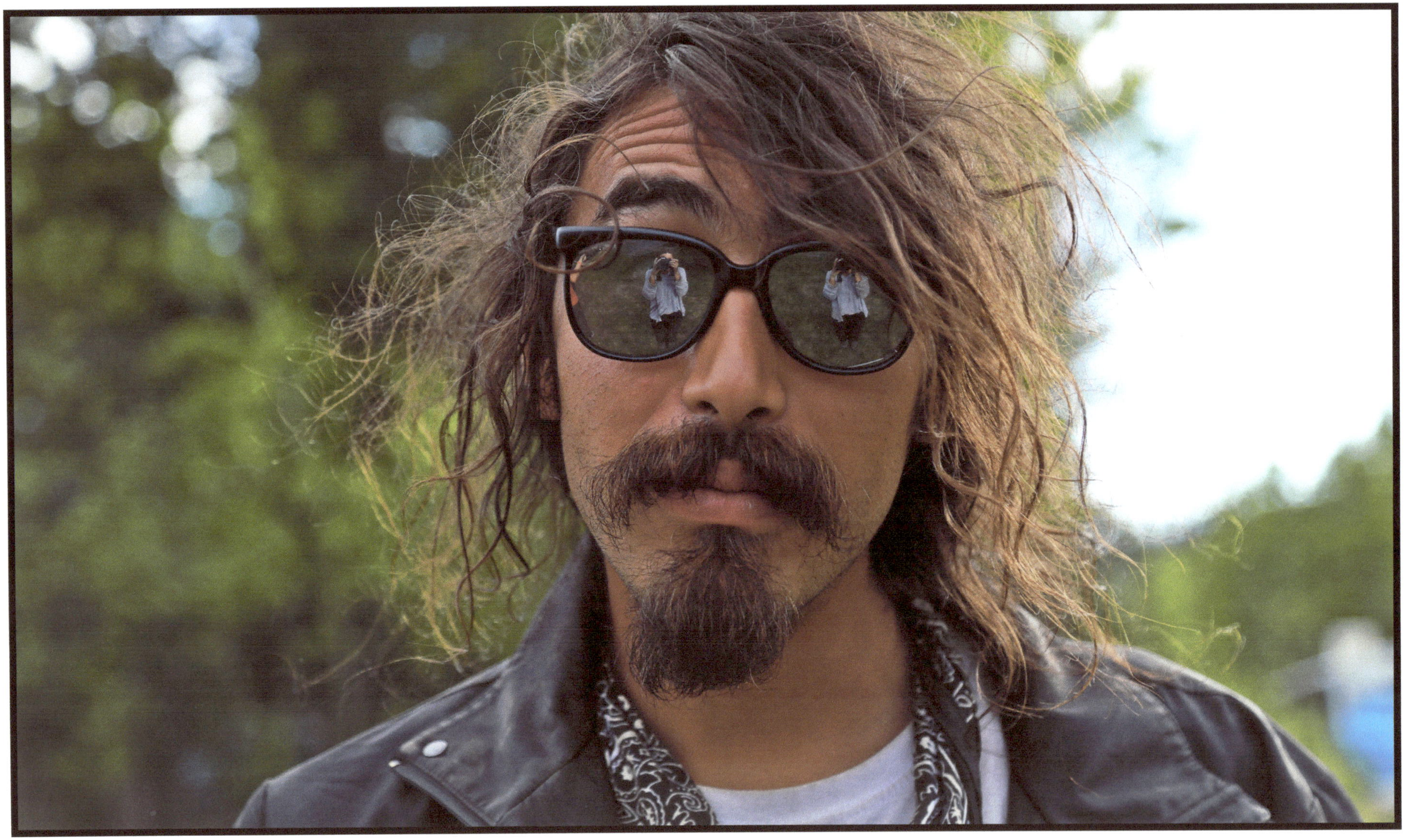

Look deep into my eyes … and you will see the photographer's reflection.

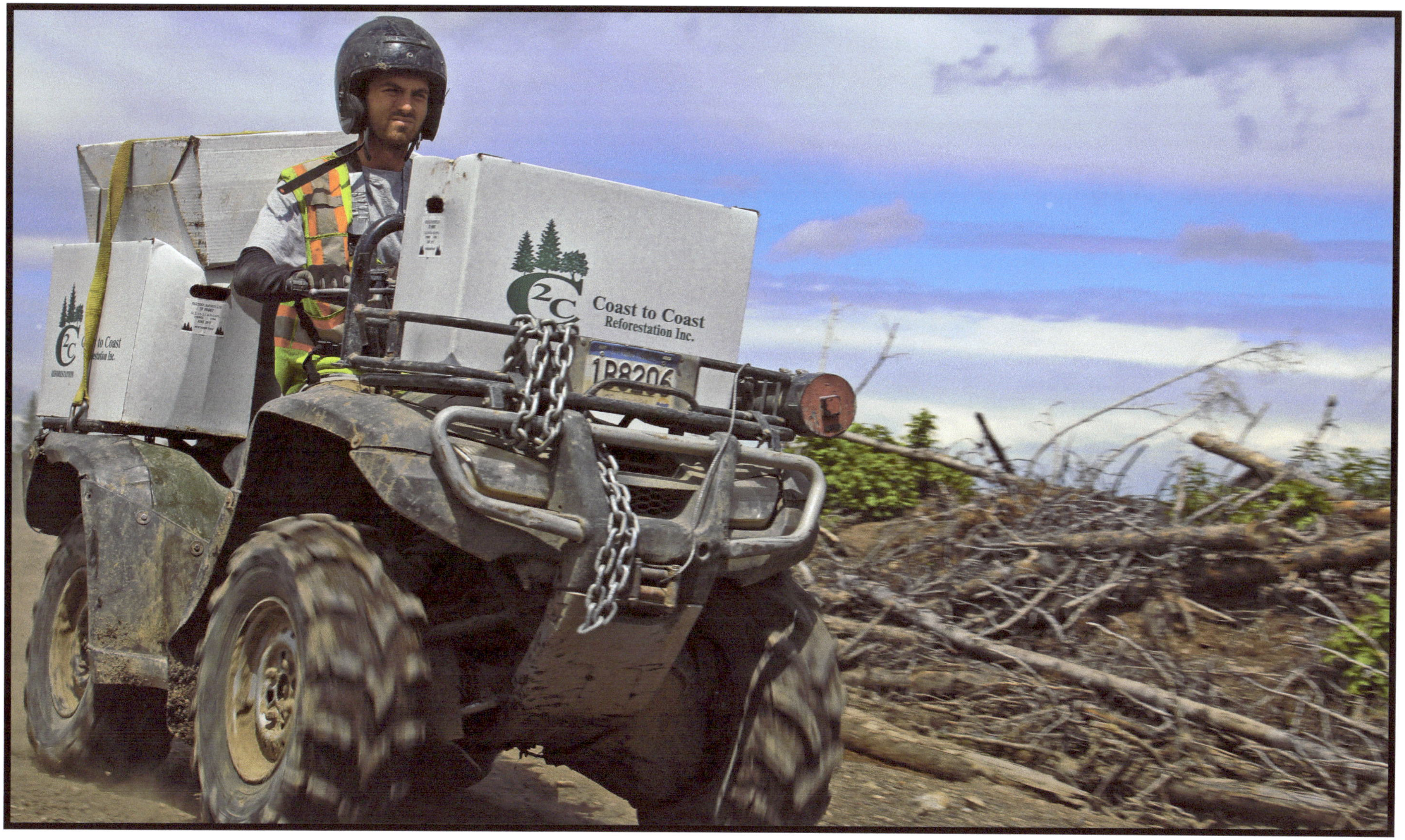

Graeme, looking very focused and down-to-business. That's always a good thing when operating a quad.

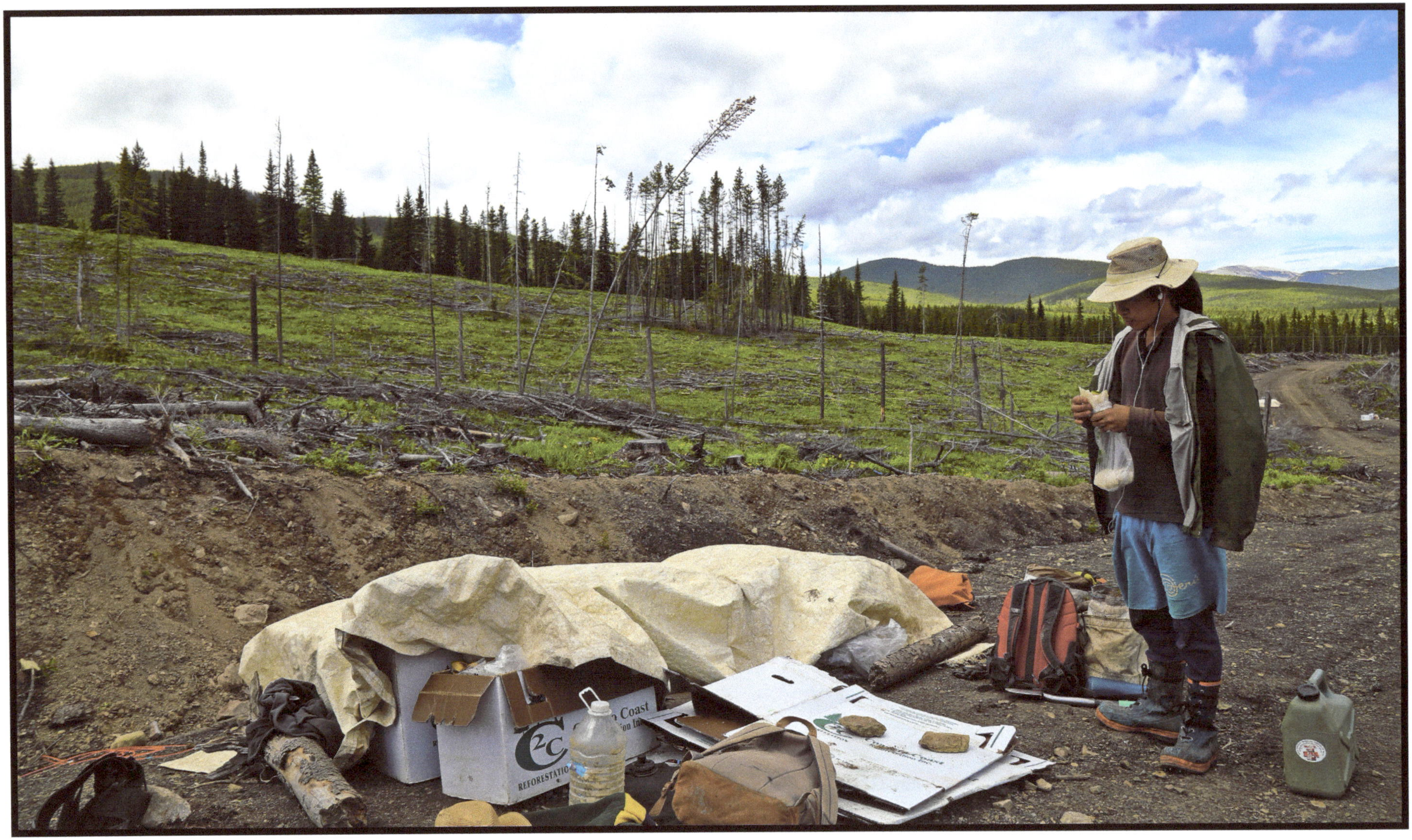

Planters don't usually spend a lot of time eating lunch during the work day. You eat when you're hungry, but many planters just take two or three minutes here and there to have frequent small snacks throughout the day.

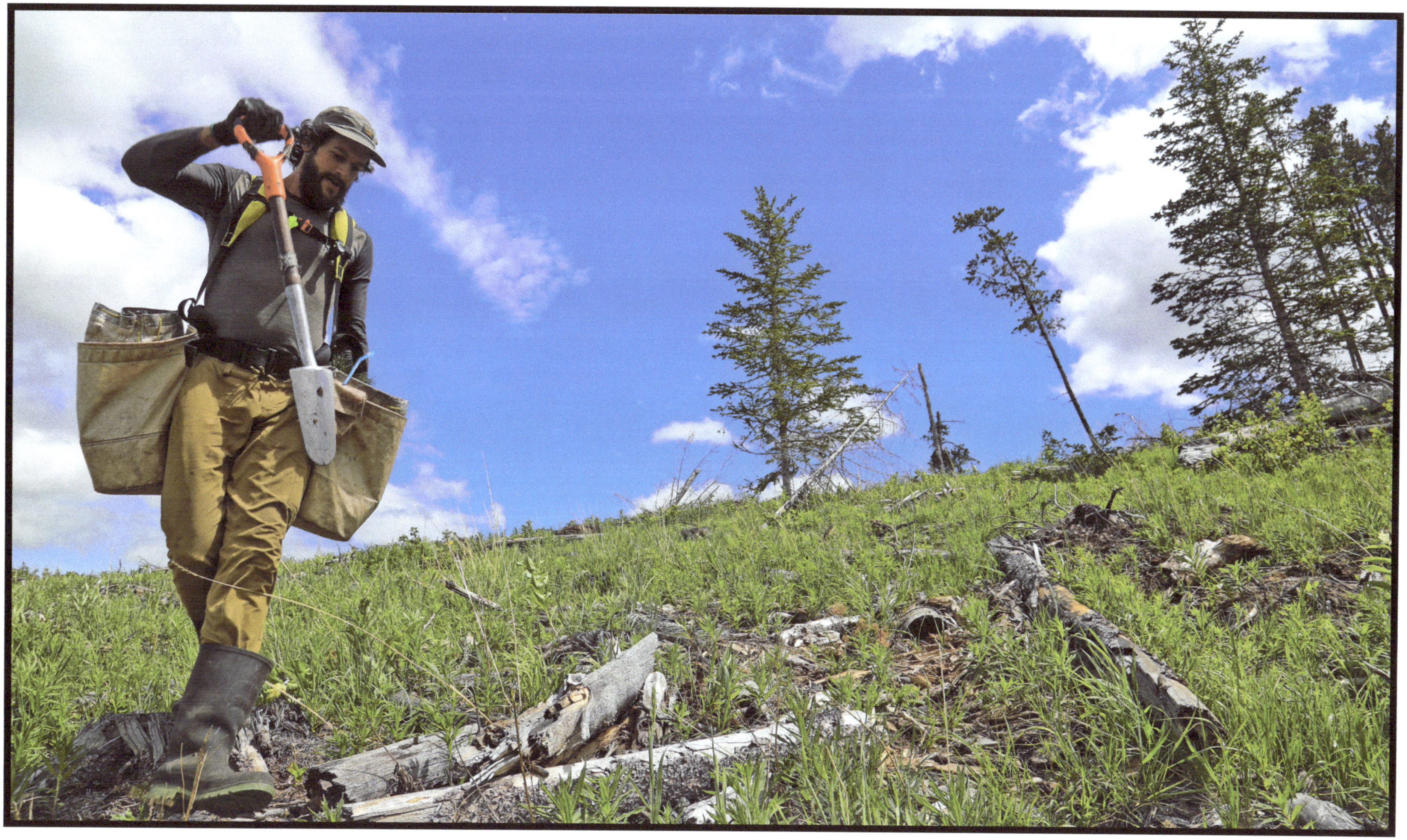

Dan has planted for a number of years, and has management experience in multiple provinces. He is also a Forest Technician. Dan thinks that camping is one of the best parts of tree planting, and something that should always be a part of the industry. Dan would like to experiment with a food forest someday, and will likely pursue a career as a botanist working in farming and agroforestry.

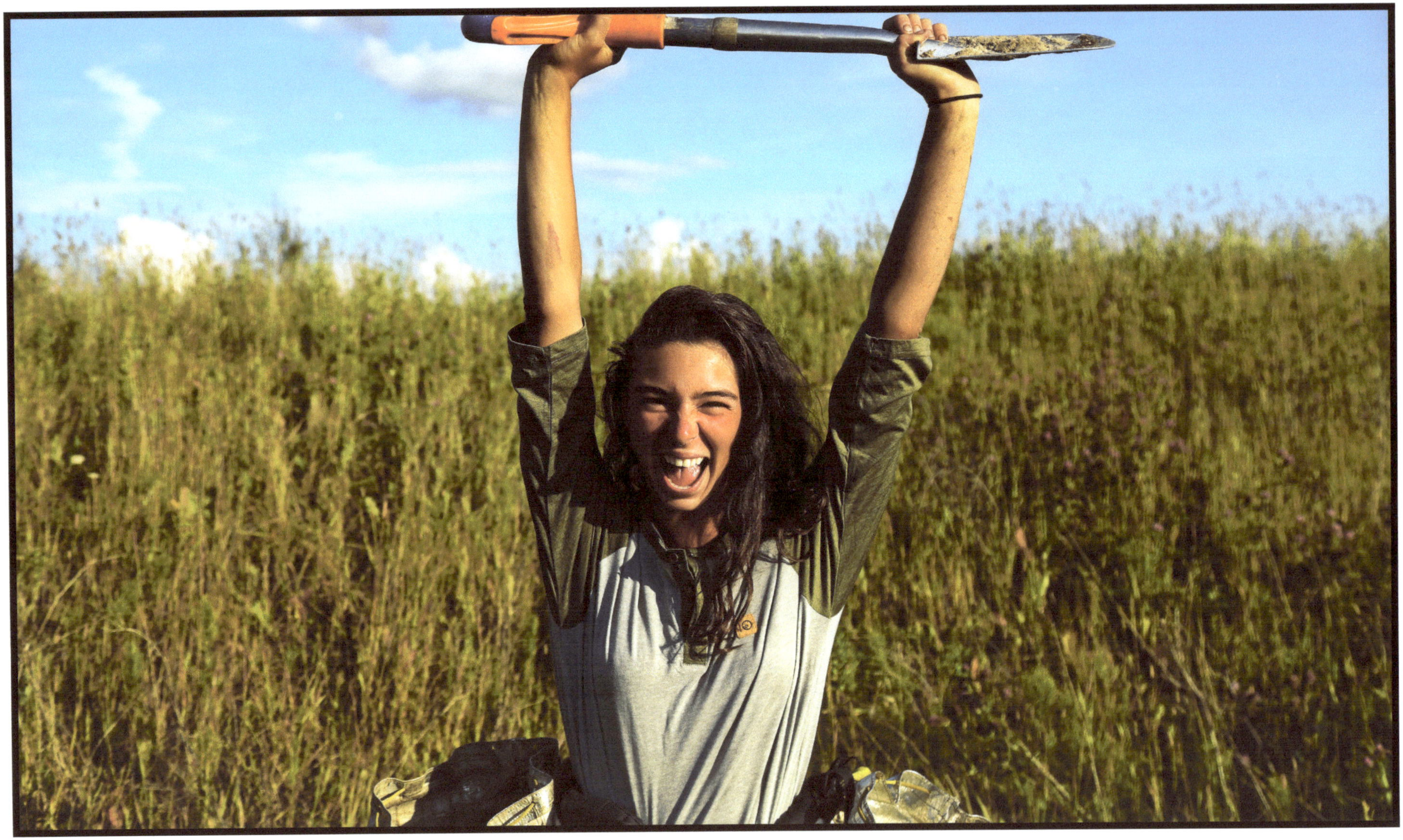

Kathleen grew up in rural Ontario, but considers Toronto to be her true home. She attended the University of Toronto, specializing in sexual diversity studies, and has also received a Social Service Worker diploma from George Brown College. When she's not planting, Kathleen works at a not-for-profit organization, working with barriered and at-risk youth, teaching them pre-employment and life skills. Her hobbies include yoga, painting, writing poetry, and thrifting.

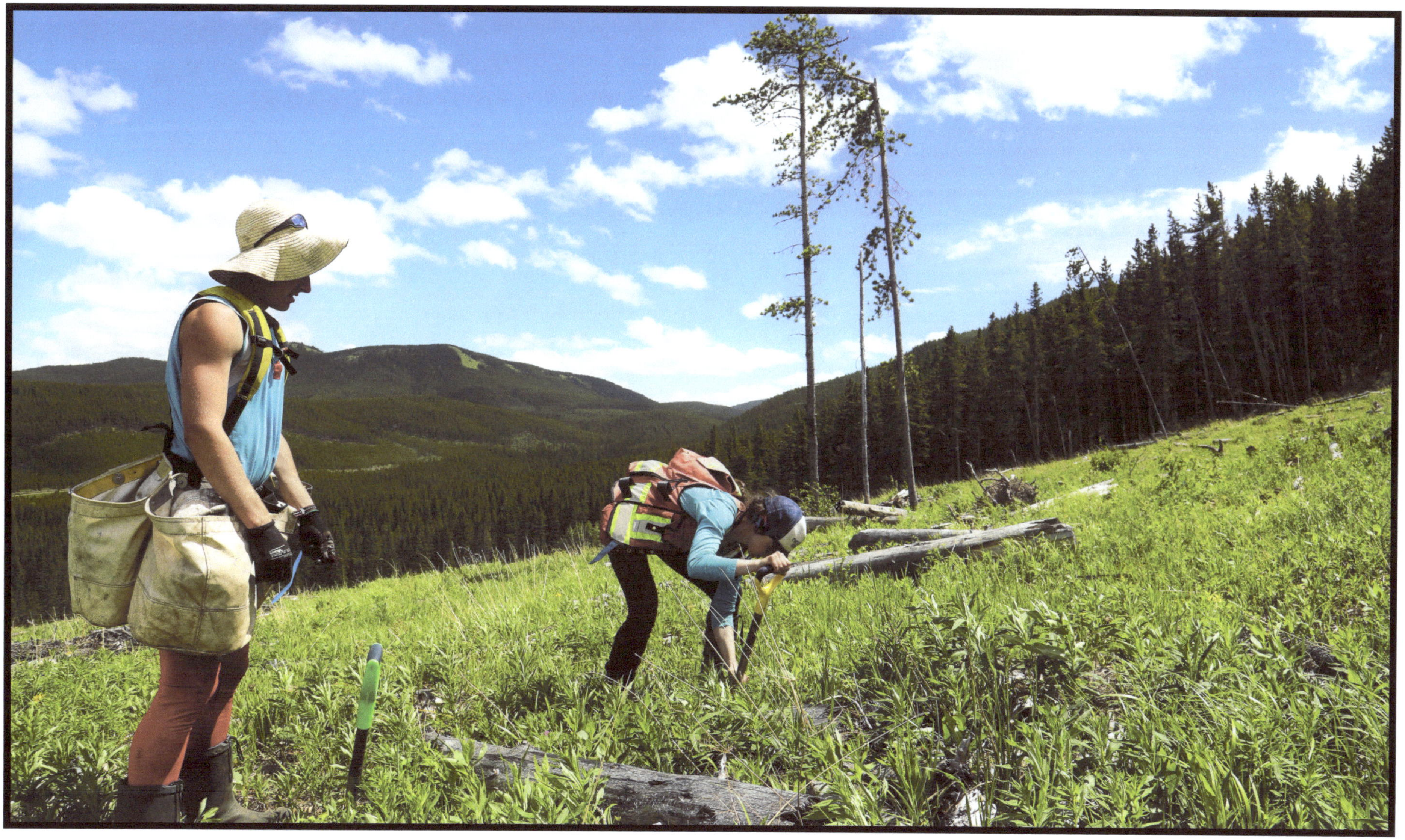

At this moment, Scott is getting some coaching from a member of our management team, and is also starting to realize the true magnitude of the job. Tree planting is a much more complex job than most people would assume! Outside of planting, Scott has supported himself by playing online poker full-time for the past couple of years, and plans to spend the next several years travelling around the world.

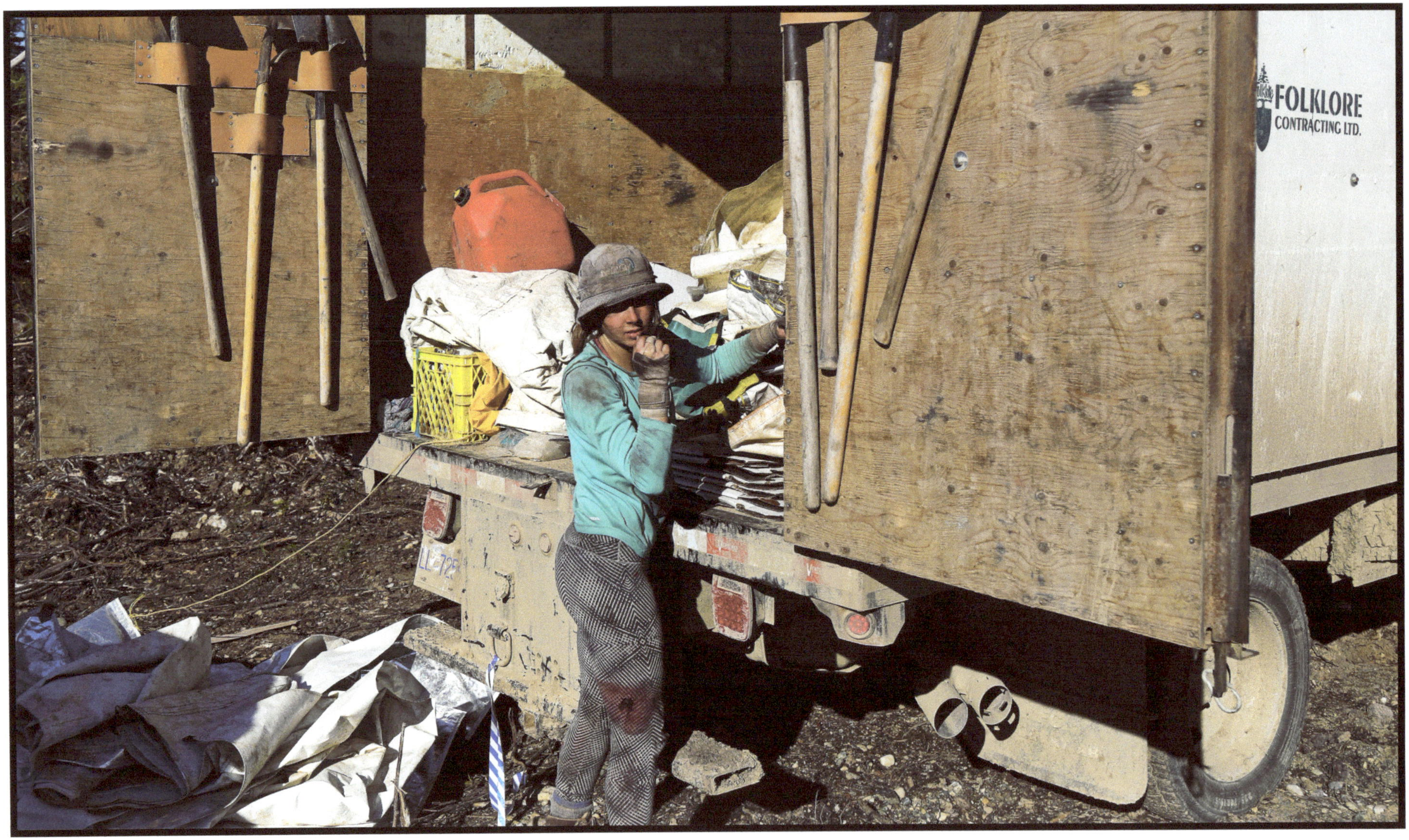

Laura seems completely unfazed by the large war-wound on her leg in this photo, and was more concerned at the time about what was for dinner. Laura studied Mechanical Engineering at Queens. She plans to someday use her degree in conjunction with a career that includes a focus on the environment, or environmental reclamation work.

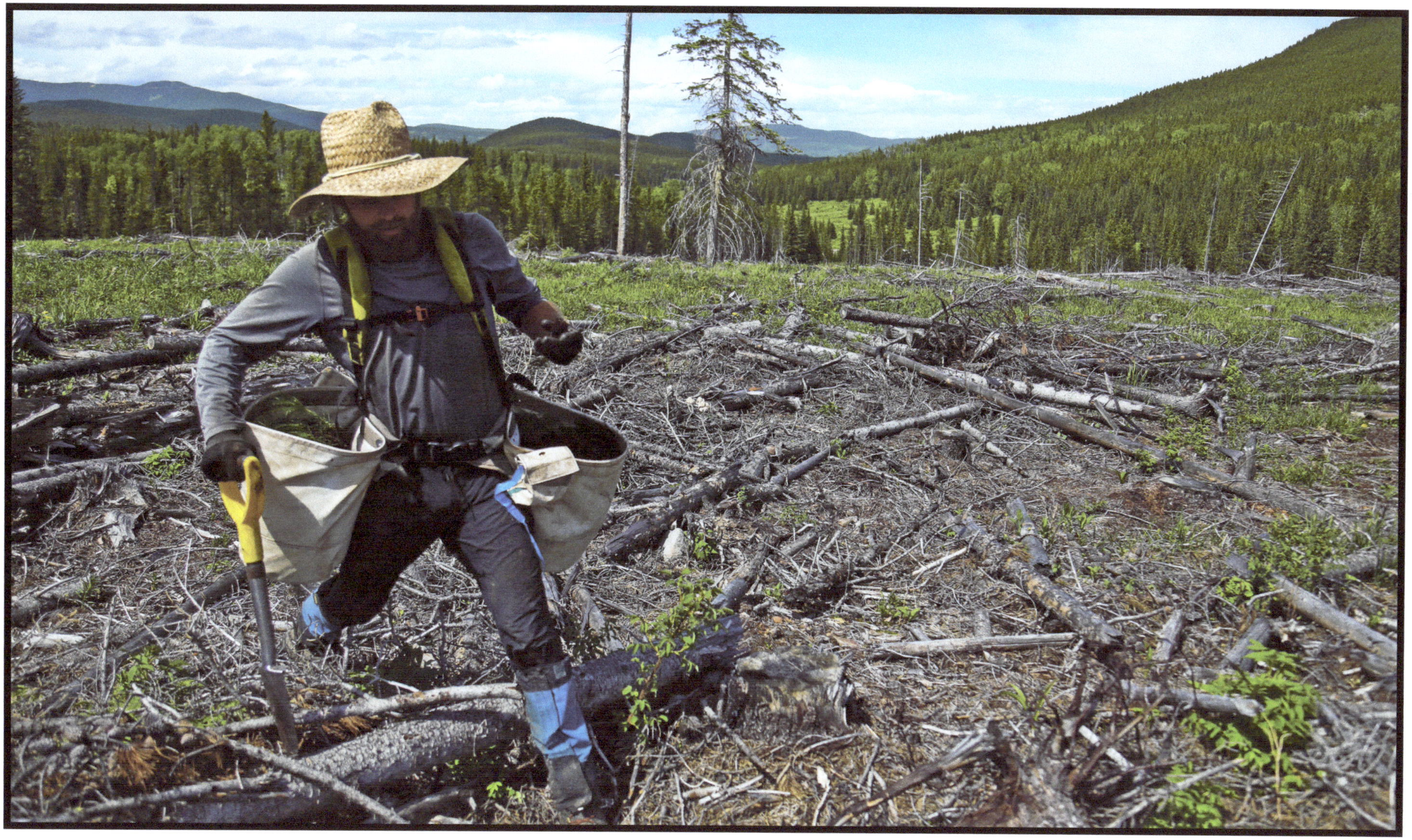

Fergo is a legend within the planting industry. Fergo started his planting career during his undergrad at the University of Guelph, where he studied business and marketing. After that, he set his eye on graduate studies at the University of Massachusetts, for an MBA/MS in Sport Management. When he's not immersed in tree planting or academic pursuits, he likes to keep active with skiing, surfing, and skateboarding.

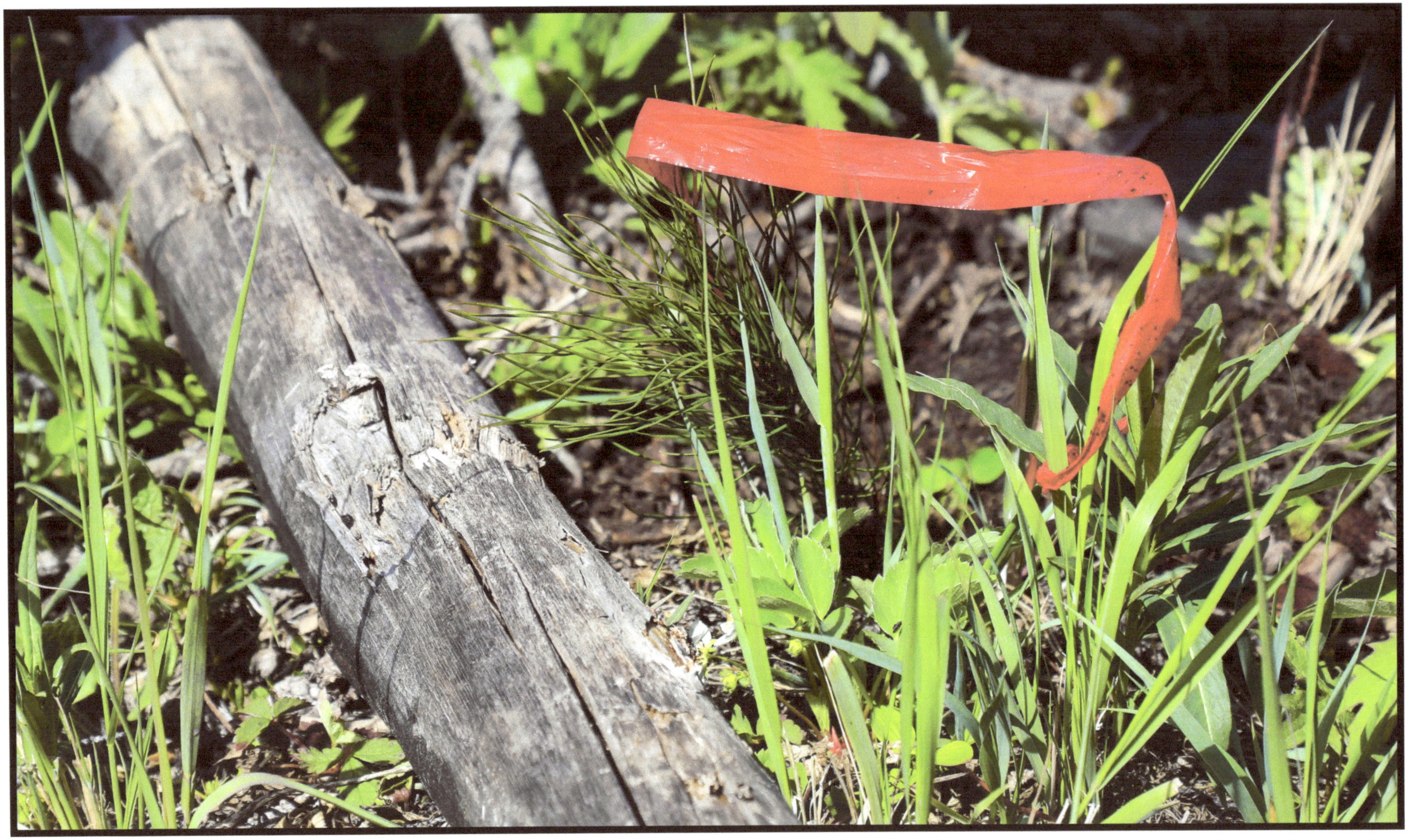

Although blue flagging tape is almost universally preferred by tree planters, red flagger has somewhat of a festive look to it. When you see how well this small amount of grass hides this pine seedling, you can see why short pieces of flagging tape can be very effective in helping to spot trees that have already been planted.

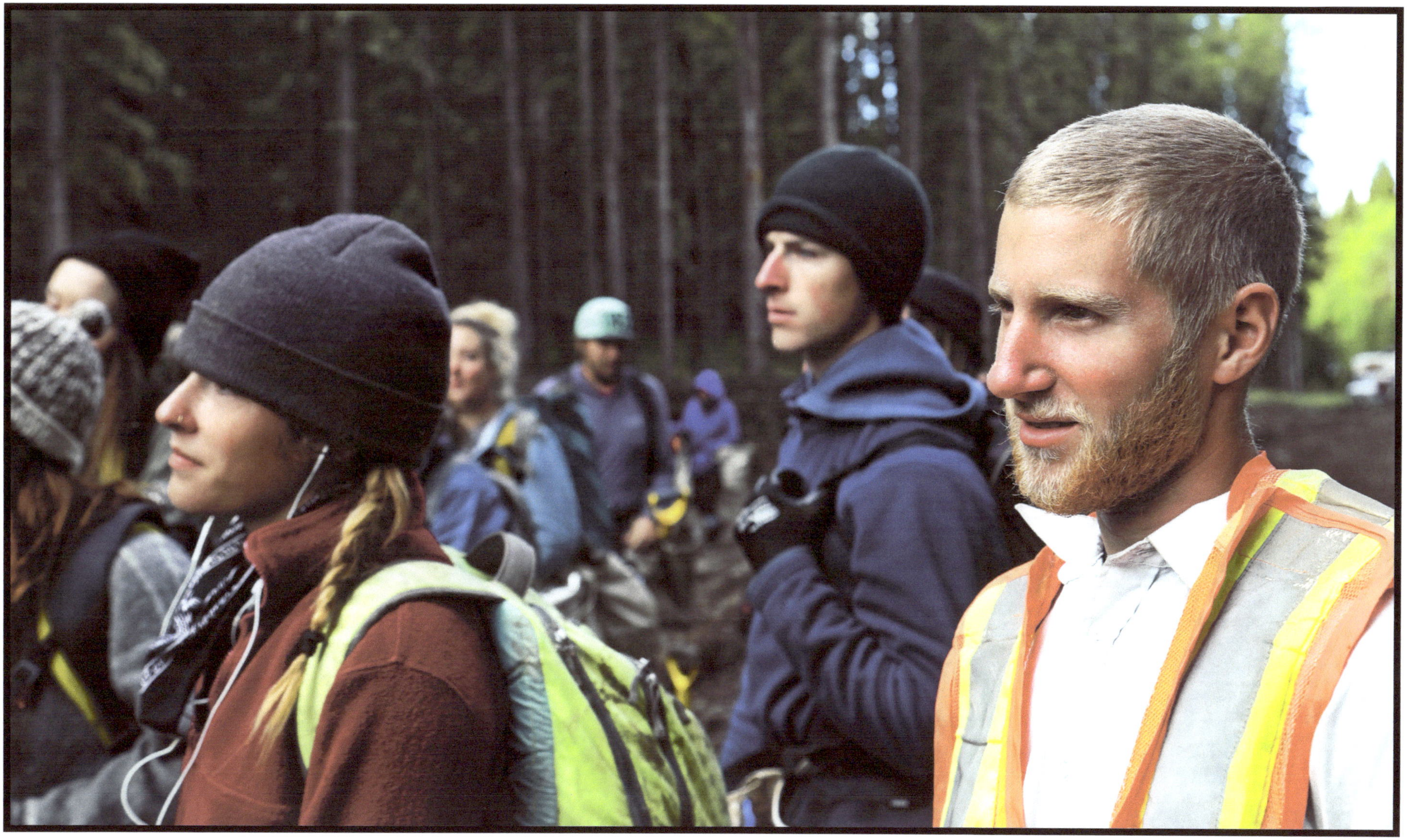

Steve, on the right side of this photo, is one of our crew leaders. Originally from Ontario, Steve took a Bachelor of Theology program at Emmanuel Bible College. Although Steve is quite happy with planting right now, he is considering pursuing a career as a helicopter pilot at some point in the future. On the rare occasions when he has free time and there's no snow on the ground, Steve enjoys golfing.

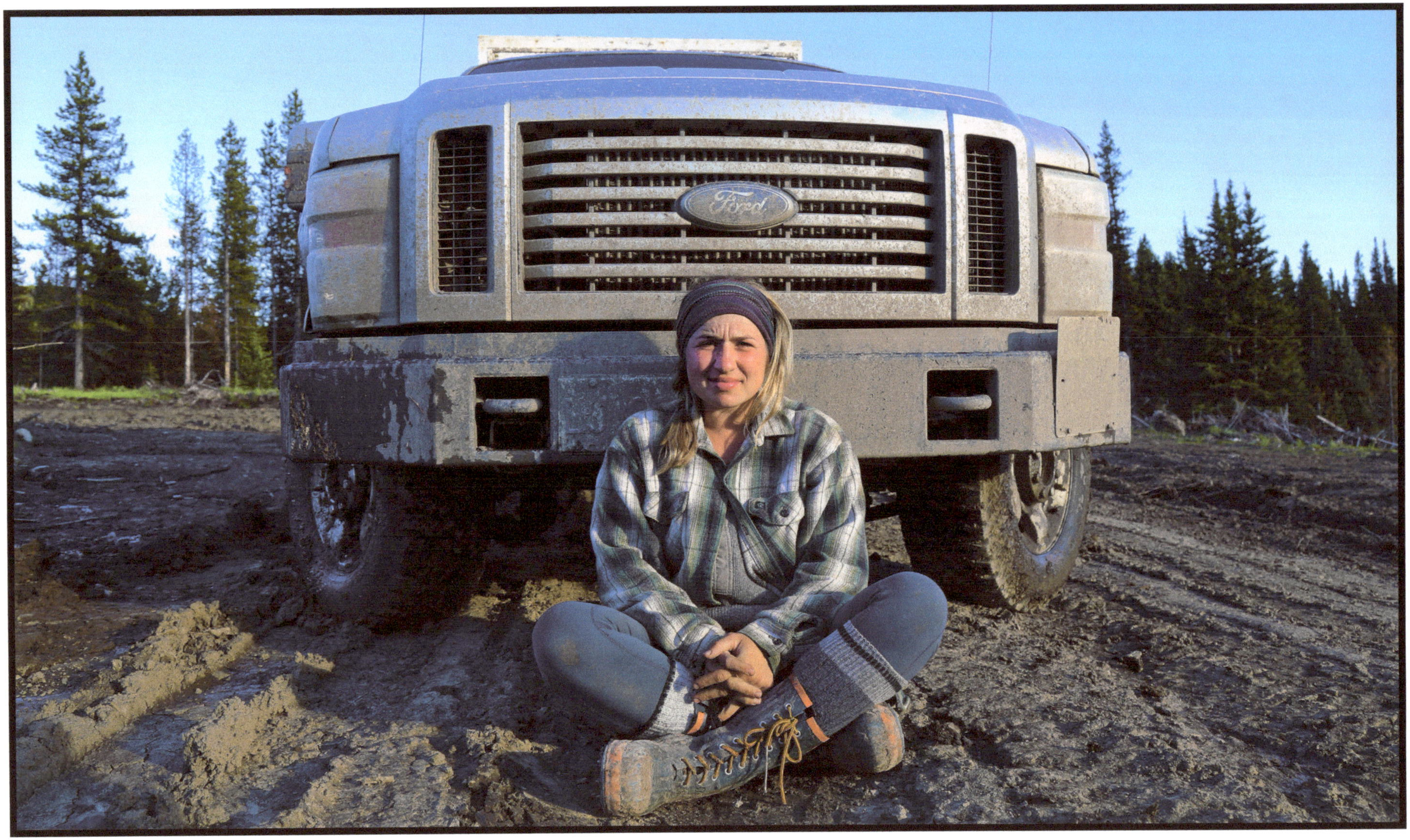

At first glance, you may think that this is intended to be an artistic photo of Ania. However, take a close look at the wheels of this truck. This photo was taken on a day when we had to go on a rescue mission with a tow truck, deep into the bush in Kananaskis Country.

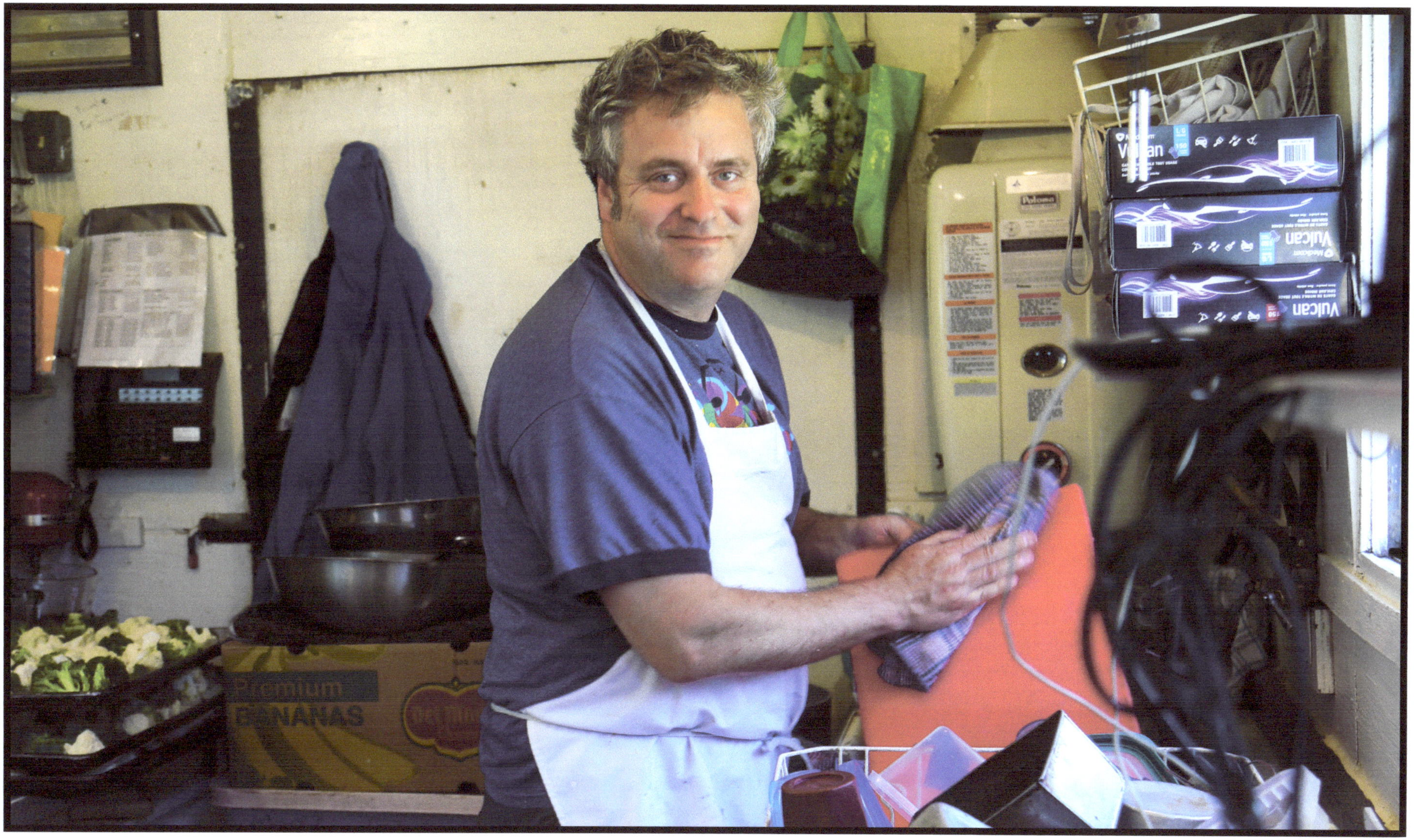

Stephane, also known as Steph the Chef, is another legend. Stephane has over twenty years of experience cooking in planting camps, many years of cooking at various high-end remote lodges and on various film sets, and he also runs the Food Services operation at the Edmonton Folk Festival. I'm quite confident that there is no better cook in the planting industry.

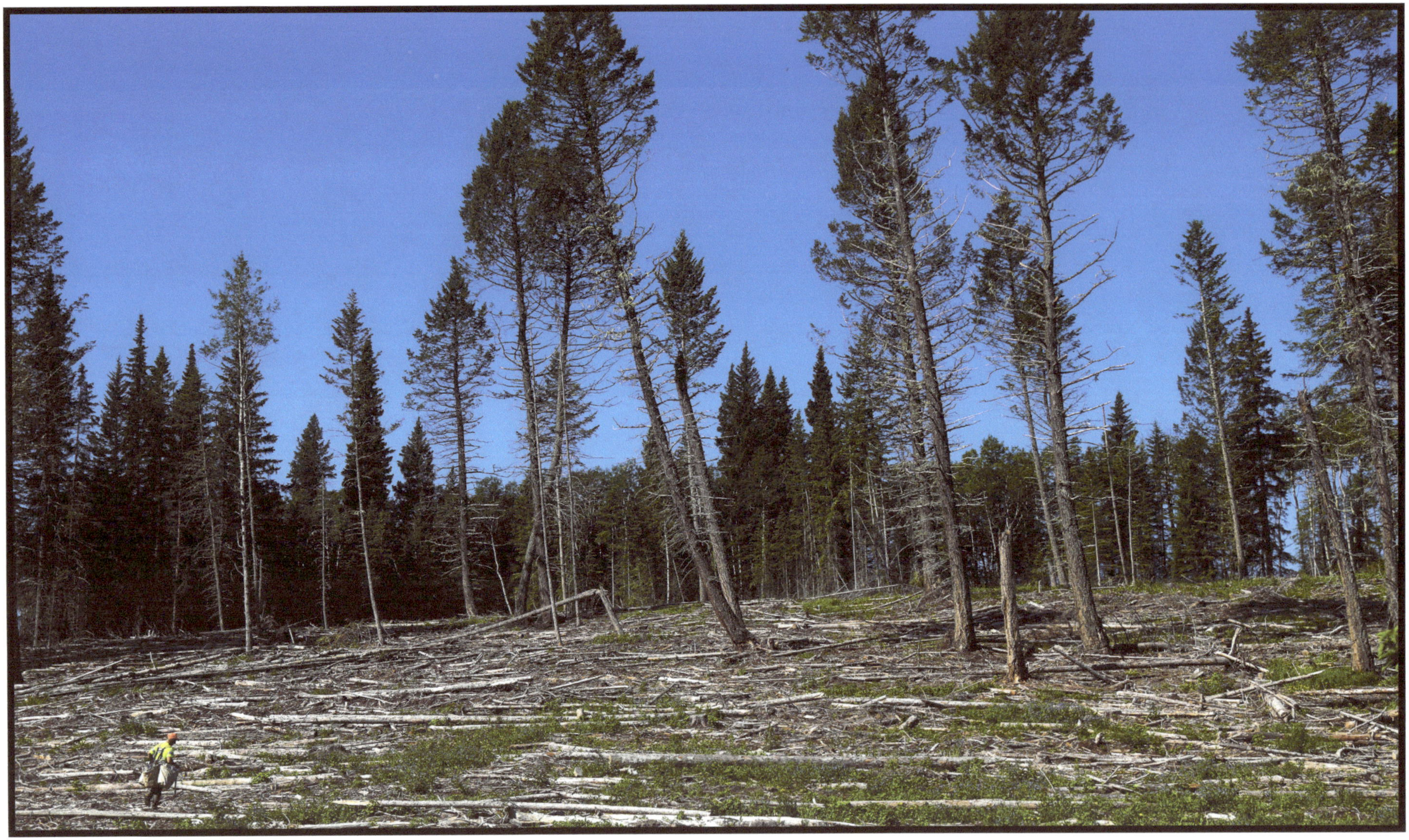

This looks like a beautiful day for planting.

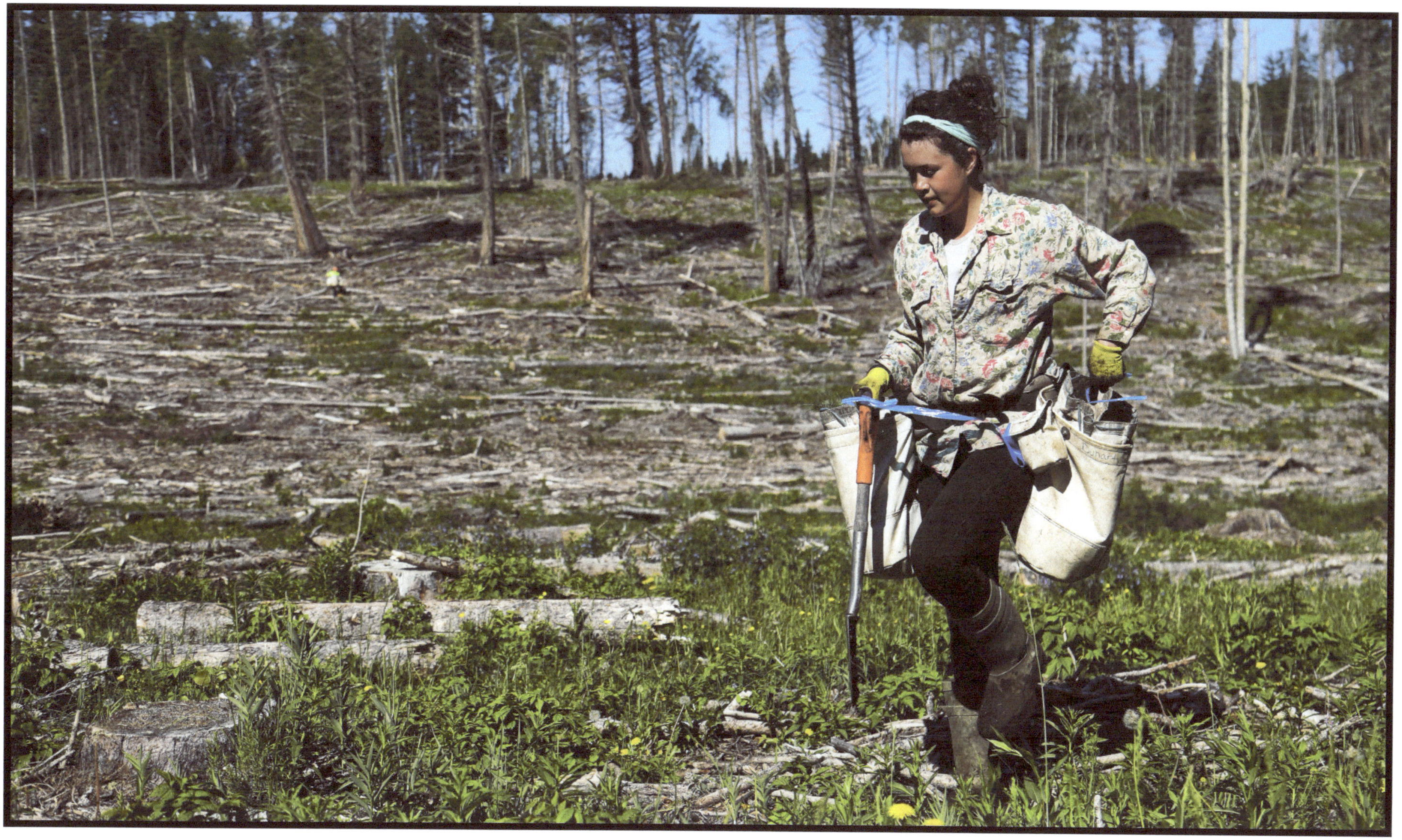

Mattea is from Glengarry, Ontario. Mattea is chipping away at a Mathematics degree at Queens, but after that, has no solid career plans in mind. Sometimes, life is more interesting if you take it one day at a time. Mattea loves travelling around the world.

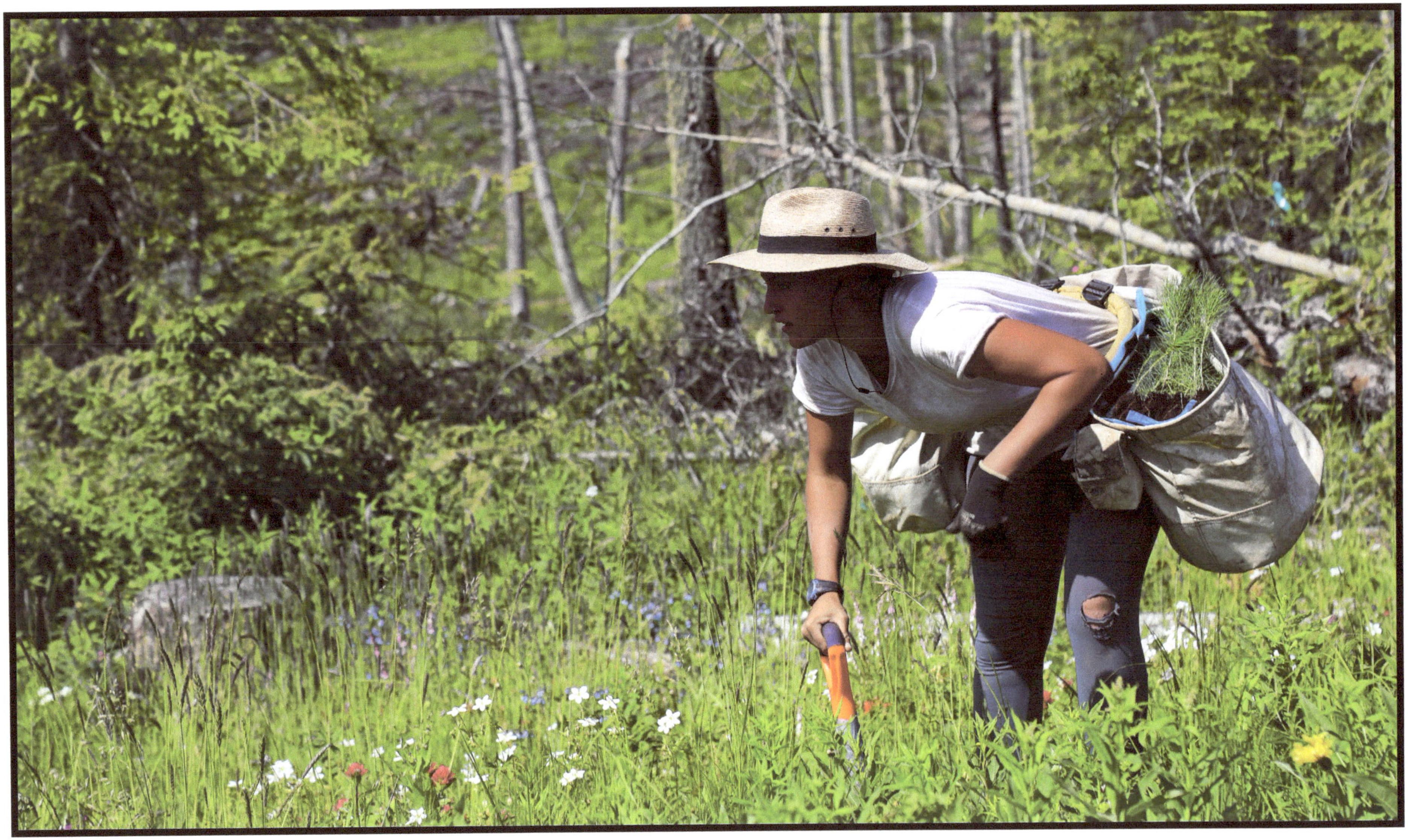

A good planter is always looking ahead and planning a path through their piece.

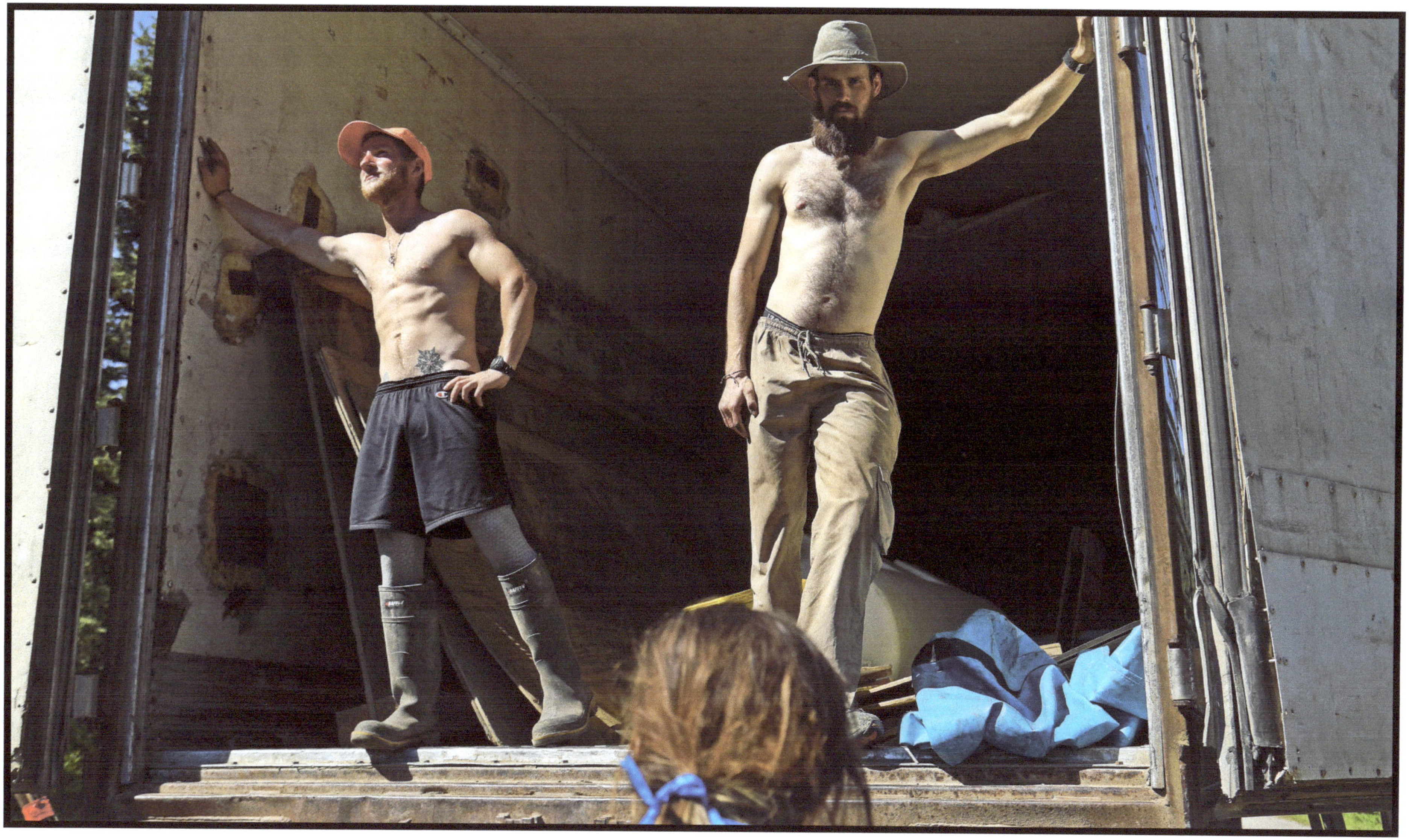

Steve and Matt, on reefer loading duty during a camp move.

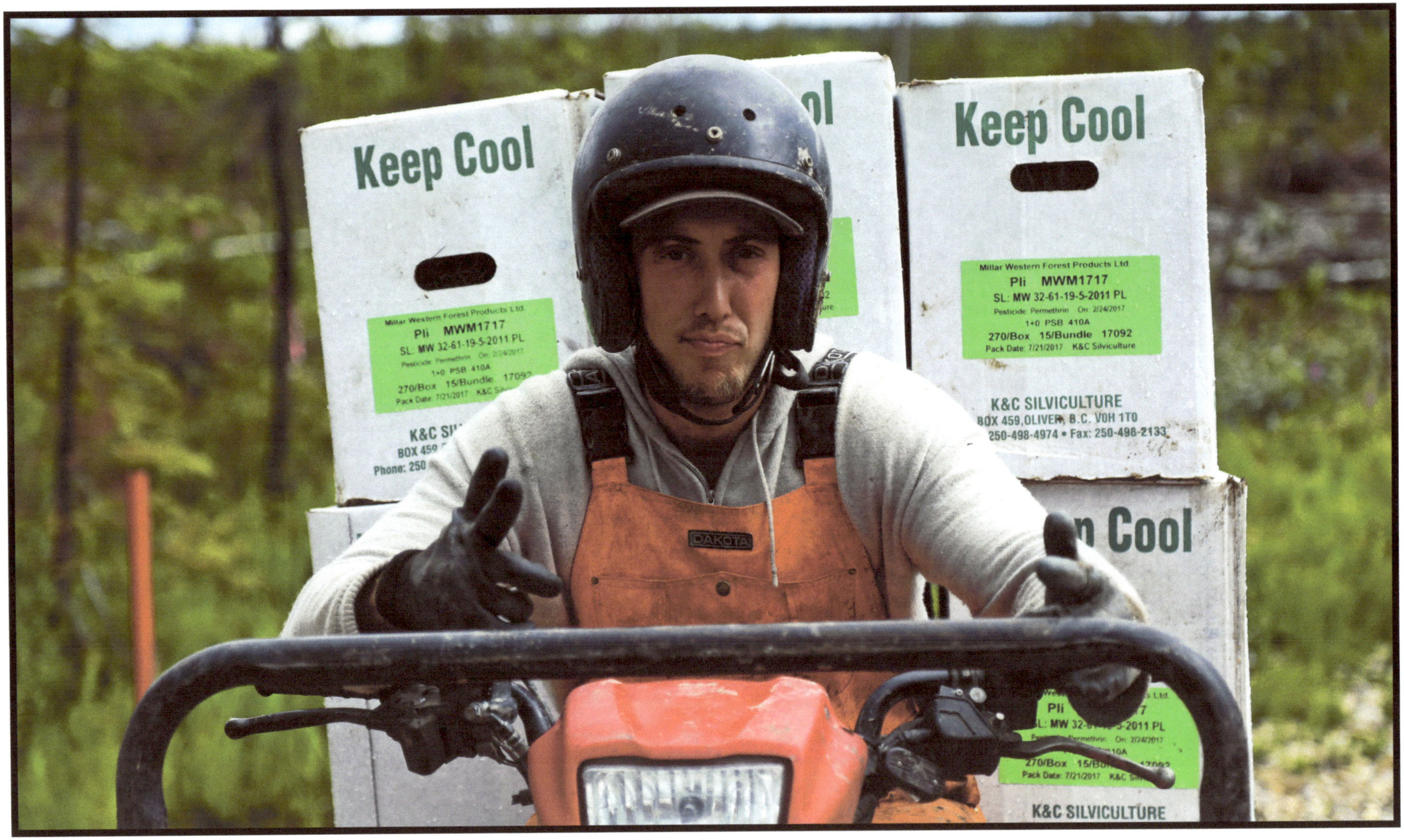

John was born on the west coast, in Vancouver, but now lives in Toronto. John went to Sheridan College for Social Service work, and Ryerson University to study working with Children and Youth. He also attended the Ontario Fire Academy. His long-term career goals include working as a child counsellor or fire fighter, and his hobbies include music, reading, and martial arts.

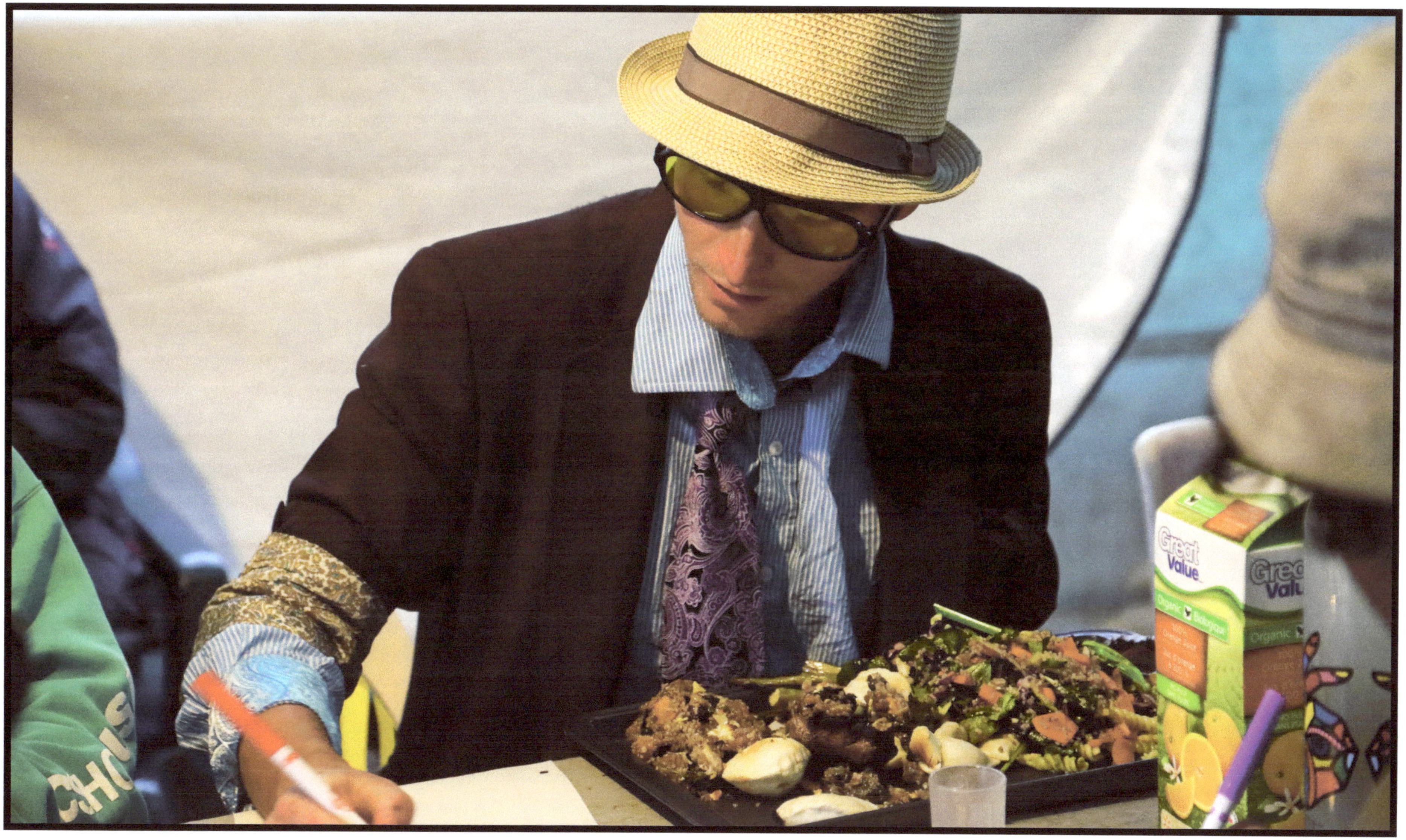

Sean, one of our management team, is responsible for quality control in the camp, as well as being our senior occupational first aid attendant. His role is essentially to act as an assistant camp supervisor, trying to help with logistics and to make sure that operations run smoothly. Sean has become heavily involved in our pre-season field training work in the recent past, which is one of the reasons why we've had such success with our first-year planters.

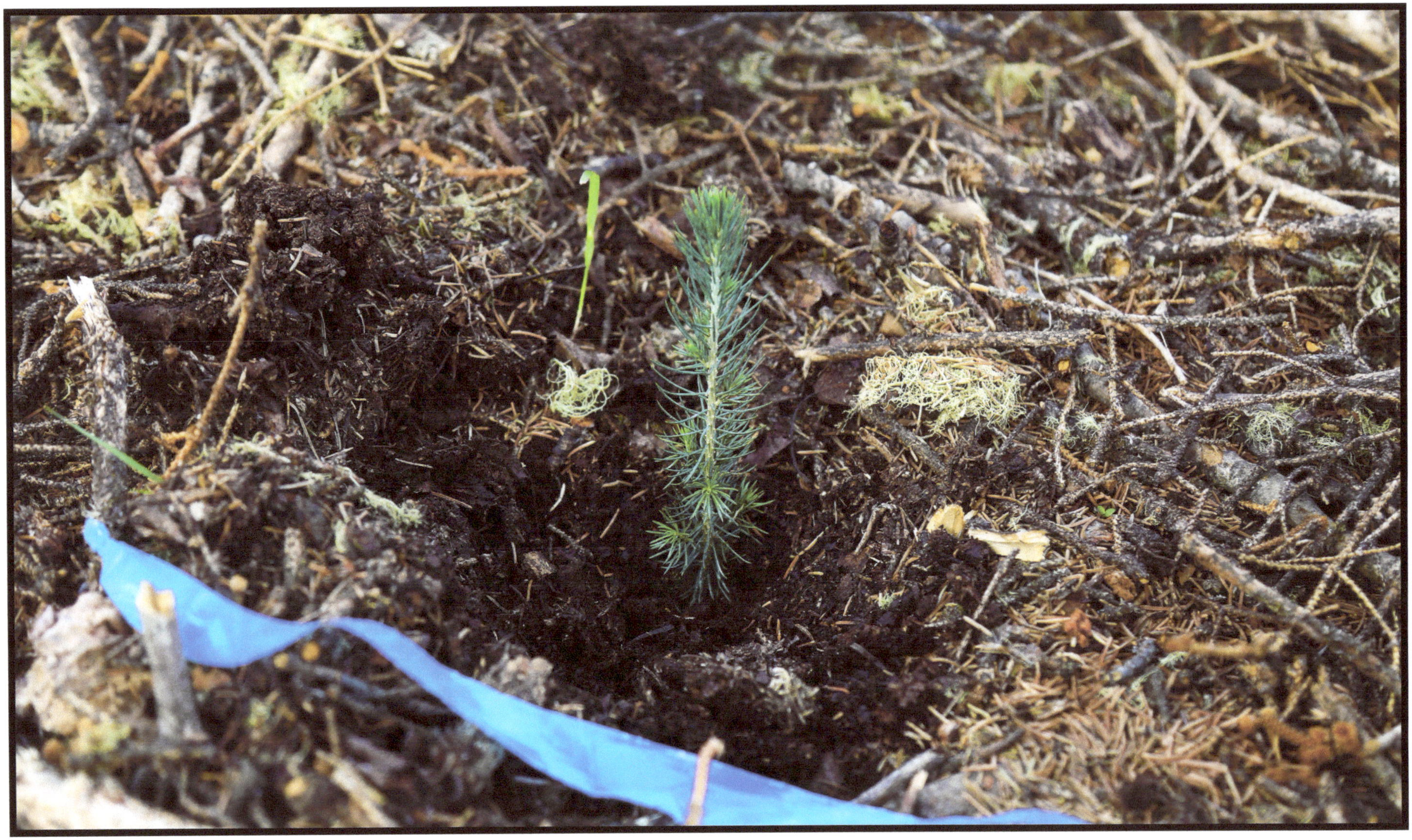

A young spruce seedling. At least half a dozen different spruce tree species are planted throughout Canada.

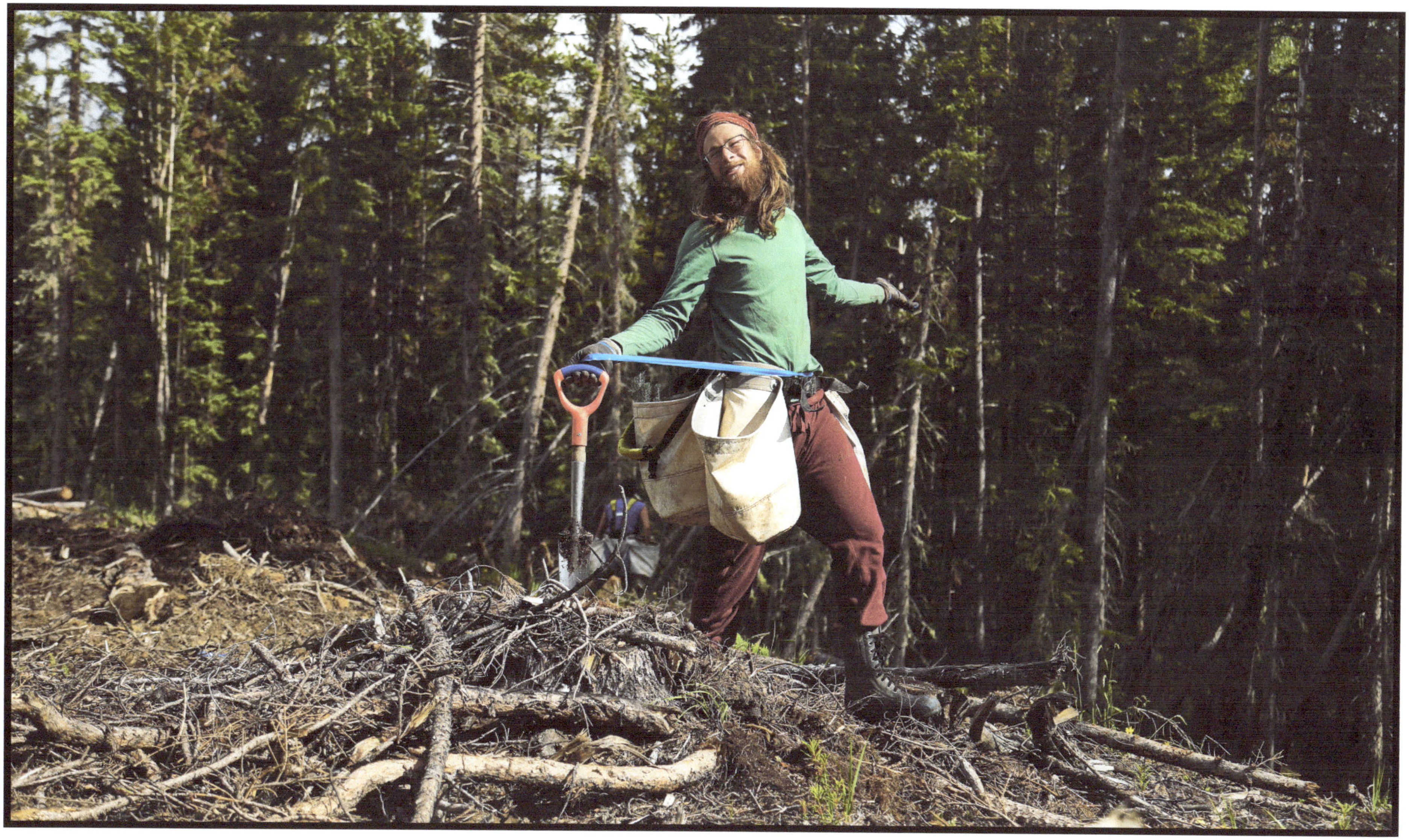

Mitch, one of our fastest planters, graduated from Trent University with a Bachelor of Science degree majoring in biology. Mitch loves planting and camping.

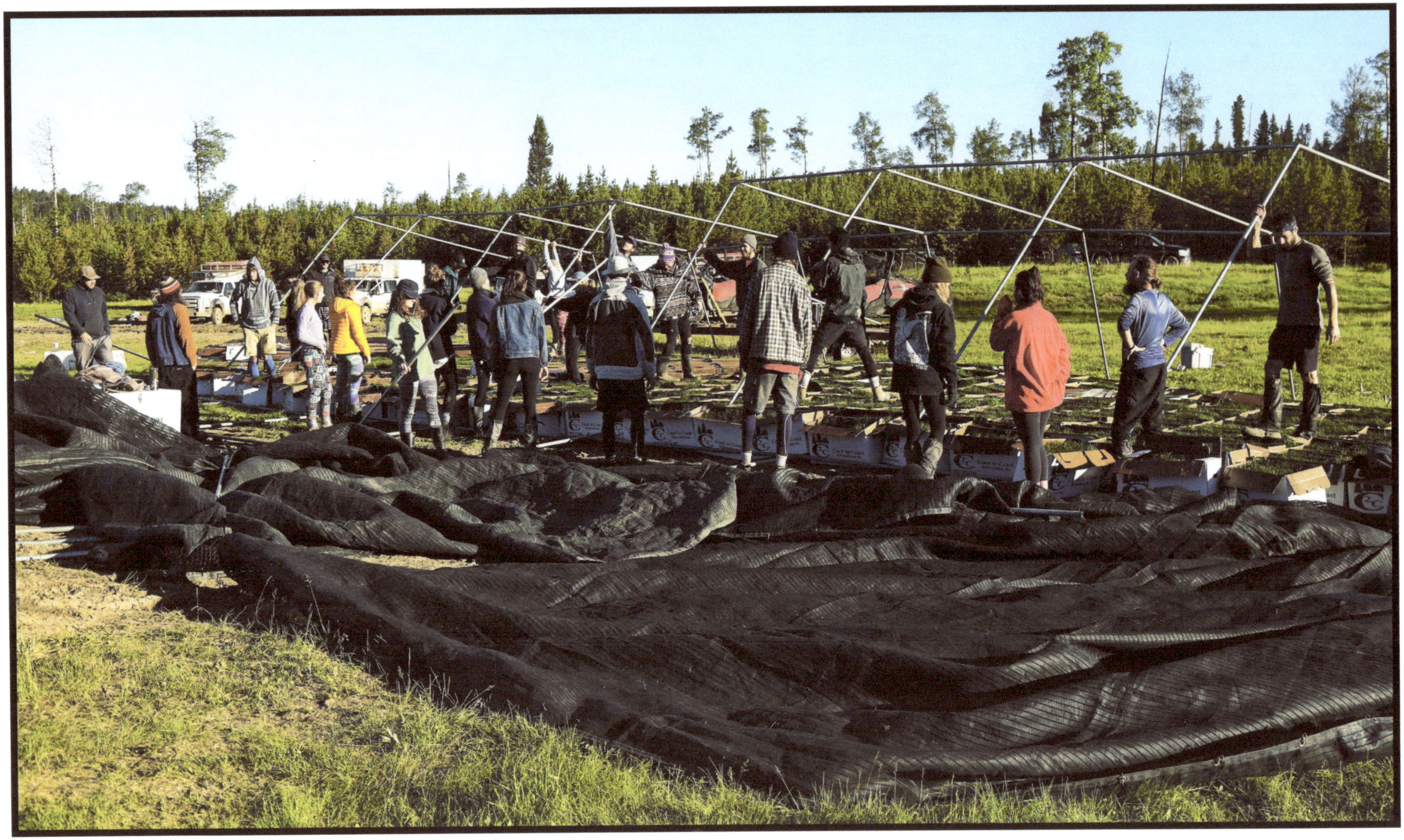

Many hands make light work. In this photo, the crew is putting the shade tent back together after another massive windstorm.

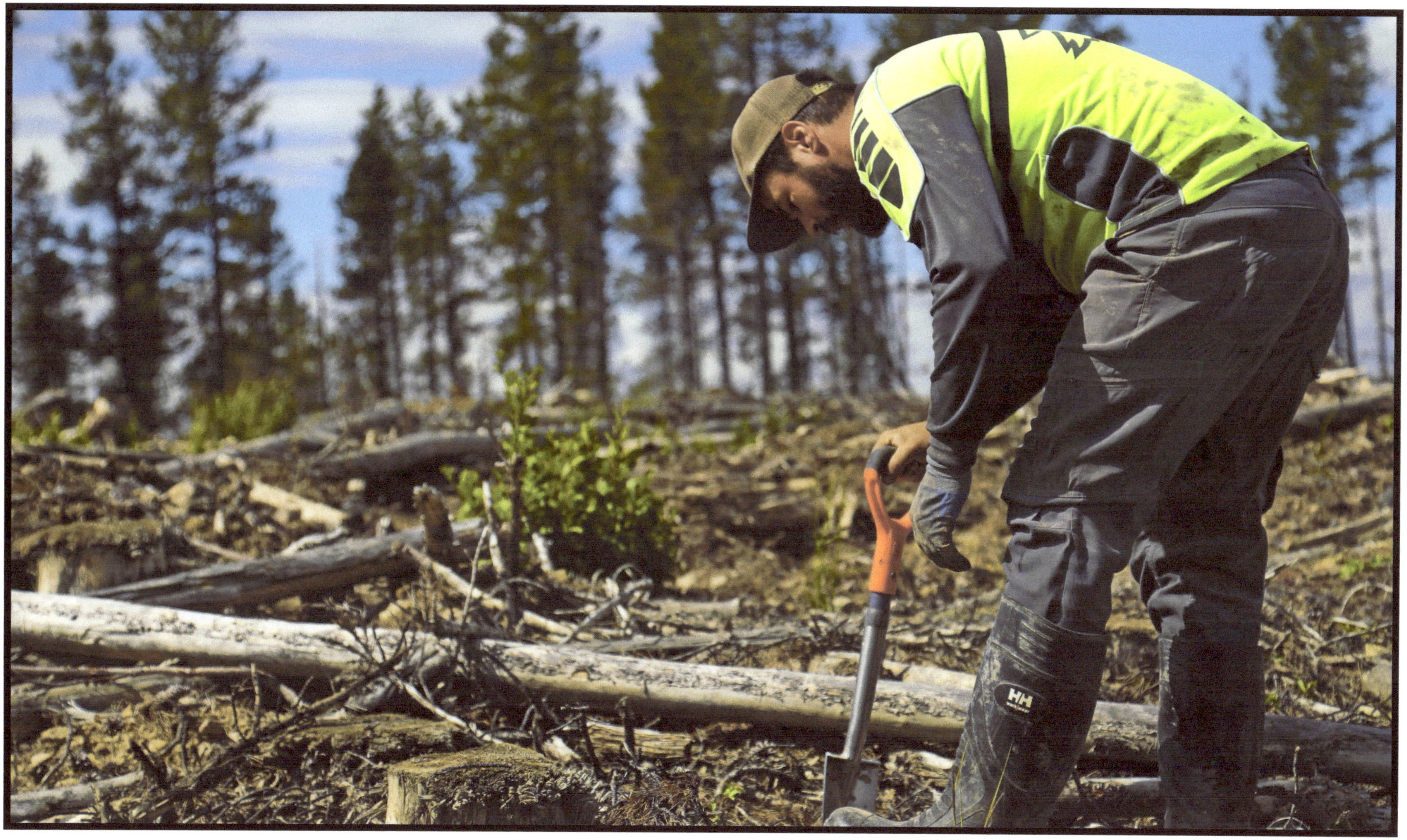

Nick is one of our crew leaders. Nick's planting experience is broad, having worked at companies in Ontario, on the west coast, and overseas.

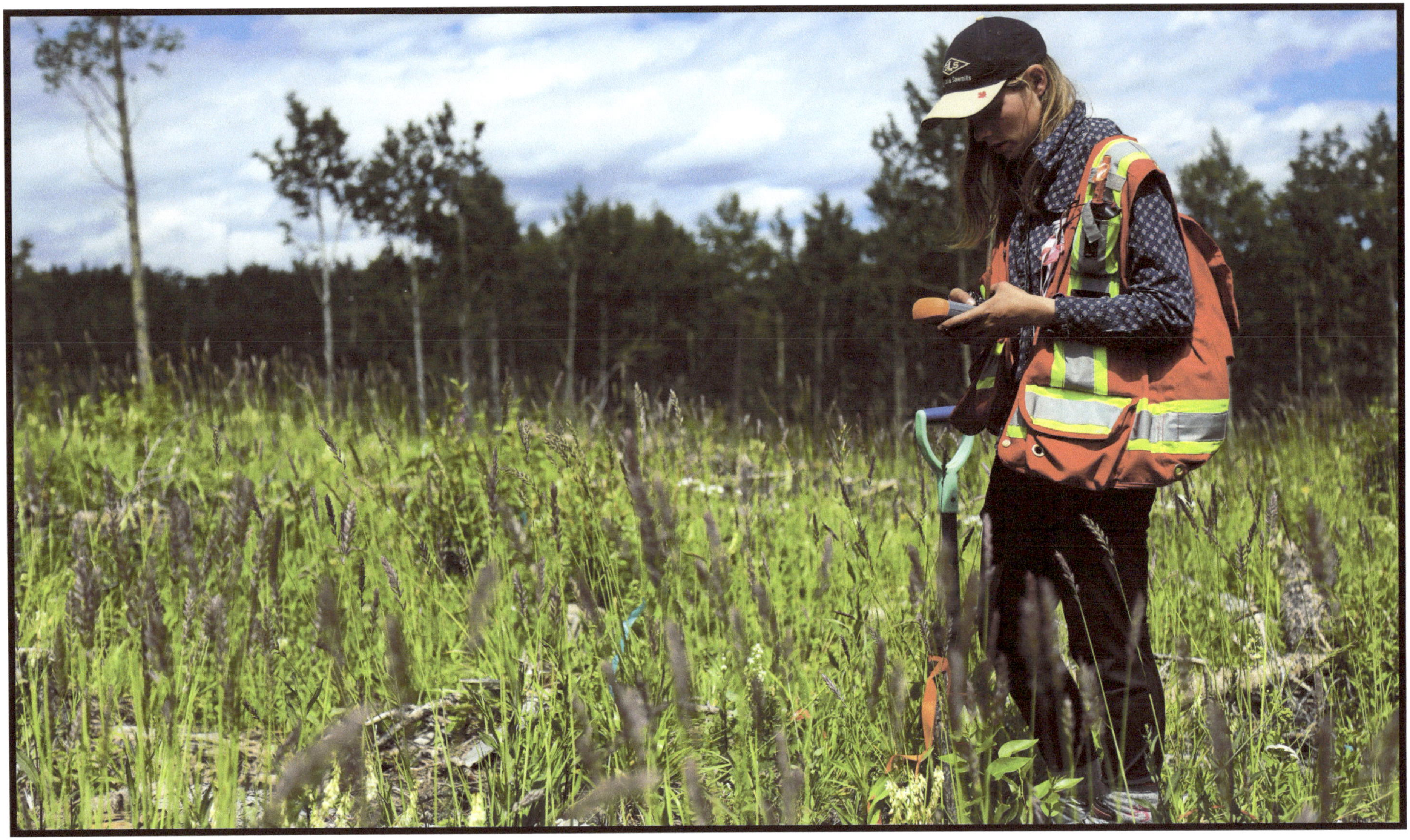

In this photo, Katherine is checking out a GPS device and recording plot results.

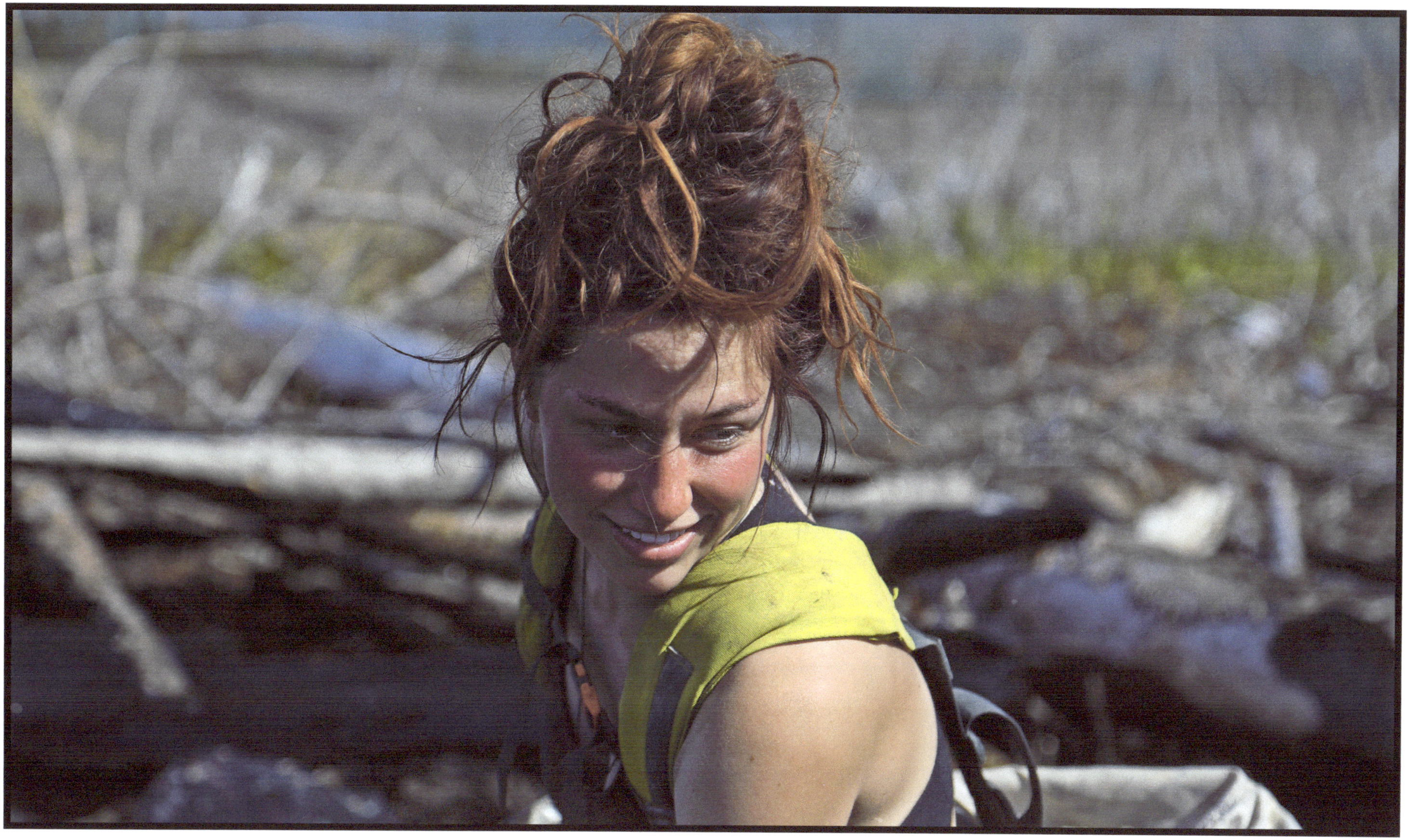

If you love your job, you'll never "work" a day in your life.

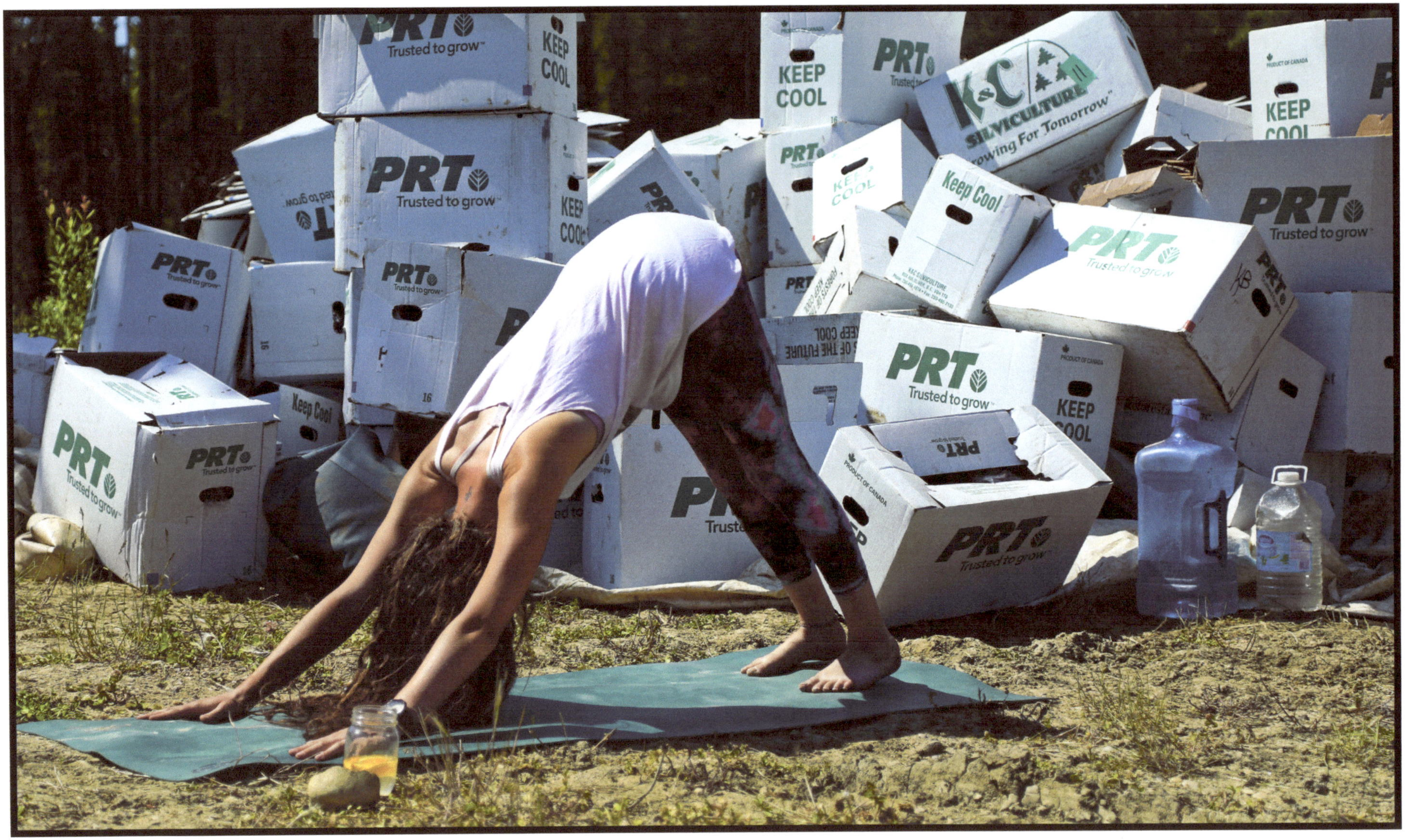

Yoga can be done almost anywhere you want to. Even in front of a large pile of garbage boxes.

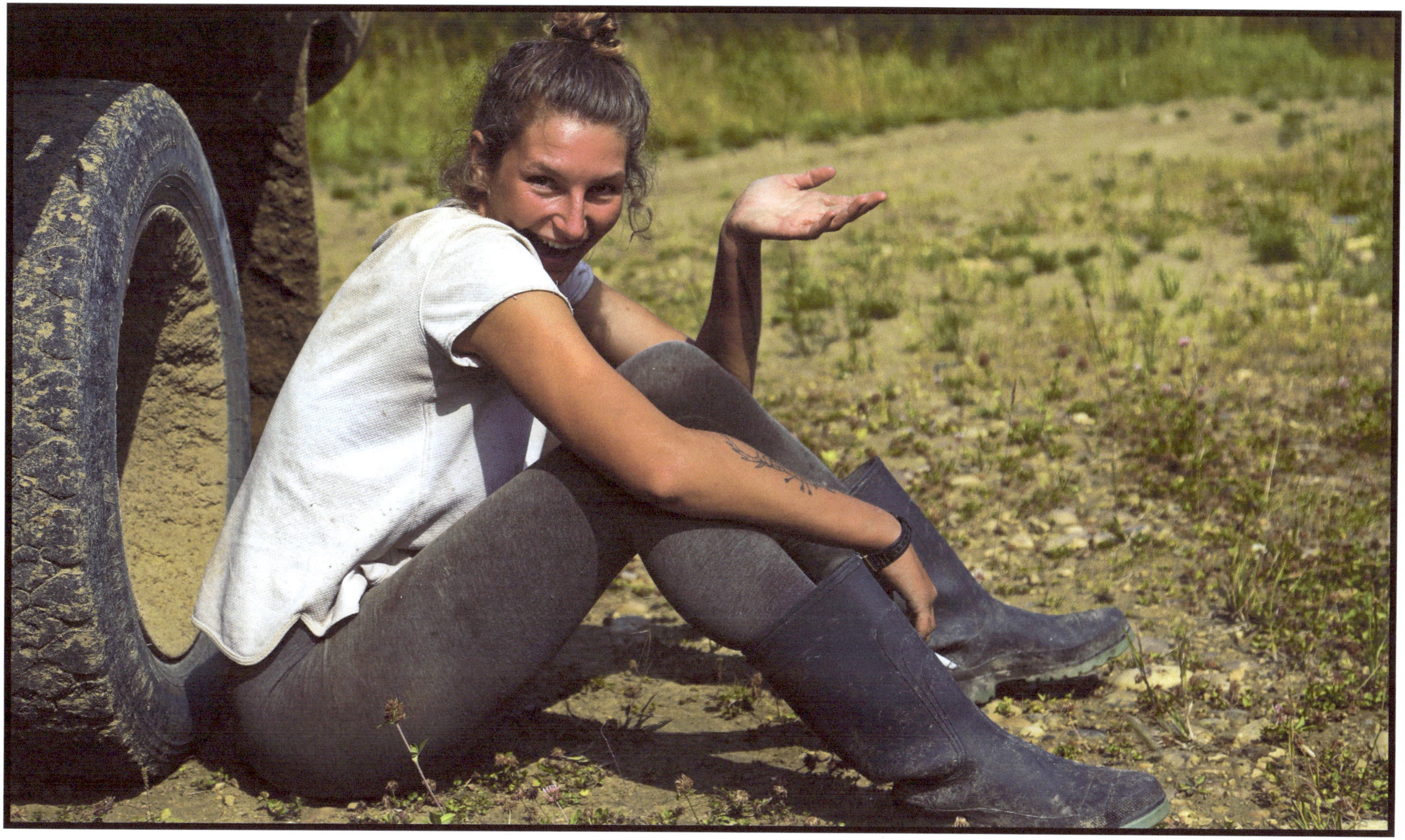

Molly, waiting patiently for the helicopter to pick her up.

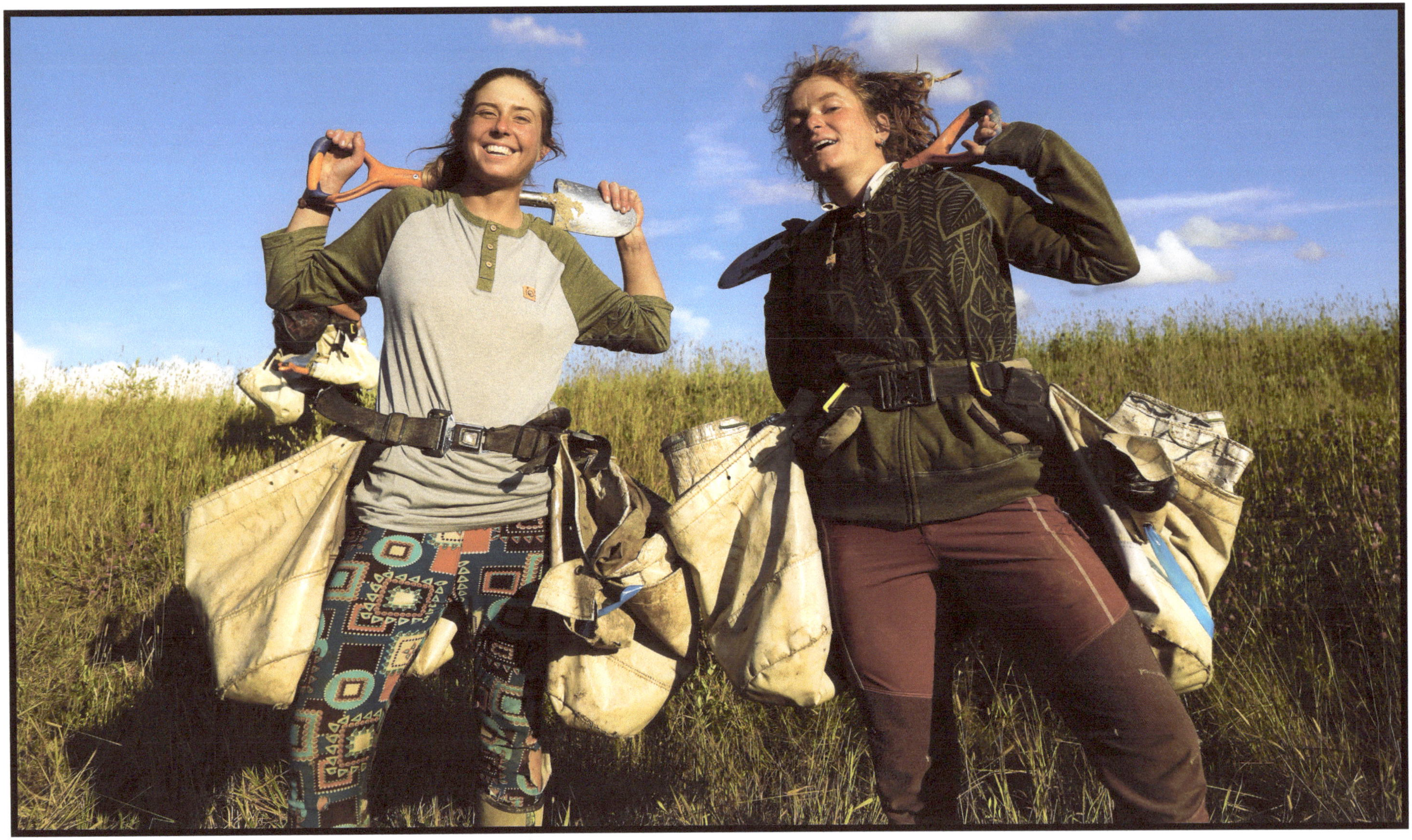

Laura and Laura.

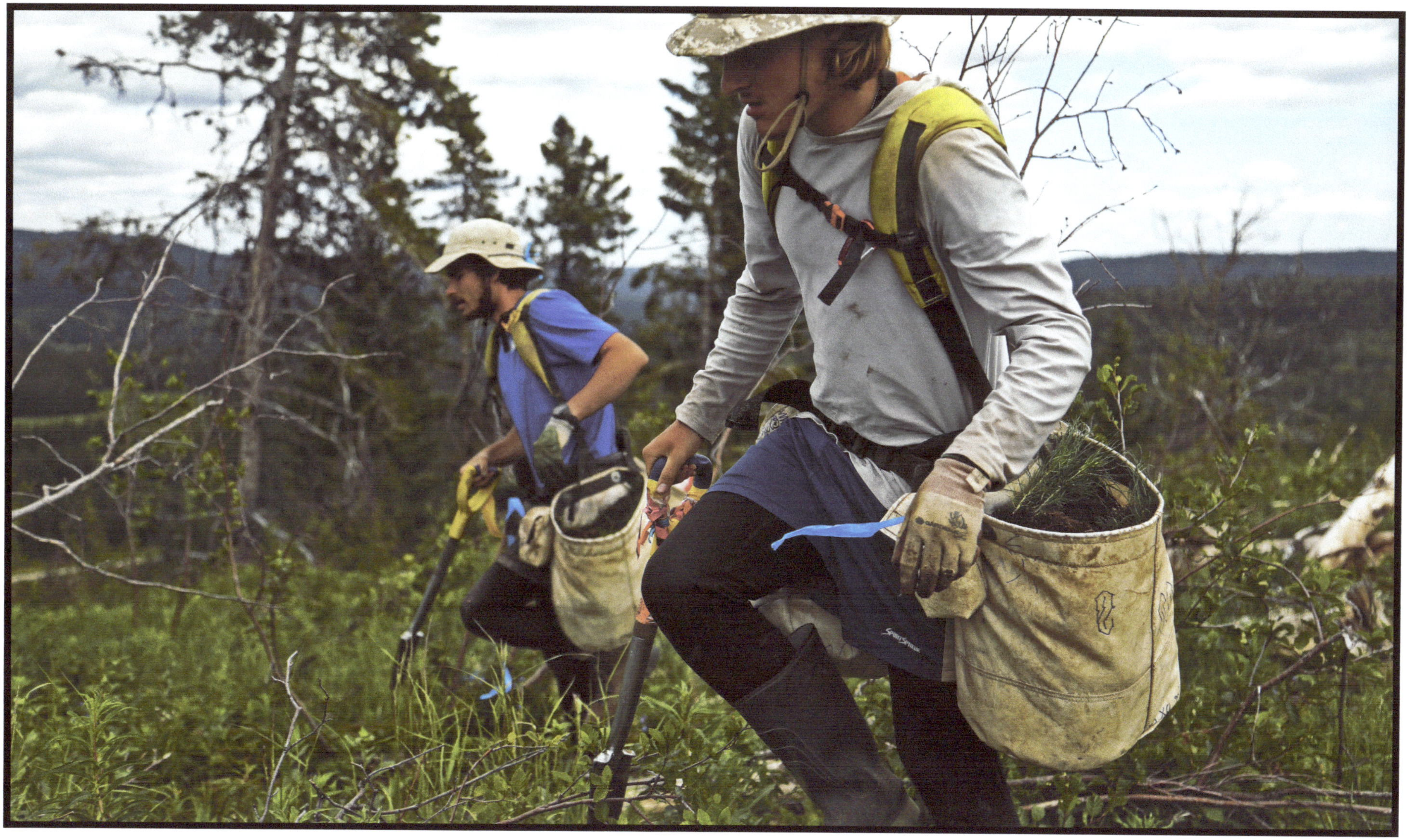

Nick, wearing the blue shirt, is an Ontario resident who studied Business at Queens, although his career aspiration is to possibly attend law school and become a lawyer. Nick's hobbies include trail running, bike-touring, and other types of wilderness endurance sports. What Nick enjoys the most about planting is that it provides the opportunity to learn about grit and work ethic. Will, in front, is a horticulture student who describes himself as blackfly bait.

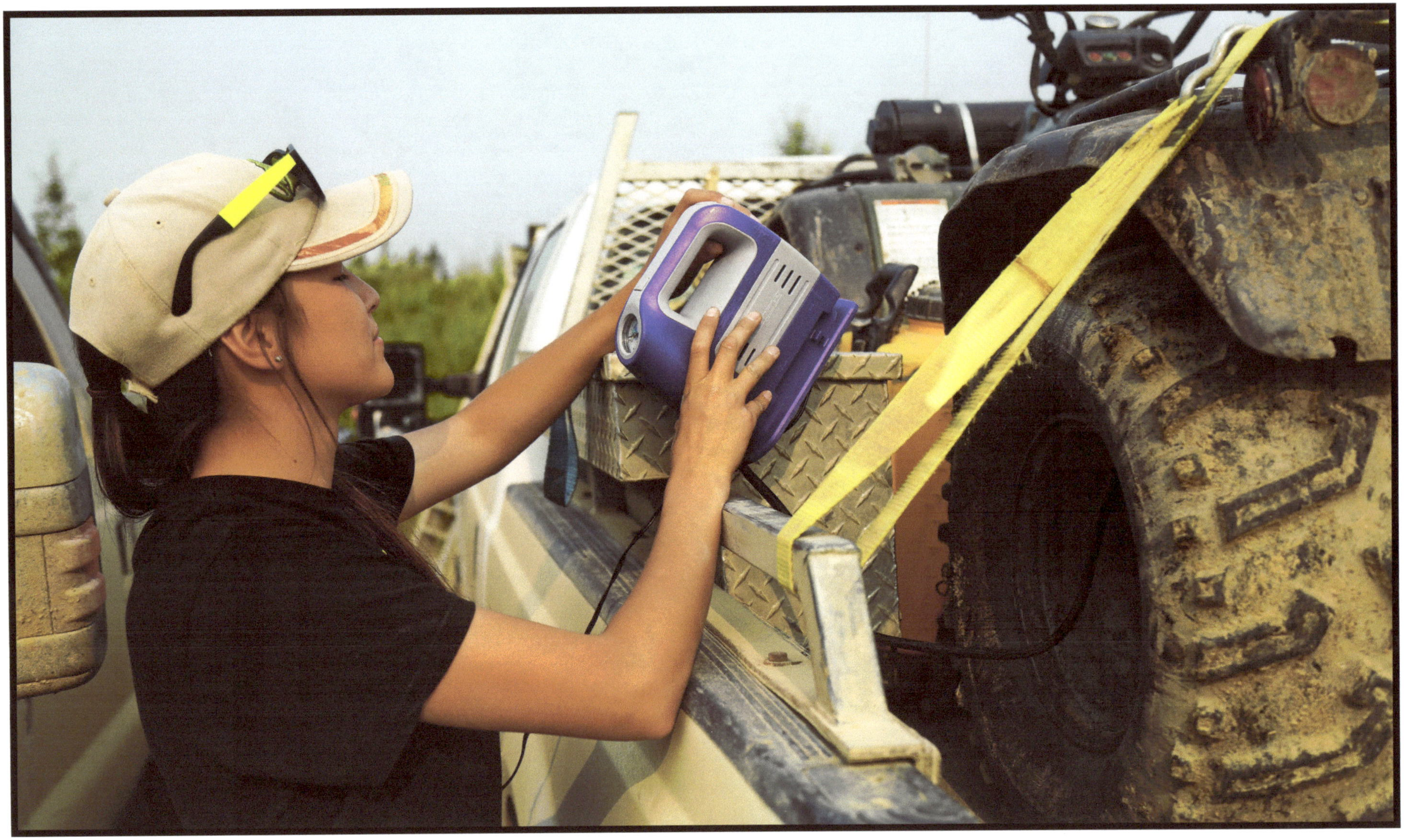

Katie knows that proper maintenance is important for keeping equipment in good working condition. Even things that may seem as minor as tire pressure can make a big difference in ensuring that equipment performs as expected.

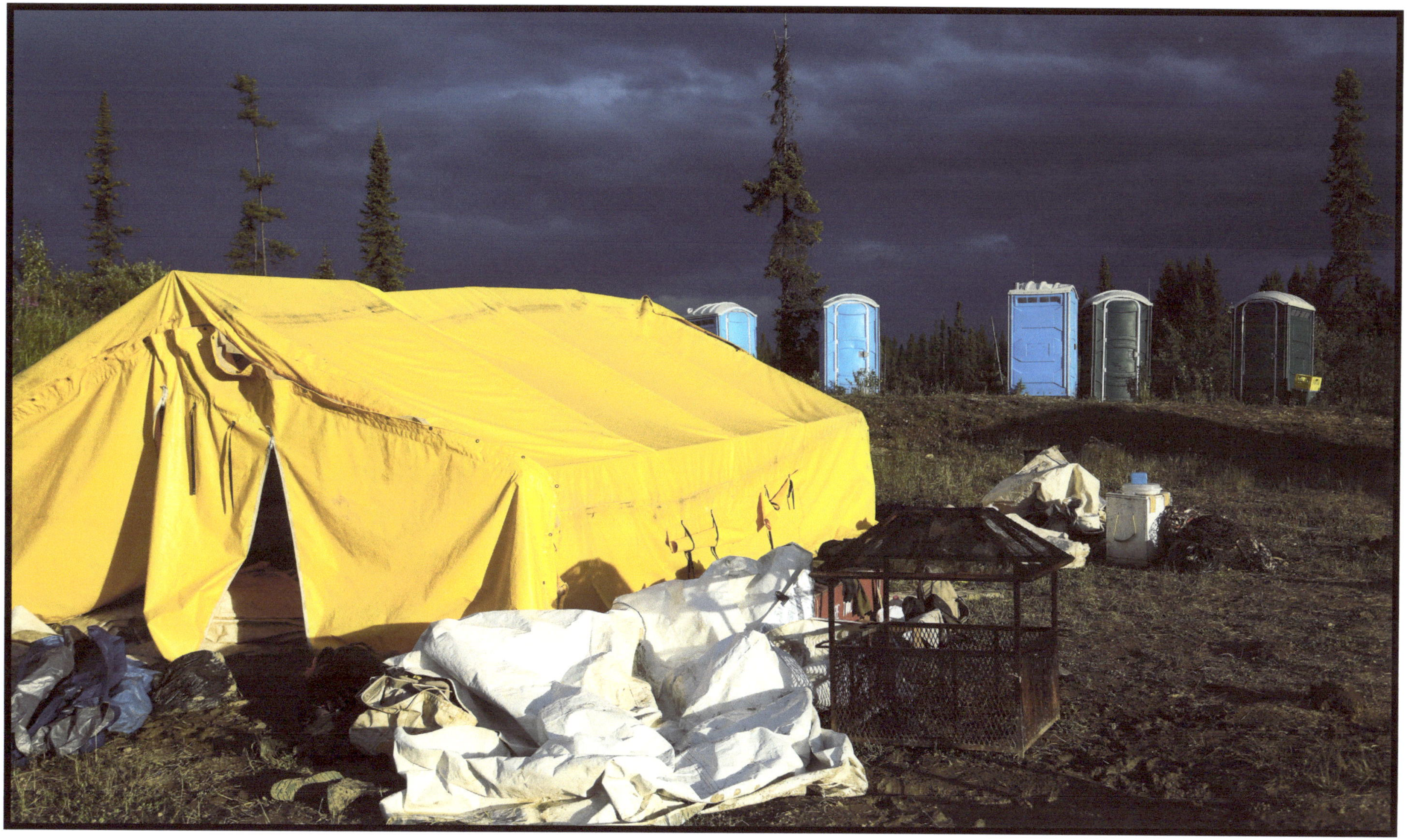

The lighting in this photo was stunning. When I took this photo, I had sunlight behind me, with a storm approaching from behind the port-a-potties to darken the sky. Come to think of it, this may have been one of several storms that tore apart the dry tent.

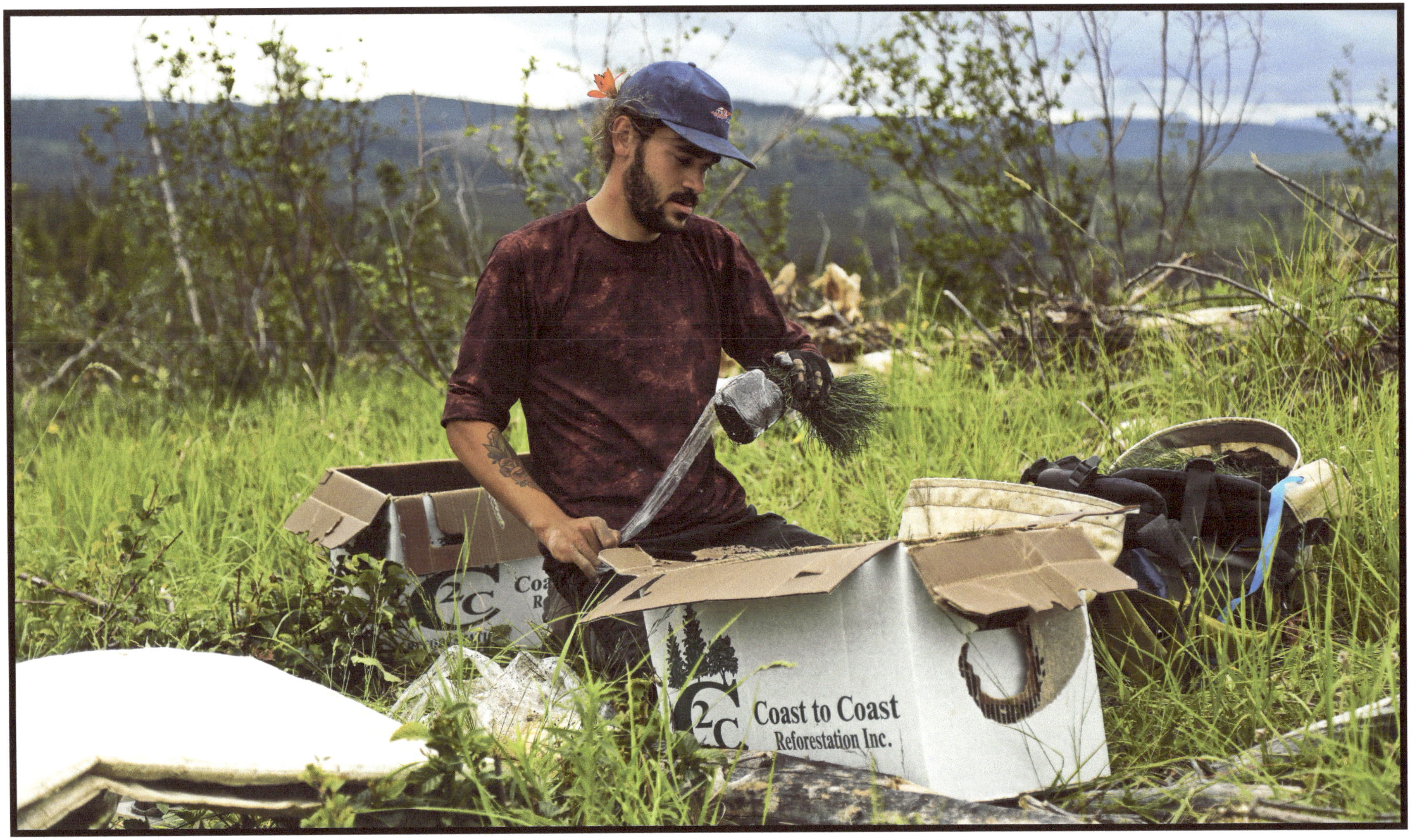

Marc-Andre is another Maritimer, coming from Caraquet in New Brunswick. Marc-Andre studies in the Therapeutic Recreation program at Dalhousie University in Halifax, and hopes to pursue a career in Community Development and Urban Planning. His hobbies include biking, bouldering, running, hiking, and skiing.

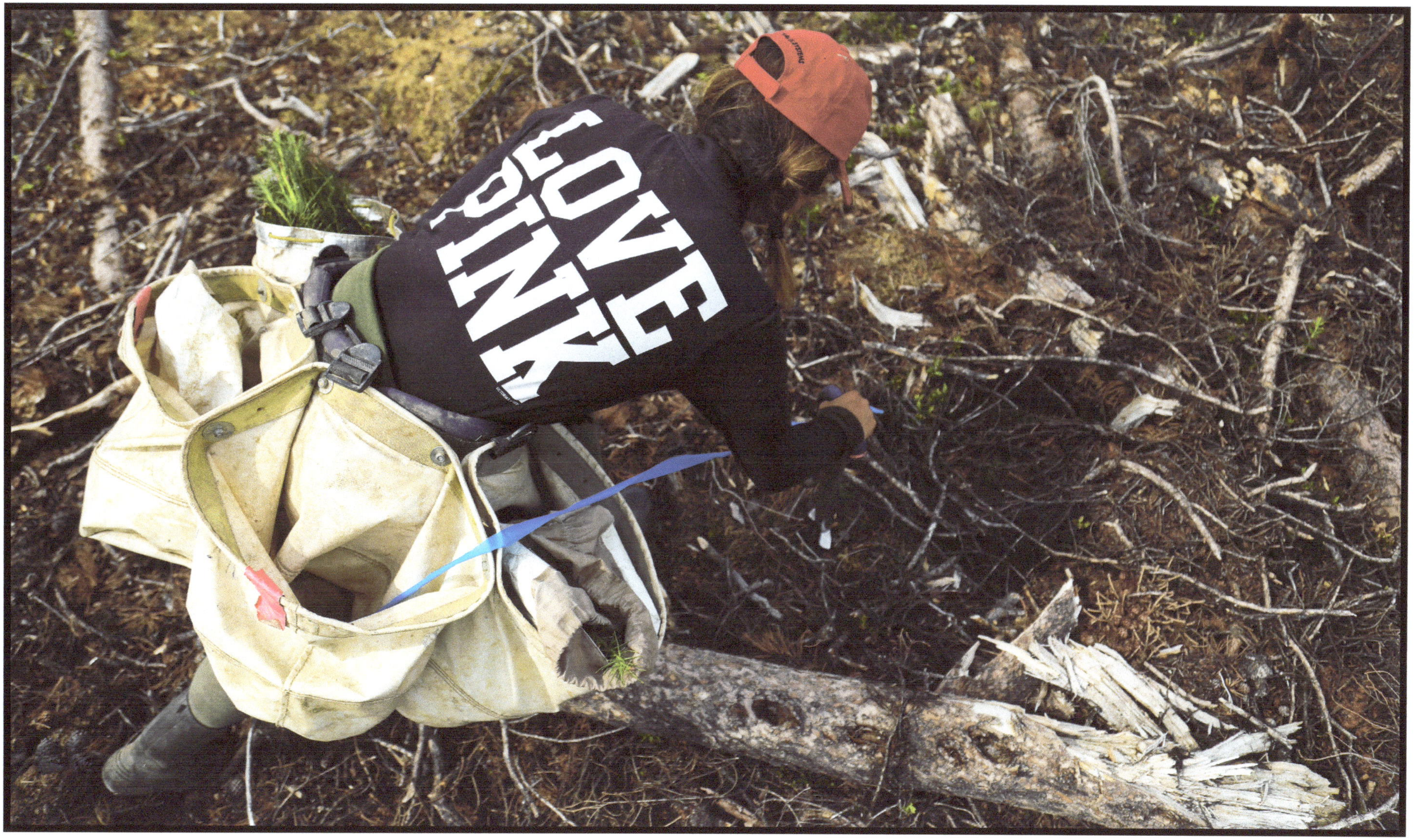

I find it fascinating to see photos looking down on a planter at work. This type of view may become more common in the future, with the increasing prevalence of drones being used professionally in silviculture and reforestation work.

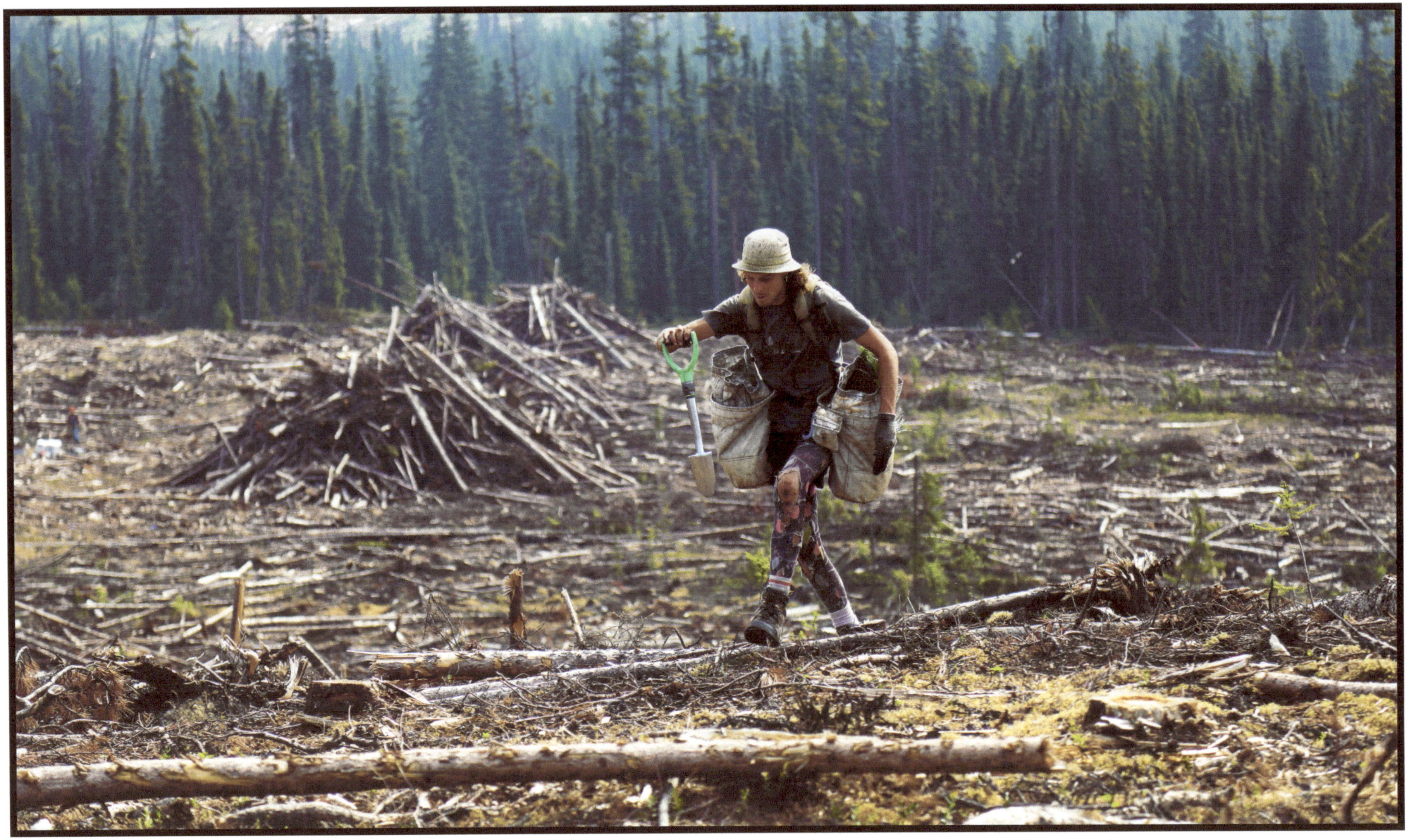

Many planters prefer to use fairly short shovels, due to their light weight. Even for a tall planter, a short shovel can be just as effective as a longer one.

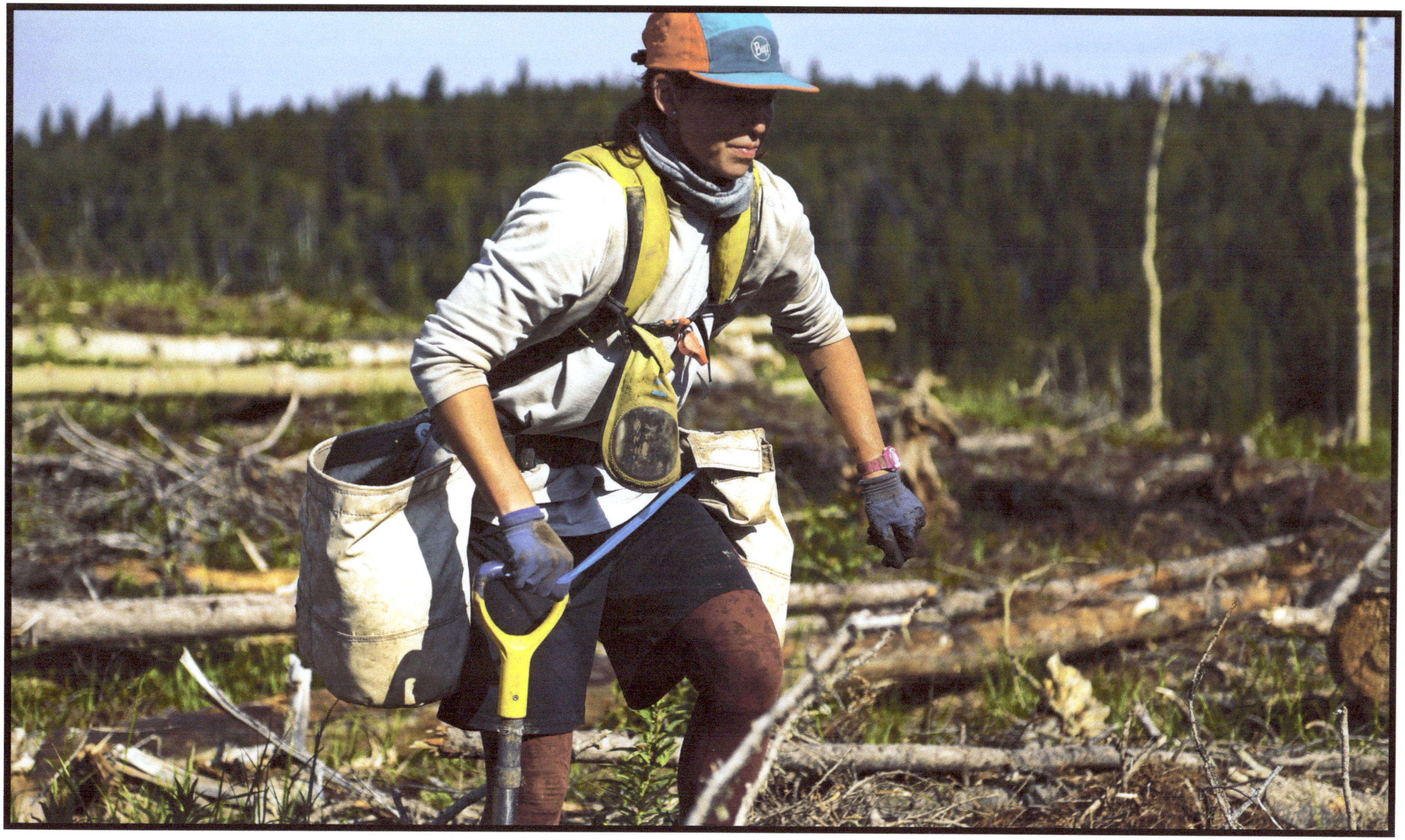

Bree studied in Ontario. Bree noted: "I'd say my favorite thing about planting is that you get to fulfill all fundamental human needs while making money. Humans innately crave physical activity, community, time spent outdoors, as well as having our body clocks match the sun. Planting facilitates all of that. Planting is challenging in so many ways, but in the end it feels great to have all of those needs met, and it feels great knowing that you earned every penny."

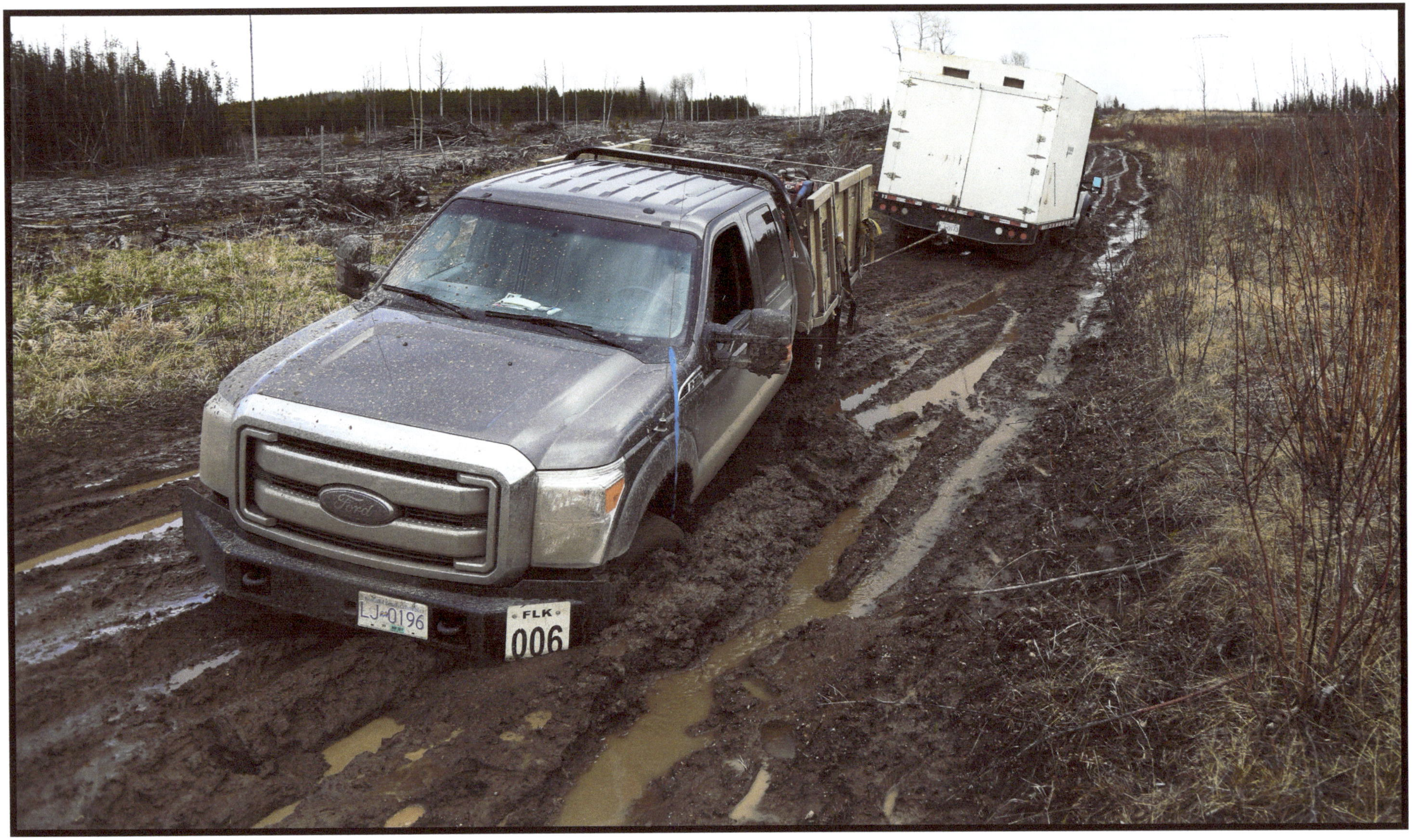

You may wonder which of these two trucks is trying to pull the other out. Sadly, they're both stuck.

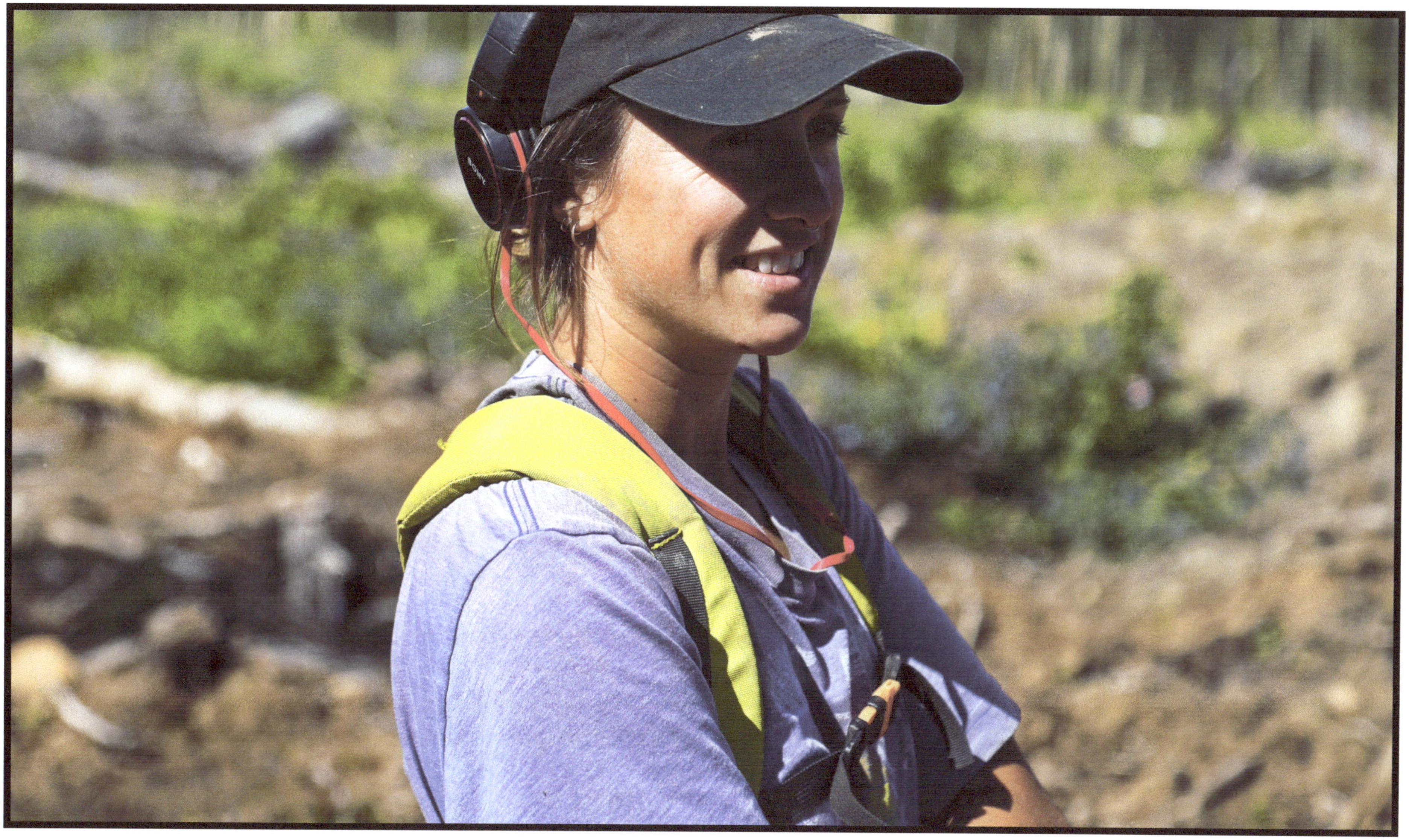

Nothing fazes Hannah.

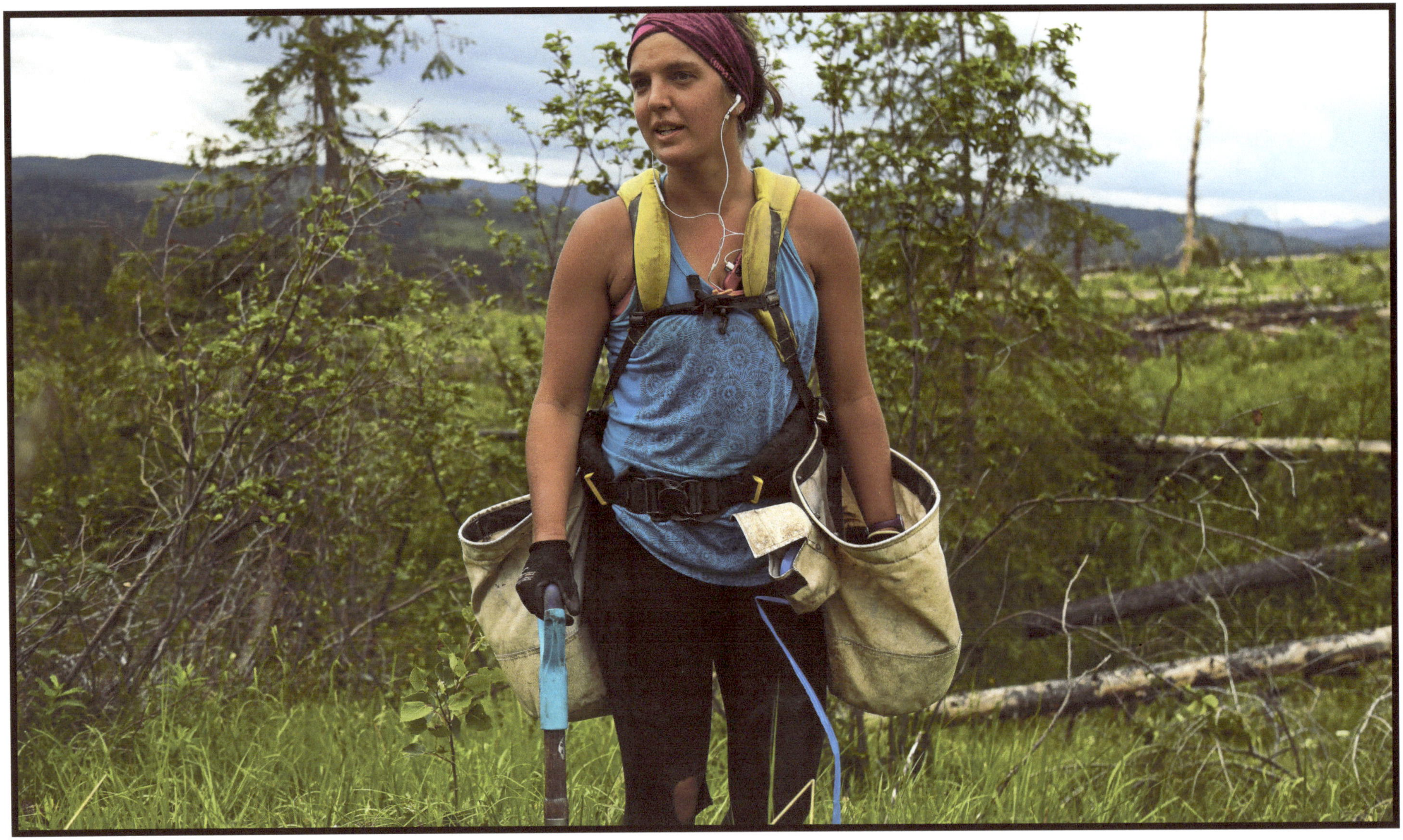

Zoe grew up in Connecticut. She completed a Bachelor of Science with Honours from Queens, specializing in Biology and Psychology, and is now working toward a Bachelor of Science in Nursing from UBC. When she isn't pursuing her studies in Nursing, she enjoys skiing, running, hiking, and spending time with friends.

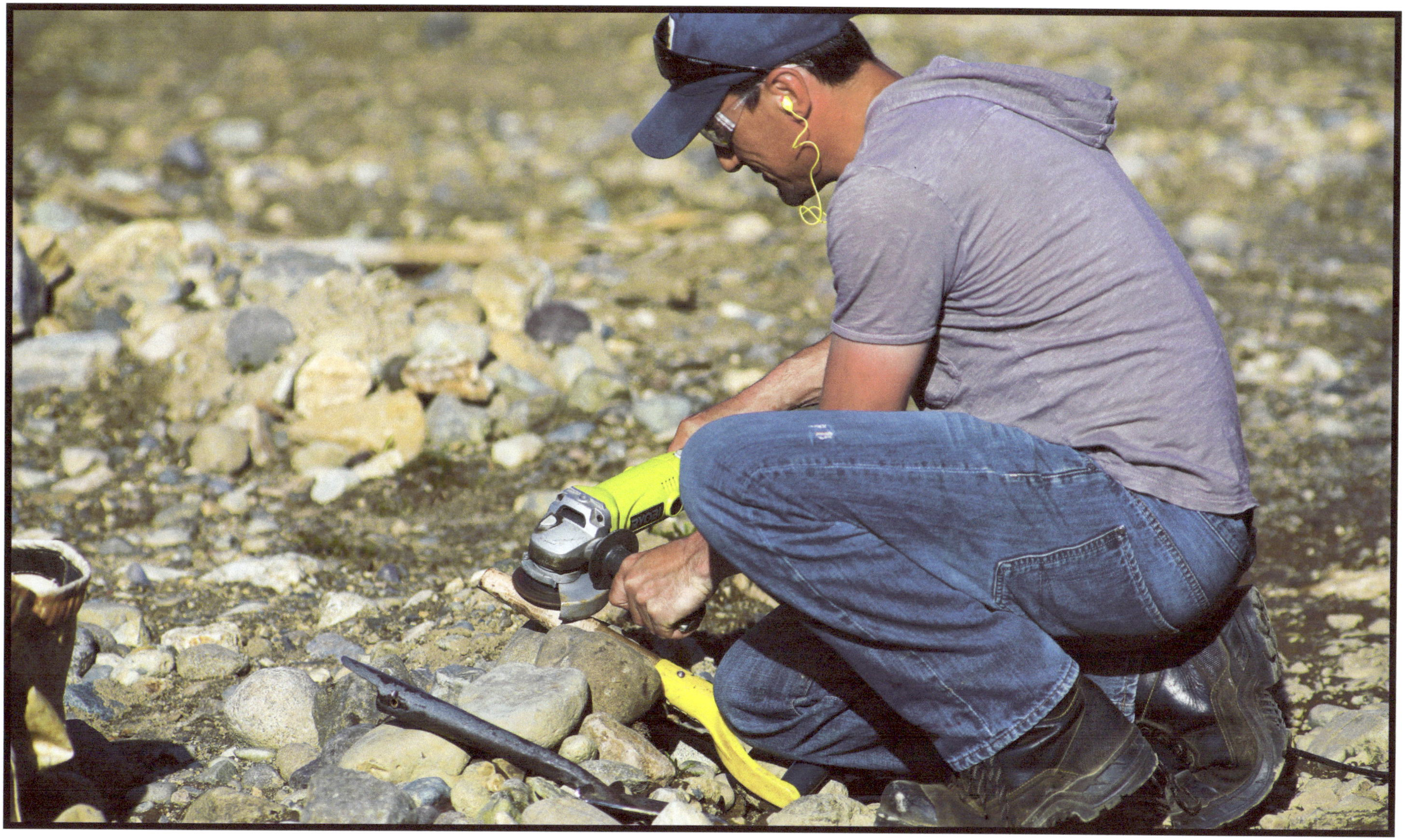

Jon has a great deal of experience with the trades, including plumbing, carpentry, and small engines. The very epitome of a versatile handyman, his skills keep the camp equipment maintained in good working order. He is also in high demand to assist with shovel modifications.

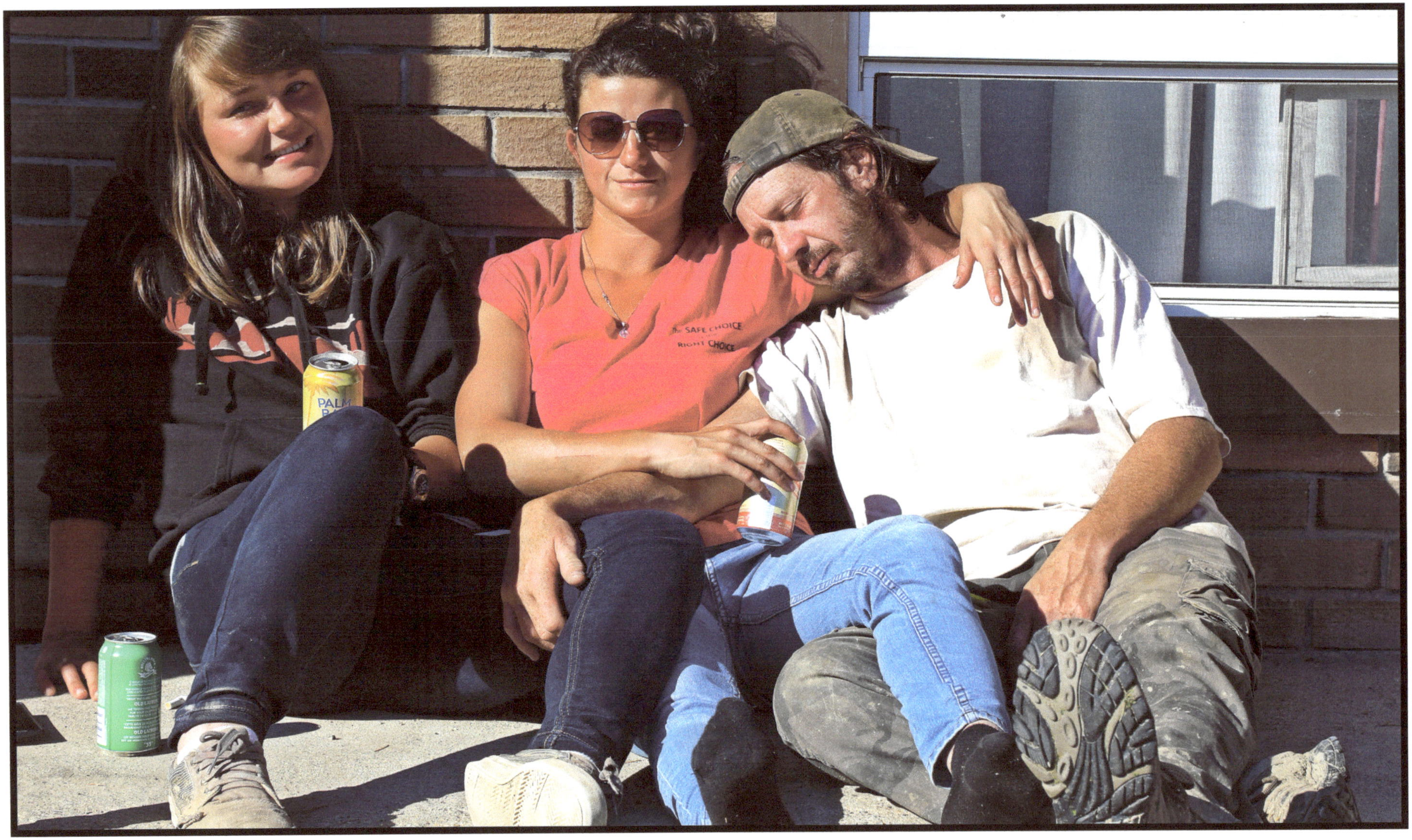

After the year-end party and a full night of socializing, it seemed that all of the planters had tons of energy. I certainly didn't.

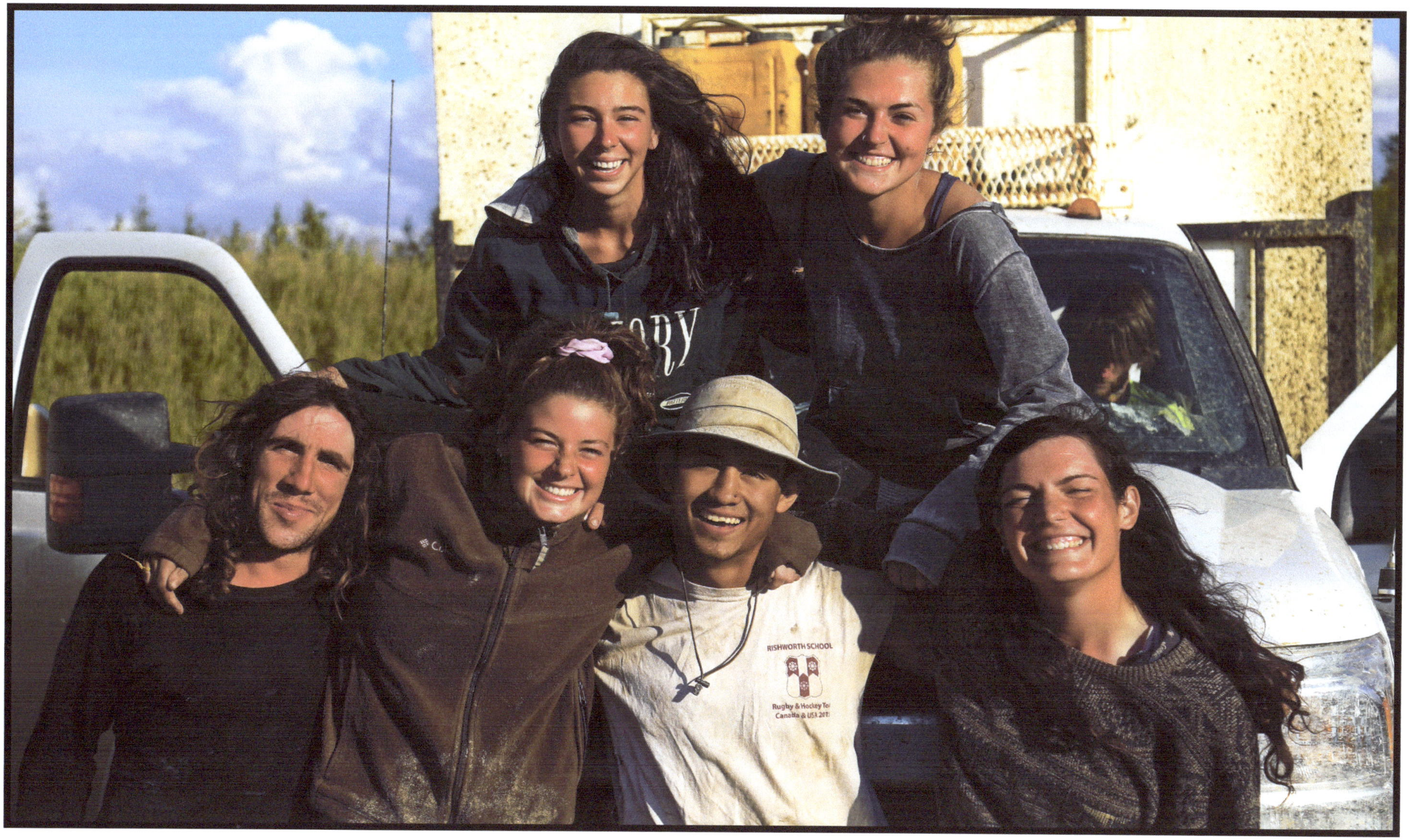

Part of Sebastian's crew. If you look carefully, Seb himself even makes a background appearance in this photo, multi-tasking as he finishes up organizing the day's production tallies.

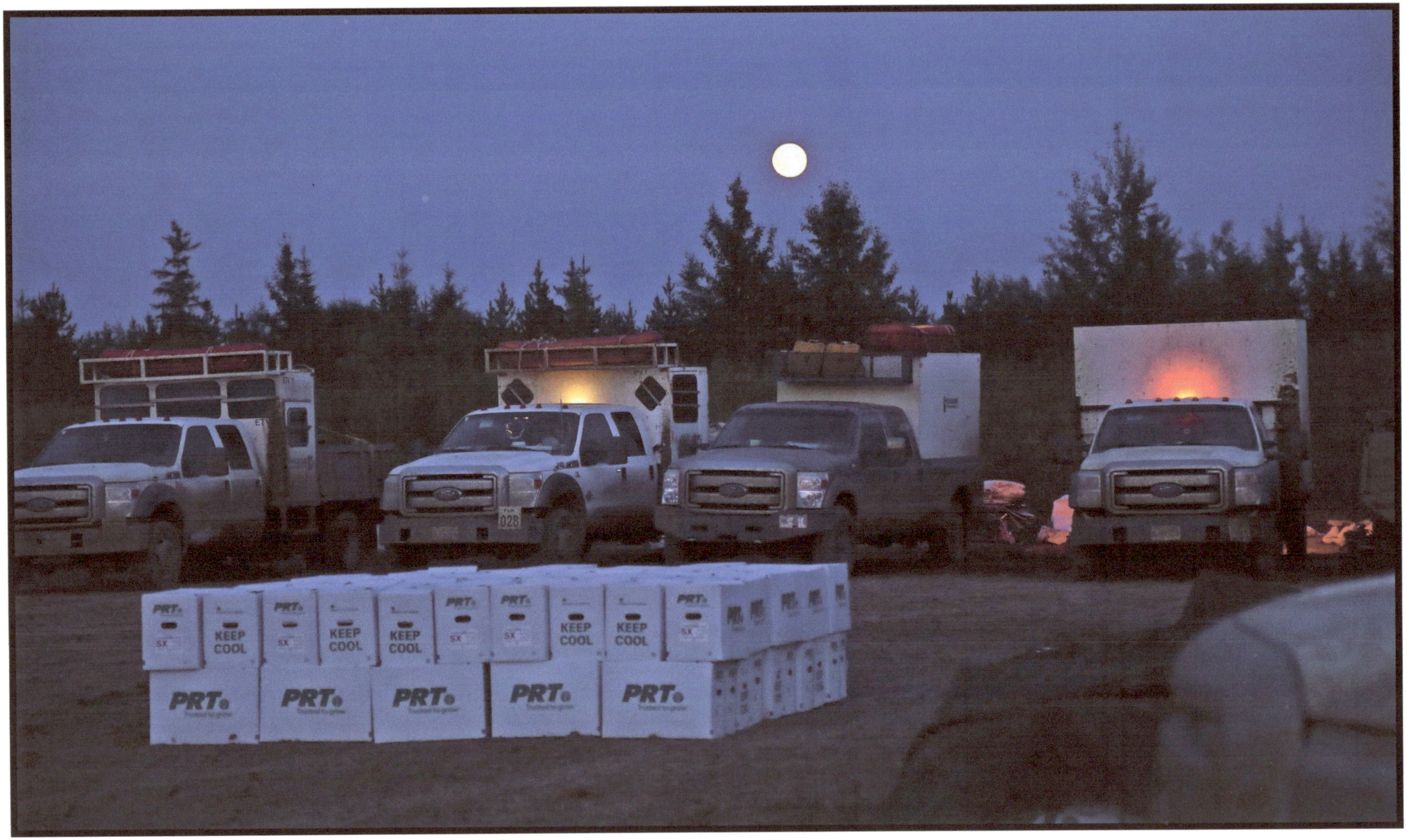

The parking lot in camp, after dark. All of the planters are probably sleeping by this point (midnight), but the management team just finished unloading a late-night reefer that showed up several hours later than it was scheduled to.

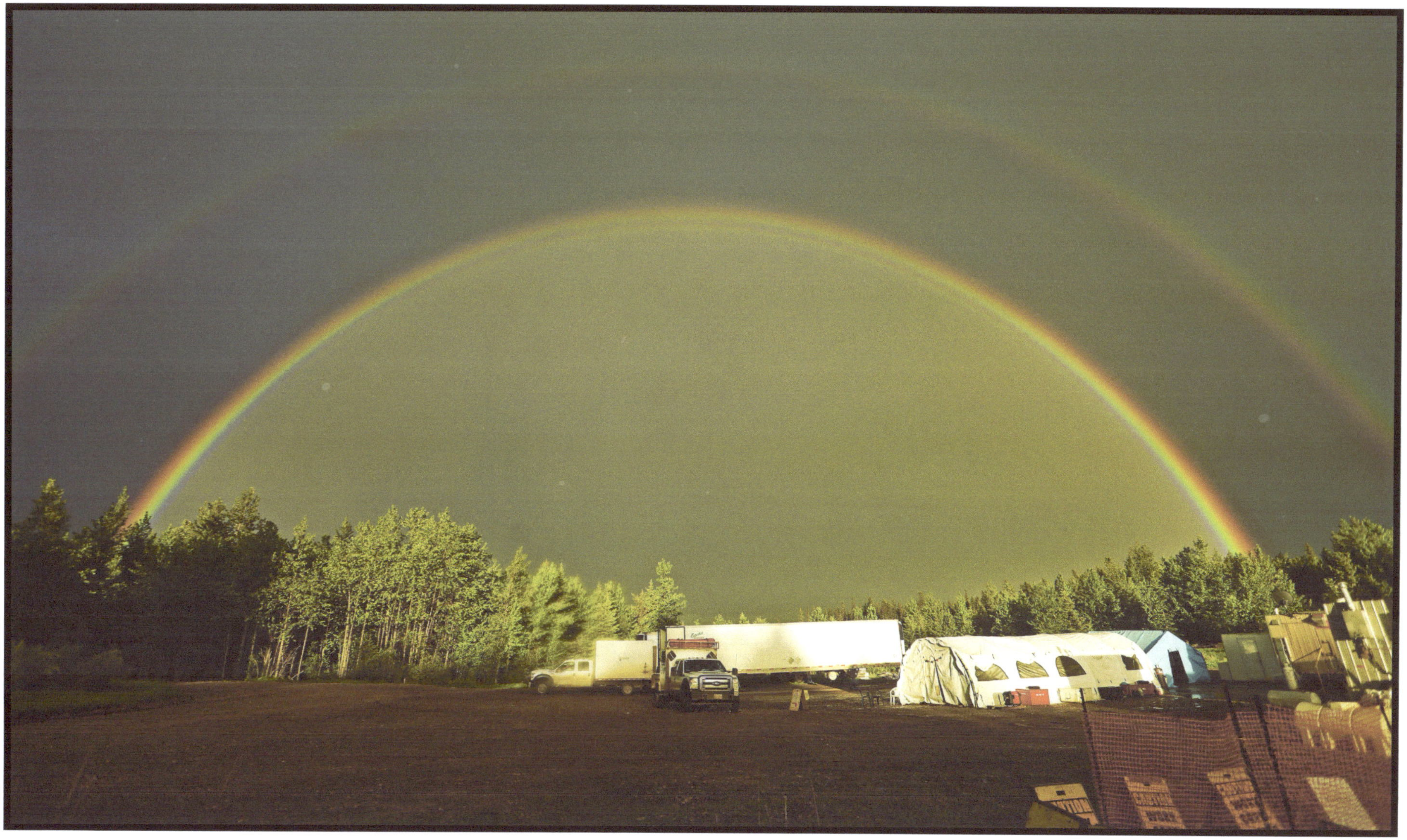

If there's one thing that you can count on while tree planting, it's that you'll see rainbows. And sometimes, you'll see double-rainbows.

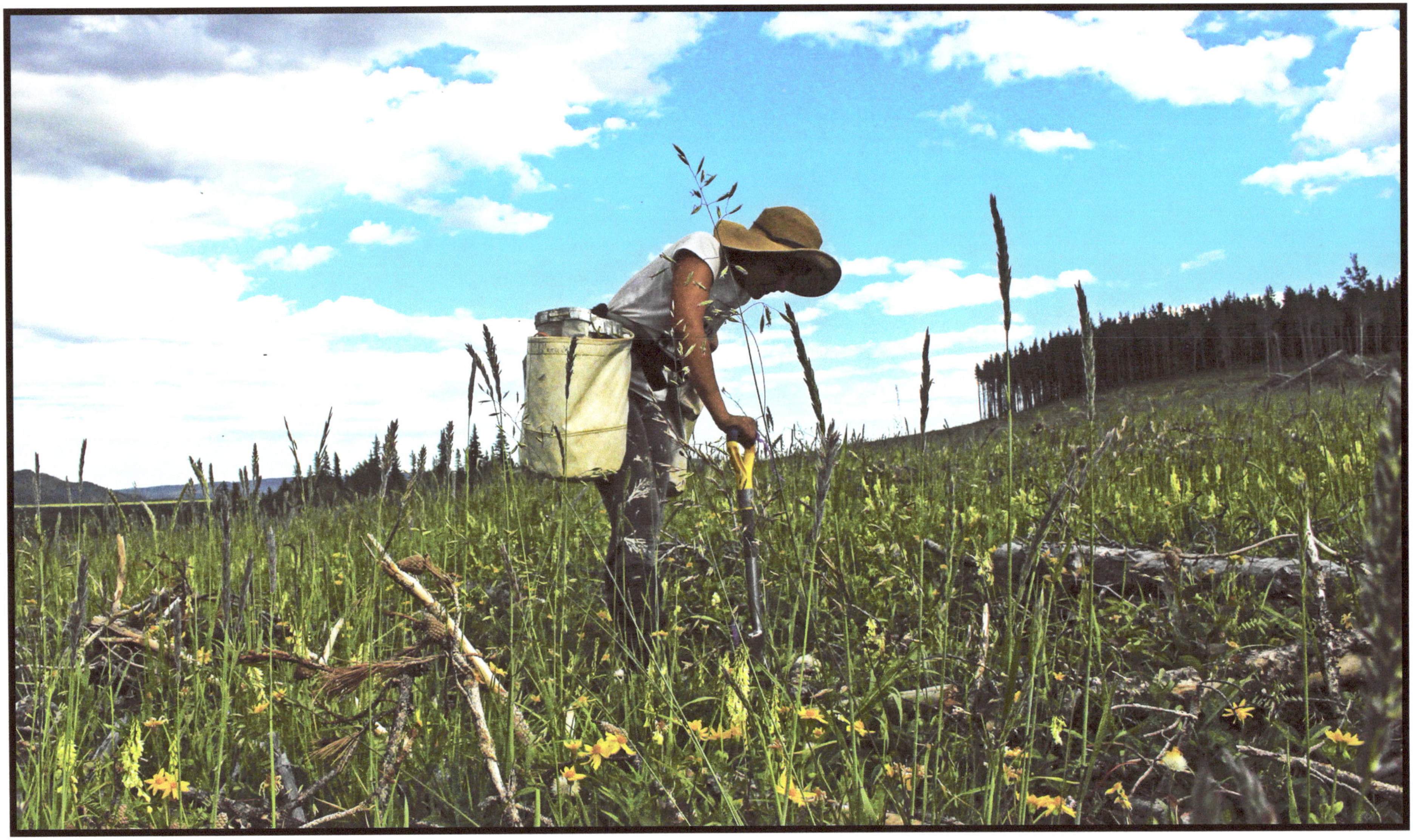

Tree planting: The job that many people love to hate, but one that most people remember fondly for the rest of their lives.

Thanks for your interest in this publication!

Jonathan "Scooter" Clark

Additional Information about Tree Planting

Books about Tree Planting: www.replant.ca/books

Photos about Tree Planting: www.replant.ca/photos

Videos about Tree Planting: www.replant.ca/videos

Music about Tree Planting: www.replant.ca/music

Replant.ca on Instagram: instagram.com/replant.ca

Other Books by Jonathan Clark

www.jonathanclark.ca

www.ingramcontent.com/pod-product-compliance
Lightning Source LLC
LaVergne TN
LVHW071514220926
848486LV00009BA/371